TISSUES AND ORGANS
a text-atlas of scanning electron microscopy

RICHARD G. KESSEL / RANDY H. KARDON

THE UNIVERSITY OF IOWA

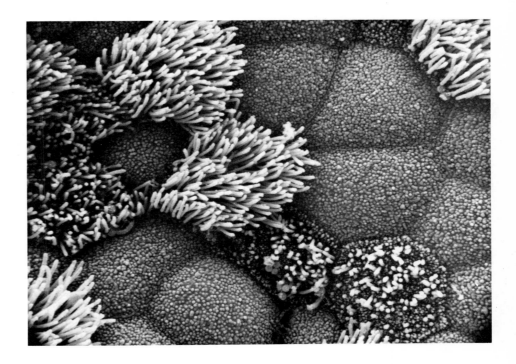

W. H. FREEMAN AND COMPANY / San Francisco

Sponsoring Editor: Arthur C. Bartlett
Manuscript Editor: Dick Johnson
Designer: Robert Ishi
Production Coordinator: Linda Jupiter
Illustration Coordinator: Batyah Janowski
Artists: Edna Indritz Steadman and Donna Salmon
Compositor: York Graphic Services
Printer and Binder: Kingsport Press

Library of Congress Cataloging in Publication Data

Kessel, Richard G 1931-
    Tissues and organs.

    Bibliography: p.
    Includes index.
    1. Histology—Atlases. 2. Ultrastructure
(Biology)—Atlases. I. Kardon, Randy H., 1954-
II. Title. [DNLM: 1. Histology. 2. Histology—
Atlases. 3. Microscopy, Electron, Scanning.
4. Microscopy, Electron, Scanning—Atlases.
QS517 K42t]
QL807.K47   611'.018   78-23886
ISBN 0-7167-0091-3
ISBN 0-7167-0090-5 pbk.

Printed in the United States of America

9 8 7 6 5 4 3

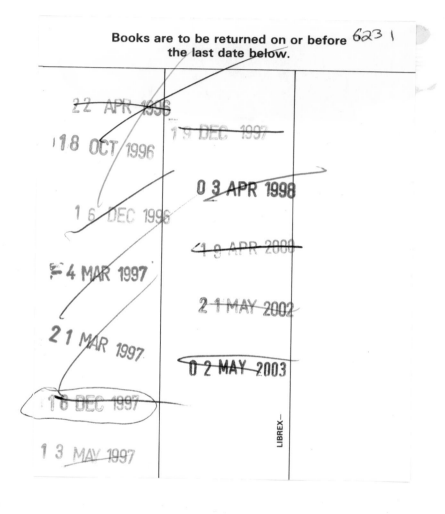

Books are to be returned on or before *6231*
the last date below.

2 2 APR 1996

19 DEC 1997

18 OCT 1996

0 3 APR 1998

1 6 DEC 1996

1 9 APR 2000

4 MAR 1997

2 1 MAY 2002

21 MAR 1997

0 2 MAY 2003

18 DEC 1997

13 MAY 1997

LIBREX—

CONTENTS

PREFACE

Over the past seven years, we have made extensive use of scanning electron micrographs of tissues and organs as an important teaching device in histology. Their use has been met with an overwhelmingly enthusiastic response among students, for they have found such images to be of great benefit to their understanding and appreciation of complex biological interrelationships. Encouraged by the response of students and the interest of colleagues, we felt compelled to prepare a text devoted to the three-dimensional organization of tissues and organs, illustrated mainly with scannng electron micrographs. In addition to its utility in upper division and graduate courses with a histological emphasis, we hope that this book will be a useful addition to the teaching of such courses as introductory biology and physiology.

The first few chapters of the book examine the basic tissue types, and subsequent chapters illustrate how these basic tissue types form the organ systems of the body. Each chapter begins with an introductory synopsis of basic structural and functional concepts about the material to be covered. The text that follows the introduction not only describes, identifies, and explains details shown in the accompanying illustrations, but, in the space available, emphasizes the relation between structural organization and functional significance. Another unique feature of the book is the wide use of microvascular casts viewed by the scanning electron microscope. These enable the student to observe examples of the vast and complex circulatory patterns in organs, especially the extensive capillary networks.

More than 700 scanning electron micrographs appear in this book. Most of them were generated during the period 1975–1977, and the bulk of the text was written during the spring and summer of 1977. Nearly all of the micrographs are original, particularly those that depict microvascular casts. Those that have been contributed by others are acknowledged in the accompanying text.

A list of suggested references appears by subject at the end of this book. These references primarily include studies with the scanning electron microscope, and have been selected from the literature prior to 1978, but a few 1978 references are included. Because of the rapid increase in publications involving scanning electron microscopy, many important references will obviously not be included. More recent reference sources can be obtained from Index Medicus, Current Contents, and the Proceedings of the Annual Scanning Electron Microscope Symposium volumes published initially by the Illinois Institute of Technology and, since 1978, by SEM Inc. The bibliographies contained within many of the selected references will yield additional articles.

Three short appendices appear at the end of the book. The first deals briefly with cell organelles and points out the potential of the scanning electron microscope for use in studying intracellular structures in three dimension. The second deals with junctional complexes, and the third describes the various methods of specimen preparation techniques that were used to obtain the various scanning electron micrographs contained within the book.

A great deal of work has gone into obtaining and preparing suitable material to be photographed so that the many structural details of the vast array of tissues and organ systems could be illustrated. Selection of illustrations proved frustrating at times, for in our course we make use of many correlative figures (diagrams, transmission electron micrographs, and light photomicrographs), not just scanning electron micrographs. Since the number of correlative figures we might like to have used ran counter to our desire to keep the size and cost of the book within reason, we decided to limit their use only to those areas or topics that we felt absolutely required them. In most cases, low-magnification scanning electron micrographs are followed by enlargements of specific areas so that transition between different magnifications can be easily followed. We welcome suggestions from readers for future improvements.

It is our hope that this book will be useful to a variety of students and investigators in their academic pursuits, and that it will serve to make these endeavors more enjoyable, complete, and meaningful. We are hopeful that the images contained in this book will provoke thought and perhaps serve to stimulate additional investigations and contributions by those who view them.

September 1978

Richard G. Kessel
Randy H. Kardon

ACKNOWLEDGMENTS

It is a pleasure to acknowledge the efforts and assistance of a number of individuals who have contributed in various ways toward the completion of this book. We are particularly indebted to Cynthia Asmussen Stiffin for her skilled and conscientious technical assistance with the extensive photographic work required for all plates used in this book. If not for the excellent quality of photographic reproduction, this endeavor would have been impossible. We also wish to express special gratitude to Dr. C. Y. Shih, Supervisor of the University of Iowa Scanning Electron Microscope and Freeze-etch Laboratory, for his expertise and dedication in maintaining the scanning electron microscope during the extensive period required to obtain the many illustrations. Dr. Shih's helpful suggestions about specimen preparation and photography were invaluable in obtaining suitable micrographs. We also appreciate the efforts of Donald Mower, who worked extensively with us to prepare, dissect, and photograph the many replicas of tissue and organ microvasculature that appear throughout the book. Mr. Mower played a major role in developing the technique so that it could be used to demonstrate the entire vasculature of tissues and organs.

We are also grateful for the invaluable assistance of many scientists who contributed micrographs or drawings for our use. Although individual acknowledgments for particular drawings and micrographs are indicated in their appropriate locations within the book, we nevertheless wish to express here our appreciation to H. Ades, J. Alvarado, M. Ashraf, R. Beeuwkes, A. Blouin, R. Bolender, M. Bonneville, J. Bonventre, A. Boyde, R. Bulger, J. Carneiro, L. Chen, F. Chlapowski, A. Contopoulos, D. Dickson, M. Dym, H. Elias, H. Engstrom, D. Fawcett, F. Feuchter, T. Fujita, R. Greep, R. Grossman, A. Ham, E. Handler, D. Hillman, E. Jaffé, S. Jones, L. Junqueira, D. Lagunoff, T. Lentz, E. Lewis, H. Lindeman, P. Lipsky, S. Luk, A. Maunsbach, J. Maynard, J. Murray, F. Netter and the Ciba Foundation, J. Revel, J. Rhodin, A. Rosenthal, J. Sherrick, G. Simon, S. Sorokin, A. Staehelin, H. Sybers, K. Tanaka, T. Tanaka, A. Weber, J. Weddell, E. Weibel, L. Weiss, W. Willis, W. Windle, and R. Young. We wish to extend special thanks to N. Azzam for providing us with several scanning electron micrographs of nervous tissue and to L. Kahn for permitting us to photograph his tissue preparation of the choroid plexus. The Departments of Zoology and Pharmacology at the University of Iowa deserve special recognition for their support, which facilitated the completion of this endeavor. Randy Kardon would also like to express thanks to the Insurance Medical Scientist Scholarship Fund and cosponsor, the Prudential Insurance Company, for their support during the period 1977–1978.

We are especially indebted to our sponsoring editor, Arthur Bartlett; our manuscript editor, Dick Johnson; and our designer, Robert Ishi, for their tireless efforts and enthusiasm in bringing this book to completion. To Ruth J. Allen of the Permissions Department and Inez Burke, Editorial Coordinator for W. H. Freeman and Company, we owe much for their constant attention to details and for the many extensive efforts made in our behalf throughout all phases of this book's preparation. To all of these individuals, as well as to members of the graphic arts and production departments of W. H. Freeman and Company, we are grateful for the many helpful suggestions and assistance that all have contributed. Many of the drawings contained in this book were redrawn, and in many cases significantly modified, by two artists, Edna Indritz Steadman and Donna Salmon. We thank these artists for their painstaking efforts, which reflect great accuracy and the highest quality of work.

Finally, we would like to express our sincere appreciation to Robin Spicher, Barbara O'Donnell, Robin Wagner, and Sharon Paul for typing the manuscript.

EPITHELIUM

Introduction

Epithelium, as a class of tissue, consists of cells arranged in sheets that cover surfaces and line cavities. In addition, cells particularly suited for secreting fluids are often organized into glandular epithelia. Since the size, shape, and arrangement of epithelial cells are directly related to their particular function, attention will be focused upon these characteristics. The five most common functions attributed to epithelial tissue are: protection (skin), absorption (intestine), secretion (glands), sensation (neuroepithelium, such as taste buds or olfactory cells), and contractility (myoepithelial cells surrounding glandular acini). One characteristic common to any type of functioning epithelium is the capacity for renewal. The replacement of cells occurs by the process of mitosis, and the rate of renewal may vary depending on the type of epithelium and its location.

The emphasis in this chapter will be upon the organization of epithelium into sheets, the form to be most frequently encountered in subsequent chapters dealing with organs. Epithelial sheets are characterized by strong cellular adhesions and a relative absence of intercellular substance. The strong cellular adhesion is accomplished by certain types of junctional complexes, which are specializations formed between plasma membranes of adjacent cells (see Appendix B for the description and diagram of junctional complexes). Epithelial sheets are anchored to underlying connective tissue via a basement membrane, and since epithelium is avascular, its source of nutrients is derived from nearby capillaries distributed within the underlying connective tissue. Epithelium can be classified by the number of cell layers and subdivided further by the cell shape (squamous, cuboidal, or columnar) in the most superficial layer. The two most common types of epithelia are simple (one cell layer) and stratified (multilayered). A third type, known as pseudostratified epithelium, consists of columnar epithelium accompanied by a basal layer of less differentiated polygonal cells, which do not usually reach the free surface. Contrary to appearance, this type of epithelial sheet is not truly stratified. Careful examination reveals that more differentiated columnar cells, which occupy the majority of space toward the free surface, still remain attached to the basement membrane. Therefore, not every cell extends through the full height of the epithelial sheet to reach the surface, but all retain a basal attachment. The variation in size, shape, and nuclear position between neighboring cells are responsible for the apparent stratification seen in this type of epithelium.

Finally, each type of epithelium can be distinguished further on the basis of surface specializations. Cilia and microvilli are the most common examples; their presence corresponds directly to specific epithelial functions. By their rhythmic beating motion, cilia are capable of moving substances across the free epithelial surface, whereas microvilli increase the surface area by extensive folding of the epithelial cell membrane. This, in turn, increases the capacity of epithelia for absorption and transport. The relative size, shape, and distribution of cilia and microvilli are best understood by the study of their appearance in the scanning electron micrographs provided throughout this book. This chapter will serve to familiarize the reader with the basic three-dimensional organization of epithelial types. Additional information will be presented in chapters dealing with the specific organs in which certain epithelial types are found.

EPITHELIUM
Simple Squamous

Simple squamous epithelium forms a thin sheet that lines the luminal surface of all blood and lymphatic vessels. Such epithelium is commonly termed endothelium because of its origin from mesoderm. Fig. 1 shows the flattened endothelial cells that line the interior of a blood vessel; blood leukocytes (**Le**) often remain attached to their surface. Simple squamous epithelium lines the ventricles and atria of the heart, making up part of the endocardium (Fig. 2). Surface folds (**arrows**) are frequently visible between adjoining cells. Dome-shaped areas (✻) of each cell project above the plane of the epithelial sheet, indicating the locations of underlying nuclei. The pericardial, pleural, and peritoneal cavities are also lined with simple squamous epithelium. In addition, many major organs are covered externally by a single layer of squamous cells, collectively referred to as mesothelium. Mesothelial cells (**MC**) from the visceral peritoneum of the spleen are shown in Fig. 3. A dense population of finger-like projections, termed microvilli, extend from their free surface into the body cavity. These surface specializations are shown at higher magnification in Fig. 4. The density of microvilli varies on mesothelial surfaces of different organs. Microvilli also possess a surface coat of glycoproteins (glycocalyx) that appears to function in making the mesothelial cell surface slippery, thus permitting the movement of organs over each other without frictional damage. A final example of simple squamous epithelium is shown from the kidney medulla (Figs. 5 and 6). Portions of the many tubular loops of Henle possess a thin wall consisting of flattened squamous cells, which project into the lumen at the location of the cell nucleus (**Nu** in Fig. 6). It is interesting to note that some of the cells possess a single cilium (**Ci** in Figs. 5 and 6) that projects into the tubular lumen, but its functional significance is unclear.

Fig. 1, ×1780; Fig. 2, ×960; Fig. 3, ×3115; Fig. 4, ×12,460; Fig. 5, ×1200; Fig. 6, ×2990

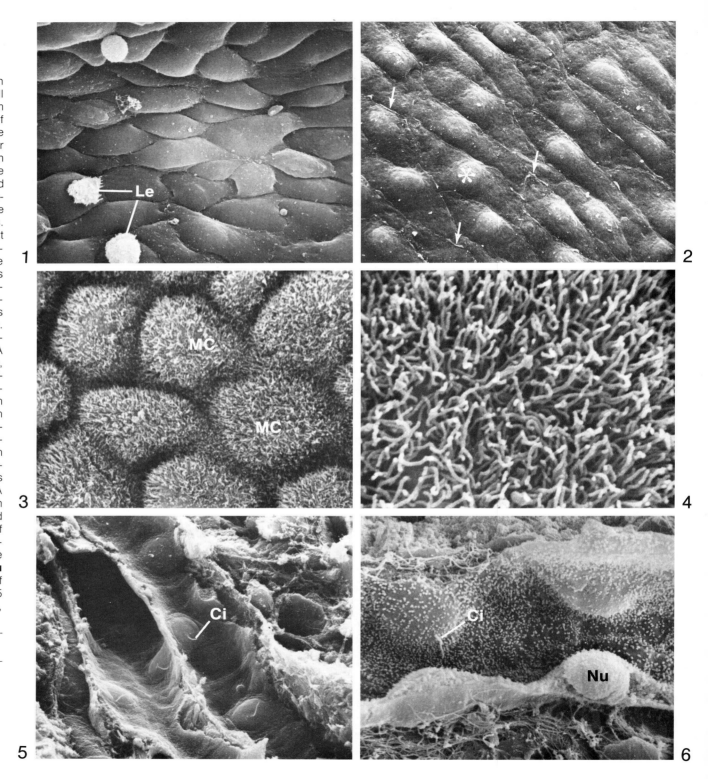

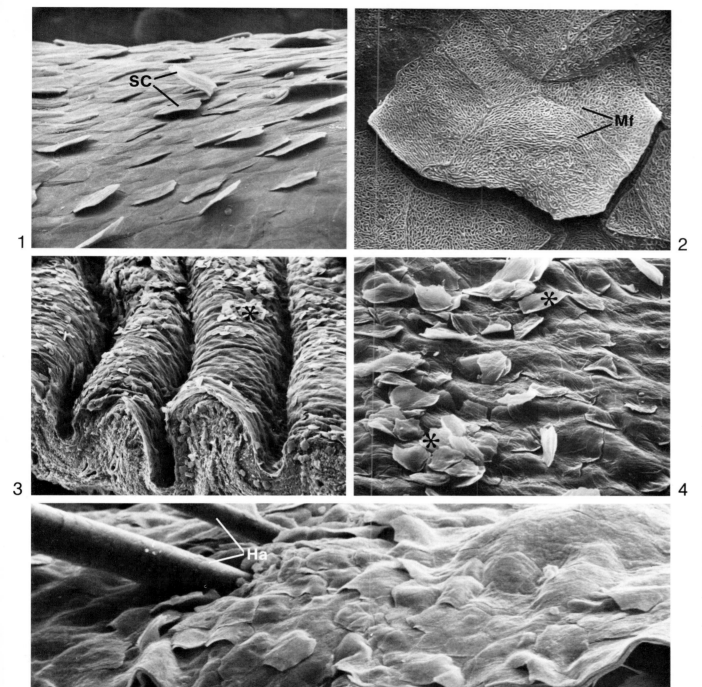

EPITHELIUM
Stratified Squamous

Stratified squamous epithelium [is character]-
ized by three or more cell layers [whose]
superficial layers consist of squa[mous c]ells.
The basal layer contains cuboidal or colum-
nar-shaped cells that are proliferative, serving
to replace older, more superficially located
cells that are extruded from the surface. By
mitotic division, newly formed cells move to
superficial layers and, in the process, undergo changes in shape (becoming more
flattened), composition, and function. Eventually, the cells become so far displaced from
the underlying vascular supply that the cells
die and are sloughed from the surface by a
process termed desquamation. Stratified
squamous epithelium is located in areas subject to wear and tear, such as the epidermis
of skin and the linings of the oral cavity,
vagina, anterior cornea, and portions of the
urethra. An area of the hard palate, which
forms the roof of the oral cavity, is illustrated
in Fig. 1. Many of the superficial squamous
cells (**SC**) are in the process of desquamation. At higher magnification (Fig. 2), a pattern of microfolds (**Mf**) can be distinguished
on these cell surfaces. The folded mucosal
surface of the vagina is illustrated in Fig. 3
and at higher magnification in Fig. 4. In both
figures, the stratified squamous epithelium
characteristically possesses superficial cells
in the process of desquamation (✱). A final
example of stratified squamous epithelium is
observed in the epidermis of skin (Fig. 5). The
stratification of the epithelium is apparent in
areas sectioned transversely. A number of
the closely packed cell layers are separated,
allowing them to be easily distinguished (**arrows**). As in previous examples, many of the
superficially located cells are undergoing
desquamation. In addition, several hairs (**Ha**)
penetrate the stratified squamous epithelium.

Fig. 1, ×515; Fig. 2, ×2545; Fig. 3, ×85;
Fig. 4, ×405; Fig. 5, ×465

Simple cuboidal epithelium consists of a single layer of polyhedral cells that forms a sheet. Cuboidal cells lining the proximal tubule of the kidney are shown in Fig. 1 from both a lateral and surface view. The lateral surface of the cells is somewhat folded toward the base, and each cell resides on a basal lamina (**BL**). The surface of these cells is densely populated with microvilli (**Mv**). Cuboidal cells lining the collecting ducts (**CD**) of the kidney (Fig. 2) appear hexagonal in surface view. Short microvilli are present on the cell surface, and a centrally located cilium (**Ci**) projects outward from each cell into the tubule lumen. A number of the tubule cells have been internally exposed, revealing spherical cell nuclei (**Nu**). Ciliated cuboidal epithelium lines the terminal bronchioles of the lung and portions of the ventricles in the brain. The epithelial lining of ventricles (termed ependyma) is viewed from the surface in Fig. 3. Groups of long cilia (**Ci**) project from each cell into the ventricle cavity, and their beating motion may assist in the movement of cerebrospinal fluid. Stratified cuboidal epithelium is rare, but can be found, for example, in the ducts of sweat glands. The germinal epithelium (**GE**) found within the seminiferous tubules of the testes (Fig. 4) is also considered by some to represent a type of stratified cuboidal epithelium. Tails (**Ta**) of developing spermatozoa lie within the tubule lumen. Transitional epithelium is also stratified, and characteristically lines the interior of the ureter and bladder. Its stratification can be observed in a transverse section through the ureter (Fig. 5), and the surface of the superficial cell layer can be viewed from the bladder lumen (Fig. 6). This type of epithelium is termed transitional because its most superficial cells undergo transitions in shape and size depending upon the degree of distension of the tube or cavity that it lines. These cells may range in shape from flattened to rounded or balloon-like (**arrows**).

Fig. 1, ×2690; Fig. 2, ×1485; Fig. 3, ×3010; Fig. 4, ×340; Fig. 5, ×480; Fig. 6, ×290

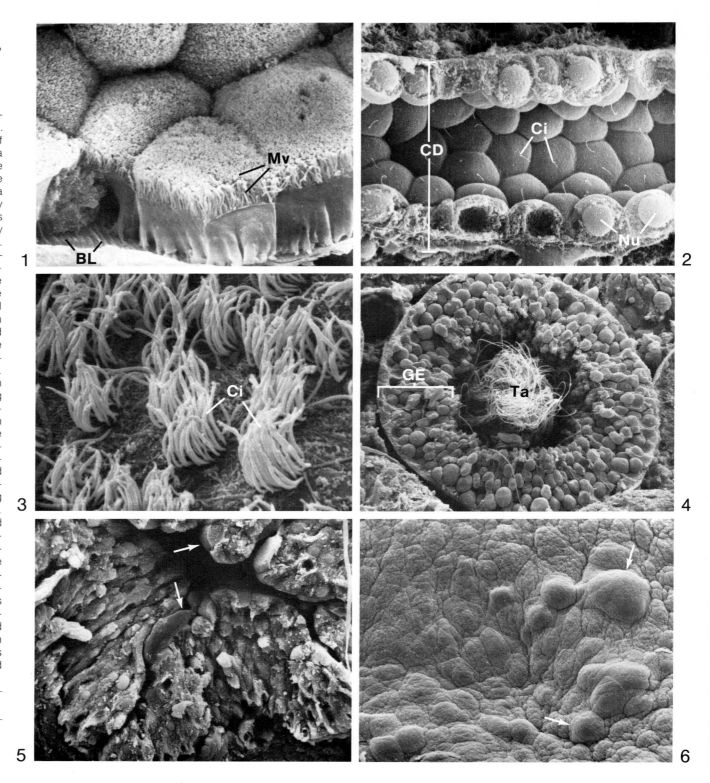

EPITHELIUM
Simple Columnar

Different aspects of simple columnar epithelium are illustrated in these figures. The columnar shape of the cells forming this single layer of epithelium is the most characteristic feature, as can be seen in Fig. 1 (oviduct) and in Fig. 3 (gallbladder). The surface characteristics of these cells, however, may vary depending on their location within the body. For example, the cells lining the oviduct (Figs. 1 and 2) are of two types: one possesses a ciliated (**Ci**) surface; the other, a microvillous (**Mv**) surface. The columnar epithelia present in the gallbladder (Figs. 3 and 4) and the intestine (Figs. 5 and 6) characteristically possess a dense population of microvilli (**Mv**) on their apical surfaces. The microvilli of the intestinal epithelial cell (Fig. 5) are longer than those associated with the gallbladder epithelium. In some tissues, another cell type exists within the monolayer of simple columnar epithelium. This cell type, shown in Fig. 6, is frequently goblet shaped, and is appropriately termed the goblet cell (**GC**). Collectively, goblet cells secrete mucus, which forms a layer over the entire epithelial surface. In these preparations the mucus layer was removed so as to reveal the microvillous cell border. When the goblet cell is fractured open, many of the secretory packets containing mucigen are observed within the cell cytoplasm.

Fig. 1, ×1760; Fig. 2, ×720; Fig. 3, ×2000; Fig. 4, ×5125; Fig. 5, ×9210; Fig. 6, ×2930

EPITHELIUM
Pseudostratified

Pseudostratified epithelium consists of a single layer of cells that appears stratified or multilayered in sections when viewed with the light microscope. This type of epithelium makes up the lining of the trachea and large bronchi (Figs. 1 and 2), the nasal mucosa (Figs. 3 and 4), and the epididymis and vas deferens of the male reproductive tract. Although all cells of pseudostratified epithelium are in contact with the basement membrane, many become narrow toward their apical end, appearing not to reach the free surface. Goblet cells (**GC**) are often present in the cell layer, and possess short microvilli (**Mv**) on their surface, whereas the rest of the cells are usually ciliated (**Ci**). This is particularly evident in Fig. 2, which is a surface view of the tracheal pseudostratified epithelium. Fig. 5 illustrates the decrease in the distribution of microvilli on the surface of the goblet cell seen when secretory product accumulates under the apical cell membrane. When the apical cell membrane is removed (Fig. 6) underlying packets of secretory material (**arrows**) are observed.

Fig. 1, ×1250; Fig. 2, ×4110; Fig. 3, ×780; Fig. 4, ×1190; Fig. 5, ×6250; Fig. 6, ×9025

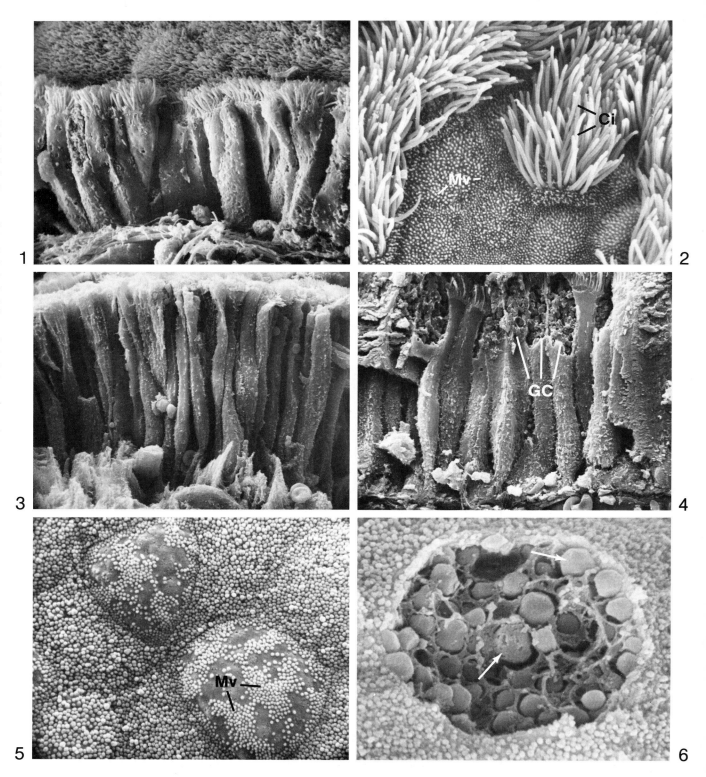

EPITHELIUM
Glandular Epithelium

Glandular epithelium develops from invaginations of epithelial sheets throughout the body. Generally, cells of this type of epithelium are specialized for the synthesis, storage, and secretion of a product. Glandular epithelium exists in either an endocrine or exocrine form. An example of the endocrine form is illustrated in Fig. 1 from a portion of the adenohypophysis of the pituitary gland. Endocrine cells (**EnC**) are polyhedral, and are collectively arranged into cords or plates that are supported by a loose network of connective tissue containing numerous capillaries (**Ca**). Endocrine cells characteristically secrete their product into the interstitium, whence it diffuses into surrounding capillaries, thereby gaining entrance into the vascular system. Exocrine glands, however, secrete their product into a duct system. Exocrine glands may exist in the form of rounded alveoli or acinar units (**AU**), as found in the pancreas (Fig. 2). Each acinar unit consists of epithelial cells arranged like a cluster of grapes around a duct leading from the secretory unit. Connective tissue (**CT**) is closely associated with the acinar surface, and is commonly organized into septa that divide the gland into lobules. In addition to the acinar form, glandular epithelium may also be organized into a tubular form, as exemplified by the mucosa of the colon (Fig. 3) and stomach (Figs. 4 to 6). The tubular glands (**TG**) are surrounded by a loose arrangement of connective tissue (**CT**). In association with the connective tissue, capillaries (**Ca**), which supply cells with raw materials for synthesis of their secretory product, are brought close to the glandular epithelium, as illustrated in Fig. 5. The relationship of the epithelial cells to the centrally located lumen (**Lu**) of the tubular gland is also shown in Figs. 3, 4, and 6. Exocrine glands are frequently classified as compound glands when there is extensive branching of the duct system to supply the many secretory units of such large glands. In addition, exocrine glands can contain both acinar (alveolar) and tubular secretory units, in which case they are termed tubulo-alveolar glands.

Fig. 1, ×500; Fig. 2, ×1070; Fig. 3, ×215; Fig. 4, ×140; Fig. 5, ×625; Fig. 6, ×670

CONNECTIVE TISSUE

Introduction

Connective tissue proper is a mixture of cells, fibers, and amorphous ground substance that forms a framework of support for epithelium and other tissues of the body, the most common being tendons, ligaments, deep and superficial fascia, as well as the internal supportive substance of most organs. Connective tissue also forms a sheath surrounding blood vessels and nerves, and when in the form of adipose tissue it has the capacity to store triglycerides. Even blood, bone, and cartilage are considered to be types of connective tissue, but because they are so specialized, they will be considered in separate chapters.

As the name implies, connective tissue "connects" other tissues together. In addition, the cells, fibers, and ground substance perform important functions, such as transport, storage, support, protection, and repair. An understanding of how each component contributes to these functions will be gained by a description of its organization. In general, the cellular component of connective tissue proper can be divided into two categories: those cells that are fixed in position and those that are free to wander throughout the tissue. The fixed cells include fibroblasts (the most common of all the cellular elements) and adipocytes. The wandering cells include lymphocytes, plasma cells, mast cells, and granular leu-

kocytes. The macrophage may be either fixed or wandering, depending upon its functional state. Wandering cells originate from different blood cell types, which, upon stimulation, migrate from the bloodstream into connective tissue. The stimulus is usually associated with an immune or inflammatory response. Therefore, the proportion of wandering cells within connective tissue varies, depending upon the prevailing functional state. Fixed cells, as their name implies, do not normally exhibit motility, but may do so when tissue is injured. For example, fibroblasts, which synthesize and secrete most of the fibers and ground substance of connective tissue, may undergo division and become migratory when repair is required.

The fibers of connective tissue are of three different types: collagen, reticular, and elastic. The fibrous elements determine mechanical properties, such as tensile strength and flexibility. When these two properties are required in areas subject to forces of stress and strain (ligaments and tendons), the relative number of fibers increases dramatically over that of cells and ground substance. The arrangement and proportion of fibers present is frequently used as a basis for classifying types of connective tissue. The distinction between loose areolar, dense irregular,

and dense regular types of connective tissue is made on just such a basis, as will become apparent from the scanning electron micrographs throughout this chapter. The structure and biochemistry of collagen, reticular, and elastic fibers is complex, and, although relatively well understood, their detailed description will be left to texts devoted more exclusively to those subjects. Here the emphasis is on the three-dimensional relationship of the fibers to the cellular component and on their distribution and orientation within connective tissue.

The amorphous ground substance, occupying the spaces between the cellular and fibrous components, has a number of unique properties that enable it to serve a variety of important functions. The term "amorphous" refers to the shapeless character of ground substance as viewed under the light microscope. Consisting chiefly of water, salts, and glycosaminoglycans, the ground substance poses a barrier to the passage of microorganisms, yet at the same time serves as a medium for the transport and storage of many soluble substances. These properties are primarily attributed to the glycosaminoglycans, a term that embraces both mucopolysaccharides and glycoproteins under one heading. The major constituents of glycosaminoglycans are hyaluronic acid and chondroitin sulfate. Hyaluronic acid consists of polymers of disaccharide units (glucuronic acid linked to *N*-acetyl-glucosamine) in which polypeptides are covalently bound throughout their length. Chondroitin sulfate has a higher molecular weight, primarily due to its larger protein backbone. Attached to this backbone are disaccharide polymers of glucuronic acid linked to galactosamine. In both hyaluronic acid and chondroitin sulfate, the polymers form a molecular "net." The openings in this net are large enough to allow passage of soluble nutrients but small enough to restrict the movement of relatively large entities, such as microorganisms. The viscosity of the matrix is related to its degree of polymerization, which can be regulated by the enzyme hyaluronidase. Of interest is the fact that some microorganisms secrete this enzyme, which allows them unrestricted passage through connective tissue. In addition, most glycosaminoglycans are negatively charged, acting as a "sink," or storehouse, for electrolytes and water. Although the ground substance is removed in most histological preparations, the remaining spaces serve as a reminder of its former presence.

In summary, it is apparent that amorphous ground substance functions as a medium for the *transport* of soluble material and is capable of *storing* electrolytes, water, and other constituents of tissue fluid. At the same time, it also acts as a barrier, offering *protection* against the penetration of microorganisms. Protection against foreign substances is also provided by the wandering cells of connective tissue, which are integral components of the inflammatory response. *Support* for connective tissue is provided by numerous fibers, which also increase the tensile strength and flexibility of tissue. Finally, connective tissue has the capacity to *repair* itself through the efforts of the fibroblast, which can replace damaged areas with new fibers and ground substance.

CONNECTIVE TISSUE
Loose Areolar

Loose areolar connective tissue forms the lamina propria and submucosa of most organs (e.g., respiratory and gastrointestinal tracts) and the subcutaneous tissue of skin. A transverse section through the wall of a bronchus (Fig. 1) illustrates the organization of the loose areolar connective tissue of the lamina propria **(LP)**, which is covered by an epithelial sheet **(ES)**. Further details of the lamina propria are shown at higher magnification in Fig. 2. The connective tissue matrix, or amorphous ground substance, is often removed during tissue preparation. Consequently, the term "areolar" refers to the small pockets or spaces (✽) that are situated between fibers and cells and which contain ground substance in the living state. Connective tissue is well vascularized in many areas, as evidenced by the presence of numerous blood vessels **(BV)**. Extracellular fibers **(Fi)** lend support to the tissue, and their arrangement here is characteristically "loose," exhibiting no distinguishable pattern of organization. They generally consist of three different types, collagenous, elastic, and reticular. Collagenous fibers, synthesized by fibroblasts, most often occur in bundles, whereas the elastic fibers occur as thin, branching strands. The cellular component **(arrows)** of loose areolar connective tissue consists of fixed cells and wandering cells. Wandering cells are predominantly spherical and frequently possess microvilli or microfolds on their surface. At times this morphology resembles that of their leukocyte **(Le)** counterparts seen in peripheral blood (Fig. 3).

Fig. 1, ×1080; Fig. 2, ×1745; Fig. 3, ×5305

CONNECTIVE TISSUE
Loose Areolar

The morphology of fibroblasts is highly variable, depending on whether they are observed in a tissue preparation (Figs. 1 to 3) or in a culture preparation (Fig. 4) and depending also on the condition of the material before preparation. The amount of available nutrients, phase of cell cycle, secretory activity, and synthesizing capabilities can all influence fibroblast morphology. In some instances (e.g., loose areolar connective tissue in the adventitia of the trachea), the fibroblast (**Fi**) may be fusiform, or spindle shaped, as illustrated in Fig. 1. For size comparison, note the presence of a nearby erythrocyte (**Er**). In contrast to the spindle shape, fibroblasts may also appear flat with a highly attenuated cytoplasm (**arrows** in Figs. 2 and 3), as shown in the lamina propria of a bronchus (Fig. 2) and in the connective tissue surrounding a tubular intestinal gland (Fig. 3). When a comparison is made with a monolayer of fibroblasts grown in culture (Fig. 4), striking similarities are apparent. In addition to fibroblasts, spherical cells (❋ in Fig. 2) possess surface properties characteristic of wandering cells. The loose, irregular network of collagen fibers (**CF**) surrounding the cellular elements in Figs. 1 to 3 is the polymerized end product of fibroblasts. Another type of fiber, the reticular fiber, is most commonly found within myeloid tissue (lymph node, spleen, bone marrow). In these locations, the reticular fiber is synthesized and secreted into the extracellular space by reticular cells. As exemplified by the medullary sinus of the lymph node (Fig. 5), the cells and fibers collectively form a dense, supportive network, or "reticulum," (**Re**) within the tissue. Processes of the reticular cells completely wrap around bundles of reticular fibers, forming a surrounding sleeve that obscures the fibers from view. When a portion of this surrounding cell sleeve is removed (Fig. 6), the reticular fibers (**RF**) become visible. Remaining portions of the reticular cell (**RC**) are also apparent. Note the smooth surface of the cell compared to that of the fibers.

Fig. 1, ×690; Fig. 2, ×1645; Fig. 3, ×3245; Fig. 4, ×1000; Fig. 5, ×635; Fig. 6, ×4600

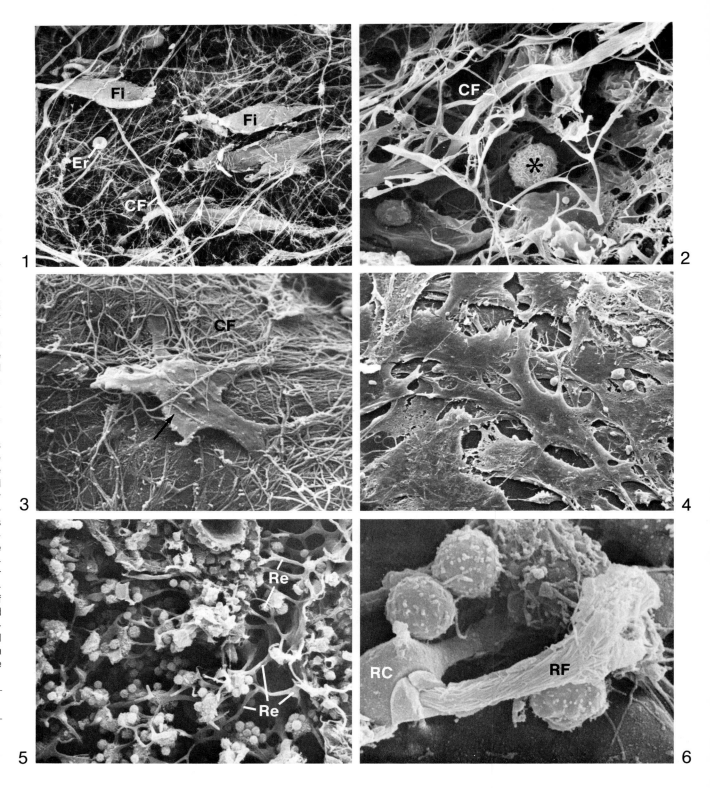

CONNECTIVE TISSUE
Loose Areolar

Phagocytic cells, such as macrophages and leukocytes, populate richly vascularized areas of connective tissue and are prominent at sites of immune reactions. Upon stimulus, macrophages are capable of ingesting foreign particles, microorganisms, and effete cells within the body. Fixed macrophages (**FM** in Fig. 1) appear flattened and possess thin, tendril-like cytoplasmic extensions, whereas the wandering type (Figs. 2 and 3) appears rounder and exhibits a ruffled surface extension, or pseudopod (**Ps**), during movement. The fixed macrophage in Fig. 1 is from the medullary sinus of a lymph node, and the wandering macrophages (Figs. 2 and 3) are from the alveolar surface of the lung. Leukocytes in circulating blood can sometimes be observed to move from the central axis of blood flow and adhere to the inner surface of the vessel (a process termed margination). Once in contact with the endothelial cells, the leukocytes (**Le** in Figs. 4 and 5) may either migrate slowly along the vessel wall or remain stationary. The process of margination may also be followed by the movement of the leukocytes through the vessel wall into the surrounding connective tissue, where they can serve a phagocytic function, particularly at sites of inflammation. The process of phagocytosis is illustrated in Fig. 6, which show a macrophage (**Ma**) in the process of ingesting two erythrocytes (**Er**). The thin cytoplasmic extensions (**arrows**) of the macrophage are observed to surround the erythrocyte surface. Although the macrophage is capable of both phagocytosis and pinocytosis, it is apparent that the former process is much more selective, particularly during an immune response, and requires a specific binding to the macrophage surface. Nonspecific recognition, however, also occurs in the lung, where the alveolar macrophage must clear the airways of airborn particulate matter. [Fig. 6 courtesy of J. P. Revel, M. Rabinovitch, and M. J. DeStefano. Reprinted from L. Weiss and R. O. Greep, *Histology*, McGraw-Hill Book Company, 1977.]

Fig. 1, ×4685; Fig. 2, ×5520; Fig. 3, ×4450; Fig. 4, ×420; Fig. 5, ×3045; Fig. 6, ×5630

CONNECTIVE TISSUE
Loose Areolar

Mast cells are typically situated in the vicinity of blood vessels, respiratory airways, and the peritoneal lining of the body cavity. They are capable of the synthesis and release of heparin, histamine, slow-reacting substance of anaphylaxis (SRS-A), and the eosinophilic chemotactic factor of anaphylaxis (ECF-A), which are stored within intracellular granules. Because of the action of these substances, mast cells are thought to play a role in the regulation of vascular permeability and airway diameter. Viewed from the mast cell surface (Fig. 1), the rounded form of underlying granules (**Gr**) can often be distinguished, and when the mast cell interior is exposed (Fig. 2), the distribution of intracellular granules becomes apparent. Corresponding transmission electron micrographs of an unstimulated mast cell and one in the process of granule release are shown in Figs. 3 and 4, respectively. The process of histamine release is thought to consist of two steps. In the first step, fusion of the plasma membrane with the granule membrane (**arrows** in Figs. 5 and 6) allows communication of the extracellular fluid with the granules, at which time a number of granules may be expelled. In the second step, extracellular cation exchange causes disruption of the protein-histamine complexes within the granule, allowing histamine to diffuse freely into the surrounding medium. In Figs. 4 to 6, granules that are in the process of being extruded (✱) from the mast cell surface are illustrated. Granules in contact with the extracellular fluid characteristically exhibit a less compact, diffuse matrix. Changes in the number of microvilli and microfolds (**Mi** in Figs. 3 to 5) on the cell surface also occur during the extrusion process when granule membrane is added to the surface. The stimulus for granule release under physiological conditions involves the binding of a specific antigen to the surface of sensitized mast cells (those with immunoglobulin E on their surface). [Fig. 3 from D. Lagunoff, "Contributions of electron microscopy to the study of mast cells," *J. Invest. Dermatol.* 58(5):296–311 (1972). Copyright © 1972 by the Williams and Wilkins Company. Reprinted from L. Weiss and R. O. Greep, *Histology*, McGraw-Hill Book Company, 1977. Figs. 4 and 6 courtesy of D. Lagunoff.]

Fig. 1, ×3205; Fig. 2, ×3130; Fig. 3, ×5520; Fig. 4, ×1070; Fig. 5, ×8455; Fig. 6, ×24,920

CONNECTIVE TISSUE
Dense Irregular, Dense Regular

Dense connective tissue, as its name implies, consists principally of fibers, with little intracellular ground substance present. Unlike loose areolar connective tissue, dense connective tissue has few types of cells. Fibroblasts, responsible for fiber synthesis, are the most commonly encountered cells. The fibers are primarily collagen, and their pattern of arrangement distinguishes two major classes, one in which the collagen fibers form a dense, irregular pattern, and one in which the fibers form a dense, regular pattern. Within dense, irregular connective tissue (Fig. 1), bundles of collagen fibers (**CF**) are extensively interwoven, and provide resistance to stress exerted in many directions. Dense, irregular connective tissue is most commonly found in the dermis of skin, deep fascia of muscles, and dura matter of the brain; it also composes the periosteum of bone and the perichondrium of cartilage. When stress is continually exerted in one direction, the collagen fibers become organized into a definite pattern, producing dense, regular connective tissue. This is particularly evident in ligaments and tendons (Fig. 2), in which all the collagen fibers (**CF**) are parallel and oriented in one direction. Because the fibers are so densely packed, the fibroblasts (**Fb**) from which they were synthesized become flattened and packed between adjacent fiber bundles. Details of the parallel arrangement of collagen fibers are illustrated at higher magnification in Fig. 3.

Fig. 1, ×1840; Fig. 2, ×690; Fig. 3, ×2890

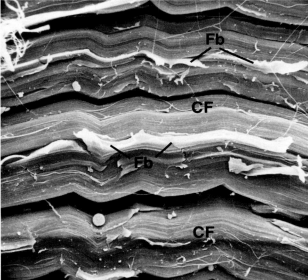

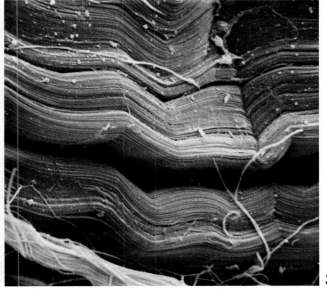

1

2

3

CONNECTIVE TISSUE
Adipose

It is not uncommon to find small groups of fat-storing cells within the ground substance of loose areolar connective tissue. In some areas of the body, however, the concentration and predominance of this cell type become great enough to warrant classifying a separate type of connective tissue, termed adipose tissue. White adipose tissue (Figs. 1 and 2) consists of large aggregations of fat-storing cells, referred to as adipocytes (**Ad**), in which septae of connective tissue fibers (**Fi**) surround and organize groups of adipocytes into lobules (**Lo**). White adipose tissue is commonly termed "unilocular" because each adipocyte contains a single large vacuole of accumulated lipid. Aggregations of adipocytes are often found within many organs. Adipose tissue commonly surrounds the kidneys and adrenal glands, and may also be associated with the mesentery, bone marrow, and omentum. Another type of adipose tissue, termed brown adipose tissue, is commonly seen in human infants and in hibernating species. Brown, or "multilocular," adipose tissue differs from white adipose tissue in that lipid droplets inside the cell do not coalesce, but instead form multiple lipid vacuoles. The brown color is due to an abundant vasculature and to the presence of intracellular lysosomes containing brown pigment. The mitochondria of brown adipocytes are larger, more numerous, and possess a complex fine structure. Since the mitochondria within brown fat are not capable of oxidative phosphorylation, they are thought to function in generating the necessary heat required of some species during hibernation. Some investigators believe brown adipose tissue is capable of transforming into white adipose tissue under certain conditions.

Fig. 1, ×55; Fig. 2, ×585

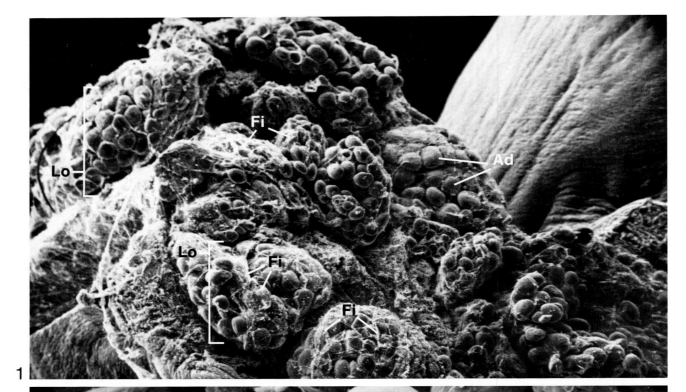

1

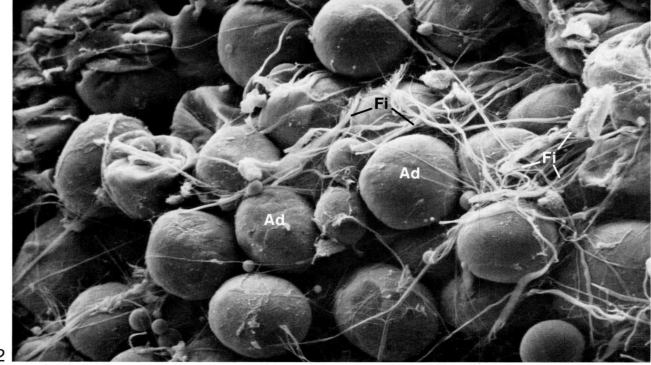

2

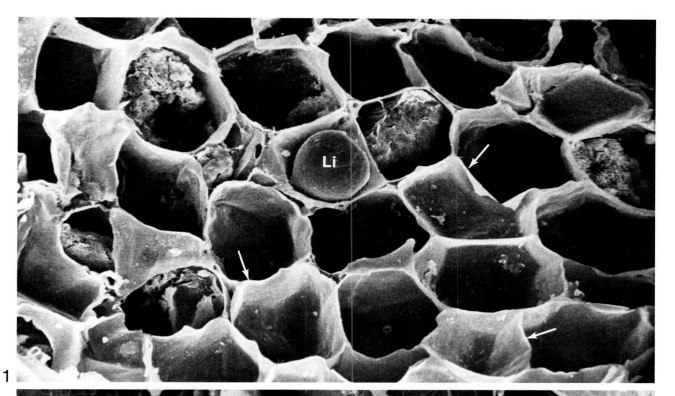

1

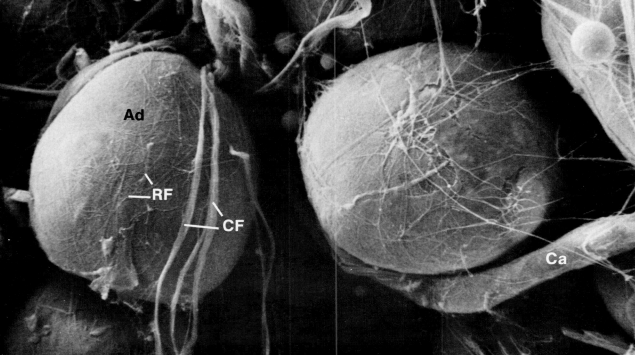

2

CONNECTIVE TISSUE
Adipose

A transverse section through adipose tissue is shown in Fig. 1. During tissue preparation much of the lipid within each adipocyte is removed; what remains is a honeycomb structure formed by the shells (**arrows**) of cytoplasm and plasma membrane. In the living state, almost the entire volume of each adipocyte is occupied by a single lipid droplet (**Li**) that consists primarily of triglycerides. Consequently, the cell nucleus is flattened and pushed to the side, occupying only one-fortieth of the cell volume. The surfaces of the two adipocytes (**Ad**) illustrated in Fig. 2 are supported by strands of collagen fibers (**CF**), and a fine network of reticular fibers (**RF**) is closely applied to the surface of each cell. Capillaries (**Ca**) course between neighboring adipocytes and are also in close association with the cell surface. Adipocytes vary in size, depending on the amount of lipid stored within them. It is important to realize that adipocytes are not merely storage centers for triglycerides; they are active metabolic sites wherein a large turnover of free fatty acids occurs. Free fatty acids are stored as triglycerides, the formation or breakdown of which is controlled hormonally. Transport of free fatty acids and substrates entering or exiting the adipocyte is also influenced by such hormones as epinephrine, norepinephrine, insulin, and glucagon. Therefore, a complex relation exists between the amount of available substrates, the presence of hormones, and the demand for energy.

Fig. 1, ×500; Fig. 2, ×1565

SKELETAL TISSUE

Introduction

Like other types of connective tissue, cartilage and bone consist of cells (chondrocytes and osteocytes, respectively) that are embedded in an intercellular matrix they have synthesized and secreted. It is the properties of this specialized matrix that make cartilage and bone ideally suited for support. The cartilage matrix consists of connective tissue fibers (collagen and/or elastic) in chemical association with proteoglycans, the combination of which produces the characteristic hardness and flexibility. The proteoglycans consist mainly of acid mucopolysaccharide molecules; chondroitin sulfate and keratin sulfate are the most common. Being avascular, cartilage relies upon the high permeability of its matrix to enable nutrients and gases to diffuse into the cells. With regard to the fibrous component, collagen fibers predominate in tissues subject to stress and strain, rendering them somewhat inflexible. But elastic fibers predominate in cartilage, where greater flexibility is required. The prevalence of one fiber type over another in the intercellular matrix provides the basis for classifying cartilage into three types: hyaline, elastic, and fibrocartilage.

Hyaline cartilage, so named because of its homogeneous, translucent matrix, is found chiefly in the walls of conducting airways (trachea and bronchi), and makes up the epiphyseal plate of long bones. A specialized form of hyaline cartilage also caps the ends of long bones, where articulation between bones occurs. The hyaline cartilage of joint areas is referred to as articular cartilage, and possesses a smooth, slippery surface that, in association with synovial fluid, facilitates movement of the joint.

Elastic cartilage is predominantly found in the auricle of the ear, eustachian tube, epiglottis, and portions of the vocal apparatus in the larynx. Its structure is very similar to that of hyaline cartilage except that the matrix contains a large proportion of elastic fibers and is more densely populated with chondrocytes.

Fibrocartilage contains numerous collagen fibers within its matrix that add properties of durability and resistance to stress. Fibrocartilage is commonly located in intervertebral disks, the symphysis pubis, and in regions where tendons and ligaments (made up of collagen fibers) insert into the ends of long bones. Details concerning the growth and development of cartilage will be discussed and illustrated later in this chapter.

Bone, unlike cartilage, possesses a calcified matrix, making it the strongest and most successful supportive tissue of the body. Because its matrix is calcified, it is highly impermeable to nutrients and gases, which are able to diffuse through cartilage. Alternatively, specialized networks of canals that permeate the calcified matrix are formed during bone development. Pathways are provided through such canals, in which nutrients can diffuse from blood vessels to cells. In areas of compact bone deposition, where the matrix is dense and thick, the networks of canals develop in an orderly arrangement to form Haversian systems, or osteons.

The formation of bone can occur by two different mechanisms, referred to as intramembranous and endochondral bone formation, or ossification. When intramembranous bone is to form in the embryo (primarily in the flat bones of the skull), mesenchyme cells aggregate and proliferate in association with numerous capillaries present in the area. A number of the mesenchyme cells differentiate into fibroblasts that produce collagen fibers so as to give the area a "membranous" appearance. Other mesenchyme cells differentiate into osteoblasts that immediately begin the synthesis and deposition of bone matrix in areas designated as ossification centers. Soon a number of osteoblasts become embedded and entrapped within the matrix they have synthesized, while others on the periphery of the ossification center continue to proliferate and secrete matrix. As the ossification centers expand, they surround preexisting blood vessels and communicate to form spongy bone, consisting of a scaffolding of bony trabeculae. A second type of intramembranous bone formation occurs concurrently with the development of trabeculae. In this type, mesenchymal cells have the capacity to form the cellular-fibrous sheath around bone, called the periosteum. Some of the mesenchymal cells differentiate into fibroblasts that produce collagen fibers, which make up the outer layer of the periosteum. Other mesenchymal cells differentiate into osteogenic cells, or osteoprogenitor cells, located in the inner layer. Some of these cells proliferate (thus maintaining their population), while others synthesize and deposit bone matrix in a lamellar form (compact bone). This type of bone deposition increases the width of bone by the addition of successive layers, or lamellae, a process known as appositional growth. Later in intramembranous bone formation, the continued deposition of lamellar bone on the surface of cancellous or spongy bone acts to fill in the spaces, thus converting much of it into compact bone.

In contrast to intramembranous bone formation, endochondral ossification is associated with a pre-existing hyaline cartilage model upon which a bony framework is built. The cartilage model is required during early development because only cartilage can accommodate the rapid growth rate of the fetus. Later, as the growth rate declines and greater structural support is required, the replace-

ment of cartilage by bone is favored. At this time, environmental factors cause the hyaline cartilage matrix to calcify, thus interfering with the necessary diffusion of nutrients and gases. Lack of adequate nutrition causes the demise of chondrocytes, which is followed by a gradual dissolution of the matrix. Concurrently, blood vessels grow into the area of calcified cartilage and carry with them osteoprogenitor cells. The osteoprogenitor cells then differentiate into osteoblasts, which begin to synthesize and deposit bone matrix on the degenerating cartilage matrix. Finally, the calcified cartilage completely disappears, leaving the newly formed bone matrix in its place.

During endochondral ossification, it is also possible for the perichondrium surrounding the cartilage model to transform into a periosteum. In such a process, the growth of blood vessels into the perichondrium exerts a profound effect upon the inner layer of chondrogenic cells. It is thought that the rise in oxygen tension, resulting from an increased vascularization of the area, stimulates the transformation of chondrogenic cells into osteogenic cells. In this way, a local conversion of the perichondrium into a periosteum takes place. Once established, the periosteum can deposit successive layers of bone matrix to increase the width of the shaft by appositional growth (actually an example of intramembranous ossification).

During adolescence, long bones increase in length by additional endochondral ossification at each end. Within each epiphyseal plate, proliferating rows of cartilage cells provide an expanding model for further bone deposition. Therefore, lengthening of long bones occurs at the epiphysis by endochondral ossification, whereas growth in width (appositional growth) results from intramembranous ossification at the periosteal region.

It is important to realize that even in the bone of adults, a continuous process of remodeling takes place. Bone is resorbed by specialized cells called osteoclasts, which demineralize bone and digest the organic components as well. At the same time, new bone deposition is also in progress at different sites. Much of the control over these remodeling mechanisms appears to be hormonal in origin. Parathyroid hormone stimulates bone resorption, thus liberating calcium into the bloodstream. Conversely, calcitonin (synthesized in the parafollicular cells of the thyroid gland) acts to stimulate bone deposition, lowering the calcium concentration in the blood. Many morphological features that characterize the extent of bone deposition and resorption can be observed in scanning electron micrographs of anorganic bone preparations (tissue from which the organic material has been removed) and organic bone preparations in this chapter.

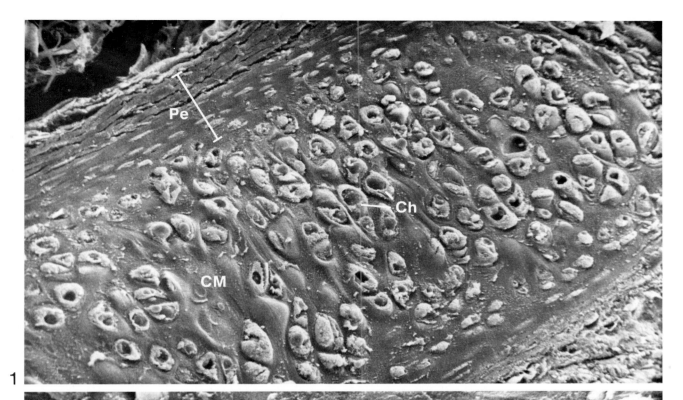

1

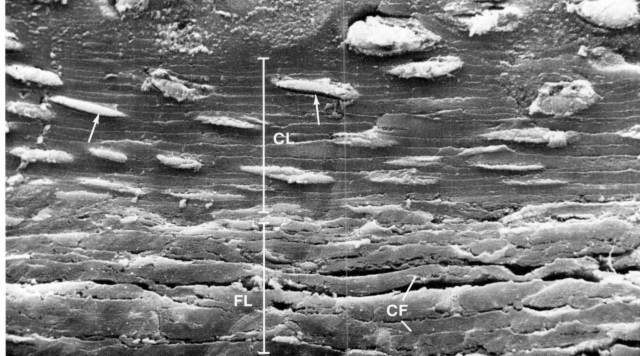

2

SKELETAL TISSUE
Hyaline Cartilage

The overall organization of a segment of hyaline cartilage from the trachea is illustrated in Fig. 1. Numerous cartilage cells, or chondrocytes (**Ch**), are embedded in a cartilage matrix (**CM**) that they have synthesized. The matrix consists of glycosaminoglycans (primarily sulfated acid mucopolysaccharides such as chondroitin sulfate), which are chemically bound to fibrils of either collagen (hyaline and fibrocartilage) or elastin (elastic cartilage). The entire outer surface, with the exception of fibrocartilage and articular cartilage, is surrounded by a layer of dense, irregular connective tissue called the perichondrium (**Pe**). The perichondrium, as shown in greater detail in Fig. 2, is organized into an outer fibrous layer (**FL**) that consists primarily of collagenic fibers (**CF**) and an inner, more cellular chondrogenic layer (**CL**). The cells (**arrows**) in the chondrogenic layer are capable of differentiation into chondroblasts, which can then synthesize and secrete new cartilage matrix. In this manner, new layers of cells and matrix are added to the cartilage surface by a process known as appositional growth. Since cartilage is avascular, its nutrition is derived by diffusion from the capillaries external to cartilage. Thus nutrients must diffuse over considerable distances through the matrix to all the cells. Since most of the chondroblasts occupy an internal location (away from the capillary bed), their need for nutrients and gases greatly limits the width attained by cartilage. Because the oxygen tension within cartilage is low, chondrocytes satisfy a large amount of their energy requirement through anaerobic glycolysis.

Fig. 1, ×525; Fig. 2 ×1870

SKELETAL TISSUE
Hyaline Cartilage

In addition to appositional growth, cartilage may also grow by the division of pre-existing cartilage cells within the matrix. This process is known as interstitial growth. Interstitial growth occurs primarily in embryonic development and continues only as long as the matrix remains sufficiently pliable to allow for further growth and division of cells. Chondroblasts (**Ch**) and cartilage matrix (**CM**), located toward the center of a segment of hyaline cartilage, are shown in Figs. 1 and 2. Small groups of cells located in areas of cell nests (**CN**), represent the progeny that have descended by mitotic division (isogenous growth) of a single cell. In Fig. 2, a cell nest containing four chondroblasts is illustrated at high magnification. As the isogenous cells mature, they synthesize and secrete matrix, which, over a period of time, causes them to become dispersed from their former nest-like configuration. Once the cells have become surrounded by matrix and are dispersed, they are called chondrocytes. The morphological distinction between chondroblasts and chondrocytes is not always clear, and chondrocytes may continue to produce new matrix for a time. The cartilage matrix has been sectioned in Figs. 1 and 2 so as to expose the chondrocytes' intracellular contents (✷), a substantial volume of which is occupied by the nucleus and large lipid droplets. A transmission electron micrograph of a single chondroblast is illustrated in Fig. 3. Intracellular constituents, such as rough-surfaced endoplasmic reticulum (**RER**), mitochondria (**Mi**), glycogen (**Gl**), and the cell nucleus (**Nu**) are present in this section. Glycosaminoglycans and precursors of collagen accumulate within the cisternae of the endoplasmic reticulum as they are synthesized and are eventually secreted to form the extracellular cartilage matrix. The cartilage matrix has a very fine, granular appearance here. The chondroblast also possesses numerous cytoplasmic projections (**arrows**) at its surface; these may function to increase the cell surface area available for absorption. [Fig. 3 courtesy of Jerry Maynard.]

Fig. 1, ×1000; Fig. 2, ×2400; Fig. 3, ×6675

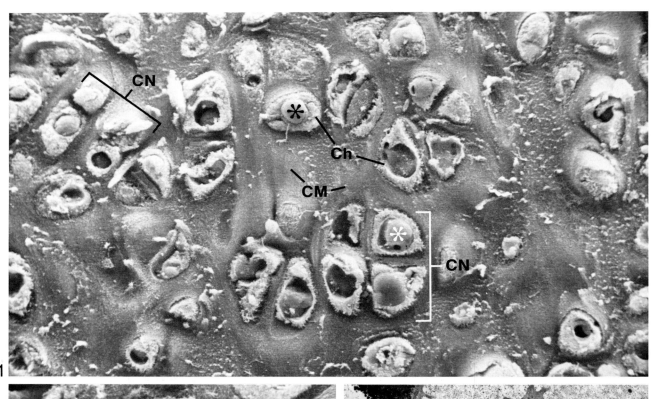

1

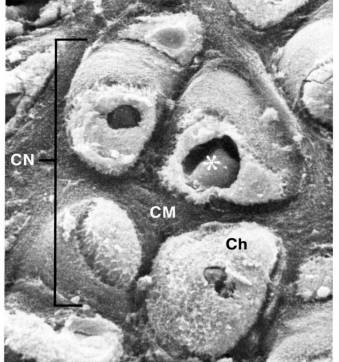

2

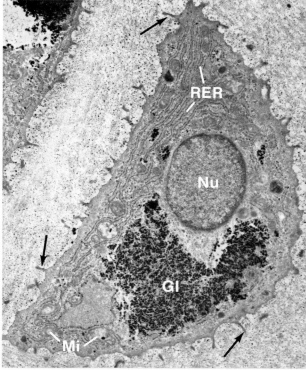

3

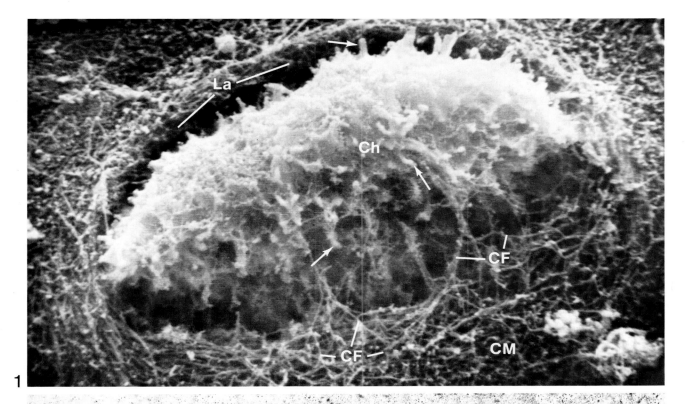

1

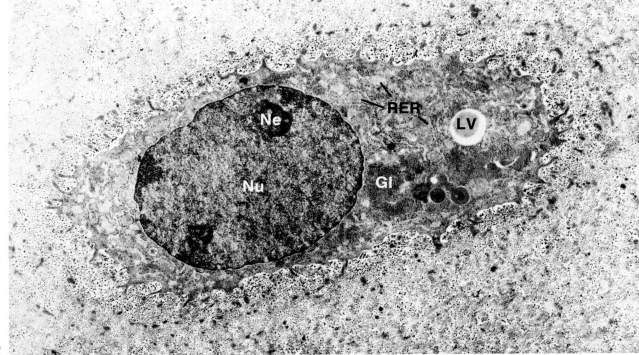

2

SKELETAL TISSUE
Hyaline Cartilage

In Fig. 1, an individual chondrocyte (**Ch**) is illustrated; a portion of the surrounding matrix (**CM**) was removed, thus allowing small projections (**arrows**) to be observed on the cell surface. Some separation between the chondrocyte and the surrounding matrix has occurred during specimen preparation, so that the lacuna (**La**), a space occupied by the chondrocyte, is apparent. Also visible is the network of delicate collagenic fibrils (**CF**) that make up a portion of the matrix. The polymerization of the fibrils is often incomplete, a condition due to the presence of a high concentration of acid mucopolysaccharide, which is thought to interfere with the polymerization process. Consequently, the collagen fibrils in cartilage matrix do not usually exhibit a periodic banding. A corresponding transmission electron micrograph of a chondrocyte is illustrated in Fig. 2. Newly formed elements of the cartilage matrix have been secreted into the extracellular space surrounding the surface projections of the chondrocyte. The cell nucleus (**Nu**), nucleolus (**Ne**), rough-surfaced endoplasmic reticulum (**RER**), lipid vacuole (**LV**), and glycogen (**Gl**) are identified. Collagen fibrils are observed within the surrounding cartilage matrix; the electron-dense particles interspersed between the fibrils are believed to be glycosaminoglycans, which are particularly concentrated around the free surface of the chondrocyte. [Fig. 2 courtesy of Jerry Maynard.]

Fig. 1, ×13,015; Fig. 2, ×15,575

SKELETAL TISSUE
Bone

Gross anatomical features characteristic of long bones in the later stages of endochondral ossification are diagrammed in Fig. 1. A typical long bone may be divided into a shaft, or diaphysis (**Di**), an intervening metaphysis (**Me**) region, and an epiphysis (**Ep**) at each end. An epiphyseal plate (**EP**), consisting of proliferating cartilage, is still present at the lower epiphysis. It is here where lengthening of the adolescent long bone takes place. Branches of the nutrient artery (**NA**) supply the bone at both the diaphyseal and epiphyseal regions. Fig. 2 shows a similar region of the rat femur in which the organic constituents were removed, thus allowing the calcified bone matrix to be independently observed. In this longitudinal plane of section, two types of bone can be distinguished; compact bone (**CB**) forms the outer portion of the diaphysis, and spongy, or cancellous, bone (**CaB**) fills the interior. Compact bone consists of a dense matrix resembling a solid mass. Small tunnels (Haversian and Volkmann canals) pass within the compact region, serving as conduits for blood vessels, nerves, and lymphatics in the living condition. Details of the canals and their surrounding cellular organization will be presented in subsequent figures. The spongy, or cancellous, region consists of a network of bony spicules, called trabeculae (**Tr**), which are thin toward the center of the diaphysis and progressively increase in thickness near the epiphysis (**Ep**). The region containing the thickest trabeculae is the metaphysis (**Me**). Cavities formed between the trabeculae are occupied by bone marrow in living tissue. [Fig. 1 from W. Bloom and D. W. Fawcett, *A Textbook of Histology* (10th ed.), W. B. Saunders Company, 1975.]

Fig. 2, ×40

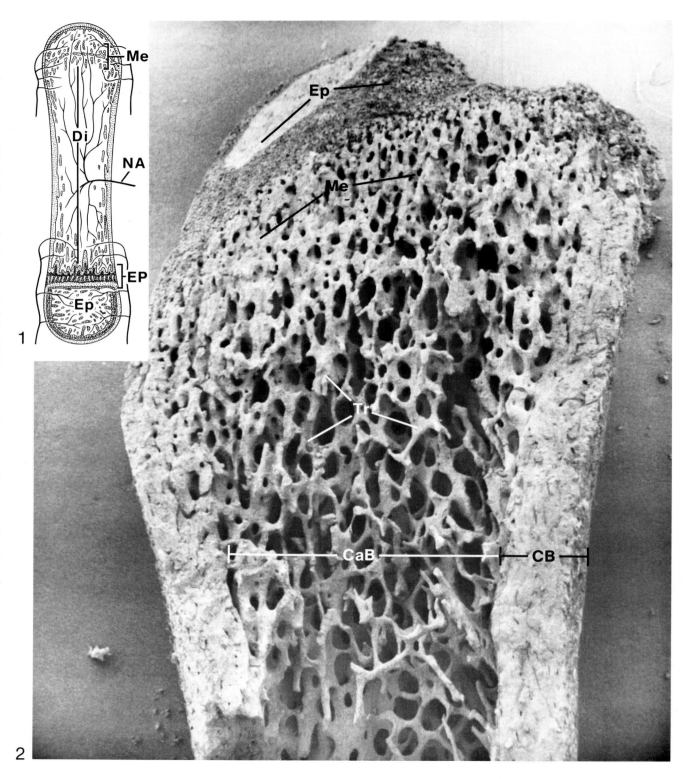

Direction of fibrils in successive lamellae of Haversian system

Internal circumferential system

Haversian systems

Direction of fibrils in successive lamellae of external circumferential system

Interstitial system

Sharpey's fibers

Blood vessels

Bony trabeculae

Haversian canal

Volkmann canal

Endosteum (thin layer of osteogenic cells and reticular fibers)

Osteocyte in lacuna

Periosteum { Fibrous layer / Osteogenic layer

Haversian system: transverse and longitudinal section

SKELETAL TISSUE
Bone

A segment of compact bone from the shaft of a long bone is diagrammed. The external surface of compact bone is continuous with a specialized connective tissue bilayer, termed periosteum. The outer layer is primarily fibrous, whereas the inner layer is more cellular and contains osteoprogenitor cells capable of differentiating into osteoblasts. Osteoblasts can then deposit bone matrix to increase the width of the shaft (appositional growth). The interior surface of compact bone is lined by endosteum, which consists of a monolayer of osteoprogenitor cells and a thin layer of reticular fibers. The openings on the endosteal surface represent the continuations of Volkmann canals, through which pass the blood vessels that supply the bone and marrow. Communicating with Volkmann canals are Haversian canals, which travel in a direction parallel to the shaft. Haversian canals contain blood vessels that supply nutrients to the majority of compact bone. During bone formation osteoblasts lay down a bone matrix that consists of collagen fibers and ground substance and which is subsequently mineralized by the formation of calcium phosphate crystals. The cells deposit this matrix in a lamellar, or sheet-like, form. Most lamellae are concentrically arranged (4 to 20 layers) around pre-existing blood vessels to form, collectively, the Haversian systems, or osteons. In contrast, those lamellae adjacent to the external and internal surfaces of compact bone follow the curvature of the shaft (external and internal circumferential systems). In the main part of the diagram, one of the Haversian systems has been drawn to show how collagen fibers within each concentric lamella are oriented in parallel and follow a helical course. The insert shows a Haversian system in section to illustrate the relationship between the osteocytes, Haversian canal, and vascular supply. [Redrawn from W. Bloom and D. W. Fawcett, *A Textbook of Histology* (10th ed.), W. B. Saunders Company, 1975; L. C. Junqueira, J. Carneiro, and A. Contopoulos, *Basic Histology,* Lange Medical Publications, 1975; and A. Benninghoff, *Lehrbuch der Anatomie des Menschen* (Vol. 17), J. F. Bergmann-Verlag, 1944 (Springer-Verlag).]

SKELETAL TISSUE
Bone

A ground, polished specimen of compact bone, whose cells and other organic constituents have been removed, is useful for studying the organization of Haversian systems. A number of Haversian canals (**HC**), formerly containing blood vessels, nerves, and a thin endosteal lining, can be observed. In transverse section, the lamellar sheets of calcified bone matrix appear as concentric rings surrounding each Haversian canal. Interposed among the lamellae are eliptical depressions, or lacunae (**La**), which represent the former location of osteocytes. In the formation of Haversian systems, osteocytes deposit lamellae of bone matrix, in succession, from the periphery toward the center. As a result of this process, Haversian canals are formed around centrally located blood vessels. The diameter of Haversian canals may vary, depending upon the number of lamellae deposited or resorbed. This, in turn, reflects the age or maturity of each Haversian system. During growth, and even in bone of adults, there is a continuous process of resorption and rebuilding of pre-existing Haversian systems. The figure shows a remnant of a first-generation Haversian system (**broken lines**) that has been replaced by a second-generation Haversian system (✱). The concentric lamellae of the older, first-generation Haversian systems now form the interstitial lamellae (**IL**), located between the newer, second-generation Haversian systems. In a number of instances, the peripheral limits of some Haversian systems are sharply demarcated by deep grooves, known as cement lines (**CL**). Cement lines appear highly refractile when ground bone is viewed under the light microscope; and they form a boundary between areas differing significantly in their collagen fiber orientation.

×490

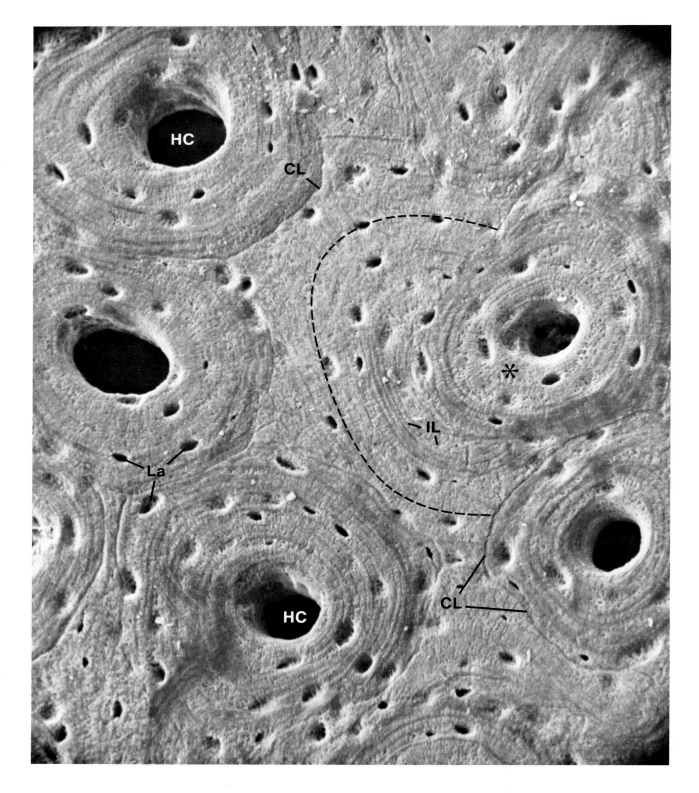

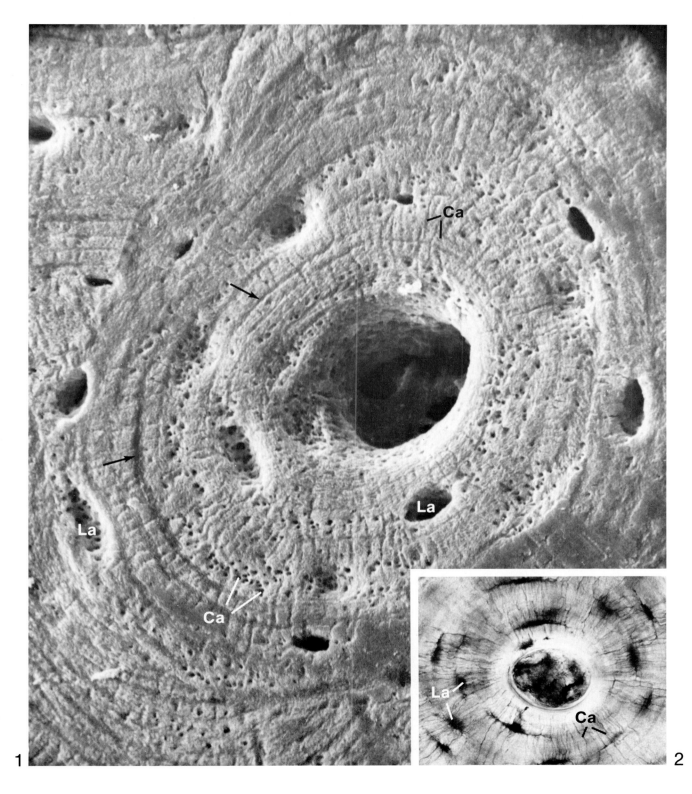

1

2

SKELETAL TISSUE
Bone

Fig. 1, an enlargement of an area from the previous figure, illustrates one Haversian system in greater detail. A light micrograph of a similar specimen of ground bone is shown in Fig. 2, and is useful for making comparisons with preparations viewed in three dimensions. In Fig. 1, the division between concentric lamellae is sharply demarcated by circular grooves (**arrows**) in the bone matrix. Lacunae (**La**), representing the spaces in which osteocytes reside in the living tissue, appear as oval depressions in Fig. 1 and as dark spider-like structures in the light micrograph. Osteocytes possess long, slender cytoplasmic processes that communicate with processes of neighboring cells to form a vast interconnecting network. During bone formation, the osteoblast (active form of osteocyte) deposits a matrix that surrounds its cell body and processes. Removal of the cellular component by acid treatment reveals thin channels called canaliculi (**Ca**), which radiate from each lacuna. Canaliculi, seen in both longitudinal and transverse section, represent the former location of osteocyte processes. The interconnection of neighboring lacunae forms an extensive network of channels that provides an effective pathway for the diffusion of nutrients and gases between Haversian canals and the cellular elements of bone.

Fig. 1, ×1345; Fig. 2, ×1110

SKELETAL TISSUE
Bone

In Fig. 1, the Haversian system is viewed at an angle that allows the internal surface of the Haversian canal (**HC**) to be observed. The numerous small openings on the canal surface are canaliculi (**Ca**). The continuity between the canaliculi and the Haversian canal provides a pathway through which exchange of nutrients and gases may take place. One of the lacunae (**La**) surrounding the Haversian system is seen in greater detail in Fig. 2. The small openings on the back wall of the lacuna mark the beginnings of canaliculi (**Ca**). The canalicular openings appear somewhat larger in the ground bone specimens because a small amount of the mineral component was removed during preparation to produce an etched appearance. For comparison, the lacuna observed in Fig. 3 was exposed by fracturing open the shaft, and the organic constituents were subsequently removed by acid treatment. The major difference is that the fractured type of preparation reveals what appears to be a pattern of collagen fibers, not observed in specimens of ground bone. Since the preparation has been made anorganic, what is actually observed is the pattern of calcium phosphate crystals laid down upon the fibers during mineralization. During mineralization of the bone matrix, it is thought that the highly ordered structure of collagen acts as a nucleation site for the formation and growth of crystals. The matrix surrounding the lacuna in Fig. 3 has the appearance of a series of short segments (**arrows**), suggesting that this is a mineralizing front that has not completely proceeded to the limits of the collagen fibers (Boyde, 1972). Consequently, removal of the organic component (collagen) leaves discontinuities between the short mineralized segments, where mineralization had not yet occurred. At the back wall of the lacuna, small canalicular (**Ca**) openings are observed. Figs. 4 and 5 illustrate the pattern of crystal formation on the inner surface of an Haversian canal (**HC**). Such a view provides information on the orientation of collagen fibers in the innermost lamella of the Haversian system.

Fig. 1, ×1040; Fig. 2, ×3740; Fig. 3, ×3725; Fig. 4, ×2670; Fig. 5, ×6320

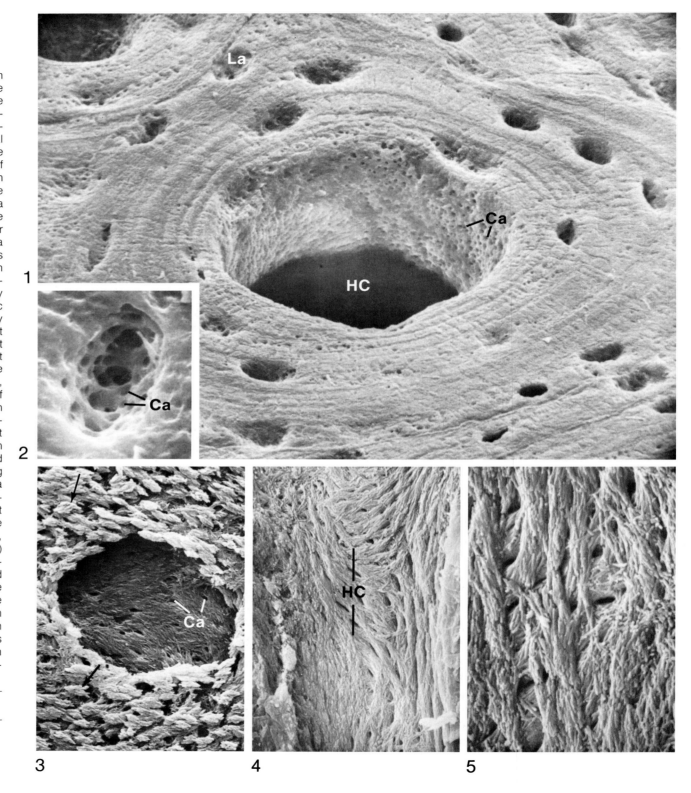

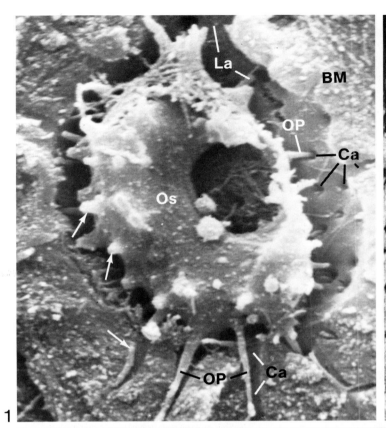

1

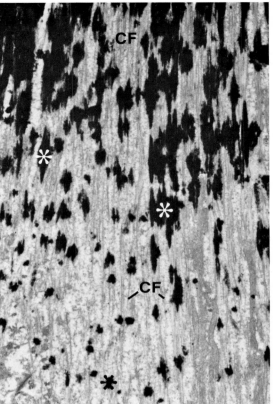

2

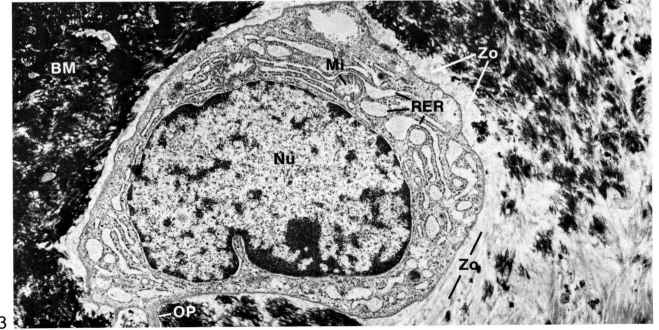

3

SKELETAL TISSUE
Bone

When osteoblasts become relatively inactive in bone formation, and surrounded by the products of their secretion, they are called osteocytes. The fractured surface of compact bone in Fig. 1 and the transmission electron micrograph of a corresponding area in Fig. 3 reveal the relationship of the osteocyte (**Os**) to the surrounding mineralized bone matrix (**BM**). The transmission electron micrograph in Fig. 2 is from an area of bone matrix that is in the process of becoming calcified. The areas of mineralization (✳) appear black and occur at foci along the surface of collagen fibers (**CF**). In Fig. 3, mineralization has occurred at the left margin of the osteocyte, but at the right margin mineralization is still in the early stages. Osteocytes are surrounded by fully mineralized bone matrix, with the exception of a zone 1–2 μm wide adjacent to the osteocyte plasma membrane. This zone (**Zo** in Fig. 3), bordering the edge of the osteocyte lacuna (**La** in Fig. 1), is most often occupied by native collagen fibers, which are devoid of calcium-phosphate crystals. In Fig. 1, a number of long, slender osteocyte processes (**OP**) extend into the bone matrix through corresponding tunnels or canaliculi (**Ca**). Many of these processes were broken (**arrows**) during fracturing of the specimen. Osteocyte processes are commonly in contact (via junctional complexes) with those extending from neighboring cells to form an interconnecting network. In some cases, branching of osteocyte processes can occur. Nutrients and metabolites are thought to circulate within a narrow space surrounding each osteocyte process, facilitating exchange between the blood vessels and the cells of bone. The transmission electron micrograph of an osteocyte in Fig. 3 illustrates the cell nucleus (**Nu**), rough-surfaced endoplasmic reticulum (**RER**), and mitochondria (**Mi**). [Figs. 2 and 3 courtesy of Jerry Maynard.]

Fig. 1, ×9160; Fig. 2, ×6950; Fig. 3, ×8075

SKELETAL TISSUE
Bone

Figs. 1 and 2 are similar low-magnification views of the diaphyseal region of the rat femur. Both are anorganic preparations and illustrate the internal cancellous bone (**CaB**) organization and external compact bone (**CB**) of the shaft. Toward the center of the shaft, the network of trabeculae (**Tr**) that make up cancellous bone become sparsely distributed. As the bone matures, the bony trabeculae in this central region are the first to be resorbed, which allows for expansion of the marrow cavity. The preparation in Fig. 1 differs in that it contains a cast of the vascular system. Numerous replicated blood vessels (**BV**) enter the marrow cavity through the openings (**arrows**) of Volkmann canals, observed from the endosteal surface.

Fig. 1, ×20; Fig. 2, ×105

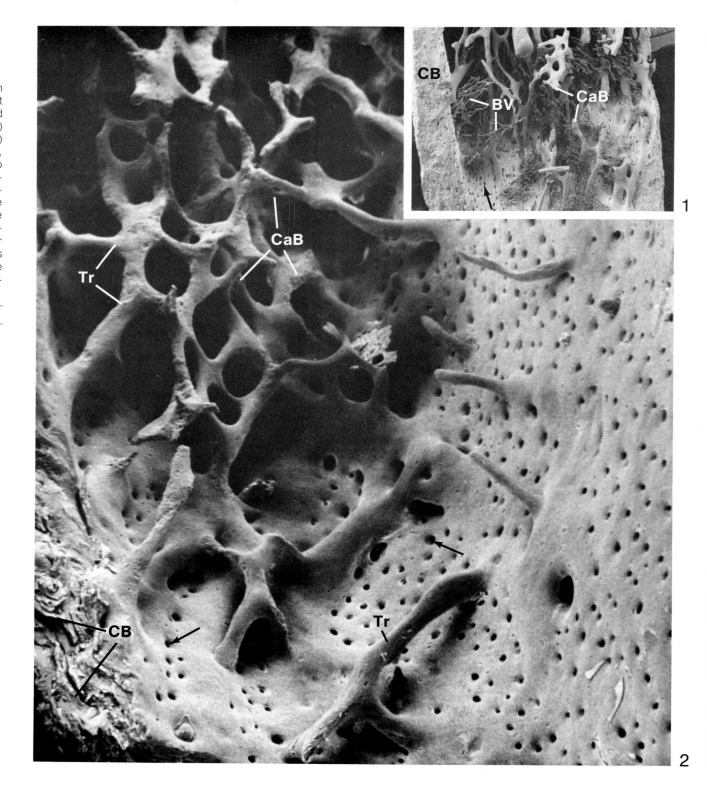

30

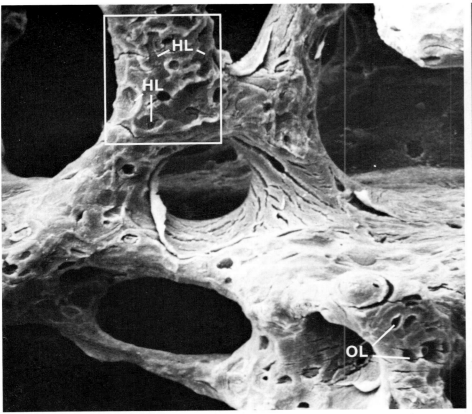

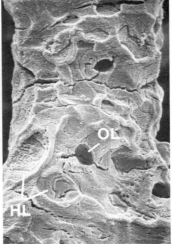

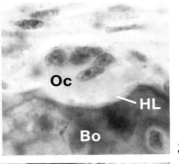

1

2

3

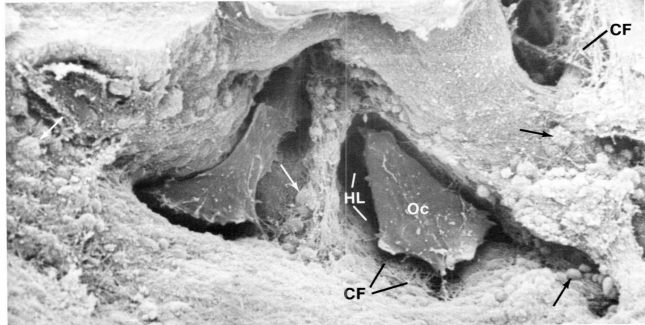

4

SKELETAL TISSUE
Bone

During growth, extensive remodeling of the trabeculae within cancellous bone may take place. This remodeling process involves the resorption of bone matrix from the trabecular surface by the action of special multinucleate cells called osteoclasts. Osteoclasts, which differentiate from osteoprogenitor cells or the fusion of osteoblasts, migrate over bony surfaces and secrete hydrolytic enzymes that take part in the solubilization and, hence, resorption of bone matrix. The anorganic preparation of cancellous bone shown in Fig. 1 illustrates the former position of osteoclasts on the trabecular surface. Such areas are characterized by surface depressions referred to as "resorption bays," or Howship's lacunae (**HL**). An enlargement of the resorption area enclosed by a rectangle in Fig. 1 is illustrated in Fig. 2. The considerable variation in the observed size and depth of Howship's lacunae (**HL**) is thought to represent loci of differing resorption activities. Howship's lacunae, which are deep and possess prominent edges, are characteristic of actively resorbing areas, in contrast with the shallow and ill-defined lacunae in areas where bone resorption is very slow or has ceased. The smaller, discrete holes on the trabecular surface in Figs. 1 and 2 represent osteocyte lacunae (**OL**). The light micrograph in Fig. 3 illustrates a multinucleate osteoclast (**Oc**) residing within a Howship's lacuna (**HL**) on the surface of bone (**Bo**). The relationship of two osteoclasts (**Oc**) to the resorbing surface of bone is shown in Fig. 4. On the surface of the two Howship's lacunae (**HL**), collagen fibers (**CF**) have been revealed by the solubilization and removal of inorganic calcium phosphate crystals by the osteoclasts. Many of the crystals are in the form of small spherules (**arrows**) because they are undergoing solubilization.

Fig. 1, ×445; Fig. 2, ×720;
Fig. 3, ×785; Fig. 4, ×5625

SKELETAL TISSUE
Bone

Shown in these figures are osteoclasts that were actively resorbing bone matrix on the endocranial aspect of young rat parietal and frontal bones. The two large osteoclasts (**OC**) in Fig. 1 both lie in Howship's lacunae and are surrounded by fimbrillated borders, the microprojections of which had been in contact with the bone surface. Such actively resorbing osteoclasts have a flat, free surface (possessing short microvilli) that is approximately flush with the surrounding bone matrix. The osteoclast in Fig. 2 has extensive branching microprojections in contact with the bone matrix. These peripheral attachment processes making up the fimbrillated border are present at the resorbing surface of osteoclasts and are hypothesized to be concerned with anchoring the resorptive apparatus to the bone matrix. The osteoclast in Fig. 3 was partly removed from the bone surface during preparation to reveal the characteristic ruffled border (**arrow**) on the undersurface of the cell. This area is further enlarged in Fig. 4. The ruffled surface exhibits microprojections, microfolds, and pits. [Figures courtesy of S. J. Jones and A. Boyde. From "Some morphological observations on osteoclasts," *Cell Tissue Res.* 185:387–397 (1977).]

Fig. 1, ×480; Figs. 2 and 3, ×1230; Fig. 4, ×5090

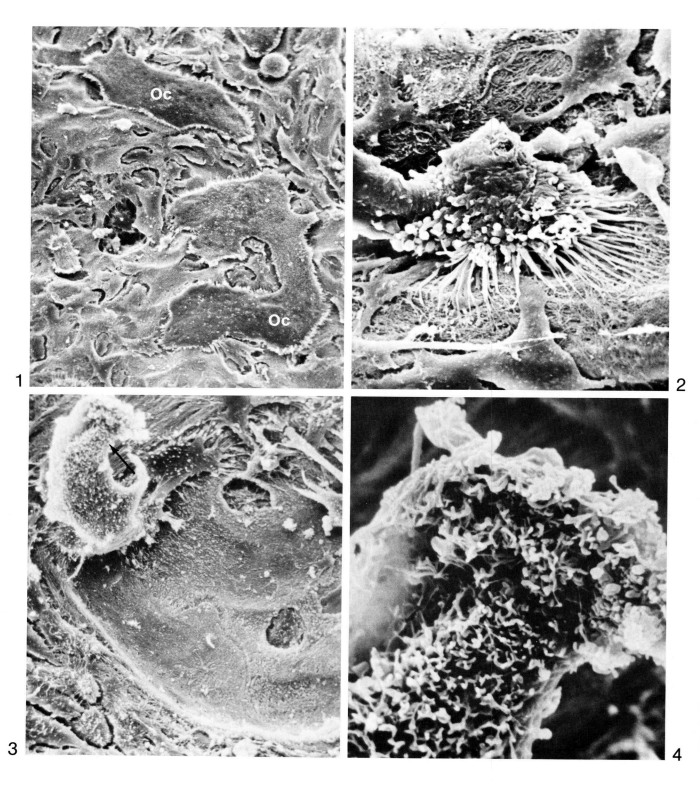

32

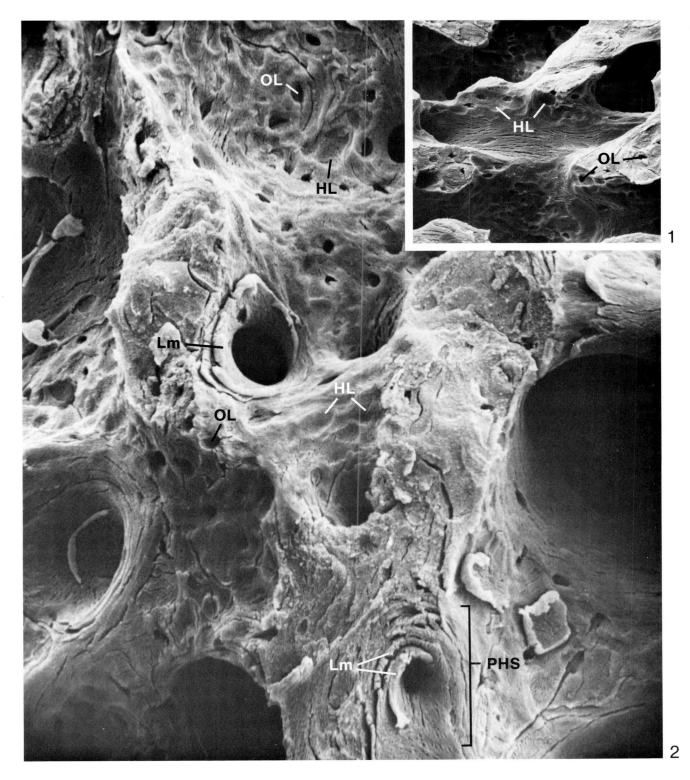

1

2

SKELETAL TISSUE
Bone

Numerous Howship's lacunae (**HL**) observed on the surfaces of the thick, bony trabeculae shown in Figs. 1 and 2 are indicative of an extensive resorption and remodeling process. The depth and area of each Howship's lacuna is highly variable and may reflect both the size of the corresponding osteoclast or the area over which a moving "wave" of resorption has taken place. Osteocyte lacuna (**OL**) are also observed from the surface. It is interesting to note that the perilacunar capsule of bone matrix often projects above surrounding areas of bone resorption to form a raised "lip" or border encircling the osteocyte lacuna. The unusually high concentration of acid mucopolysaccharide in this mineralized area is thought to bestow a resistance to resorption upon the matrix. The greater density of mineralization (close packing of calcium phosphate crystals) in this mucopolysaccharide region may act to reduce its permeability and, hence, solubility by the osteoclasts' digestive enzymes. As a result, the osteoclasts must digest around the lacuna in order to remove its contents from the surface. When greater strength and support is required of bone, much of the spongy or cancellous region is converted into compact bone by an extensive remodeling process. This process is accomplished by the deposition, progressively inward, of layers of bone matrix upon the surface of cavities within spongy bone. This deposition continues until the majority of space is filled with concentric lamellae of bone matrix, leaving only a narrow canal with a centrally located blood vessel. Fig. 2 is an example of spongy bone (anorganic preparation) in a late stage of conversion into compact bone. A number of cavities between the thick trabeculae have been filled with successive lamellae (**Lm**) of bone matrix to produce primary Haversian systems (**PHS**). Compare this to an earlier stage (Fig. 1), in which little deposition of lamellae has taken place.

Fig. 1, ×265; Fig. 2, ×585

SKELETAL TISSUE
Bone

Although much of the process of bone resorption has been attributed to the action of osteoclasts, it has also been recognized for some time that osteocytes are capable of bone resorption, and hence share responsibility for the homeostasis of calcium levels between blood and bone. The process of bone solubilization by osteocytes has been termed osteocytic osteolysis (Belanger, 1969). The process appears to be one in which bone matrix and salts lying deep within bone, away from the surface, are resorbed. Different aspects of osteocytic osteolysis, as observed in the petrosal bone of a mature guinea pig, are illustrated in Figs. 1 to 4. Osteocytic osteolysis often involves groups of mature, hypertrophied osteocytes (**Os**) that possess enlarged lacunae (**La**). In addition, these osteocytes frequently exhibit a ruffled border (**RB** in Fig. 2) at their free surface, similar to that described for the surface of active osteoclasts. The resorbing surface at the wall of the lacuna (Fig. 3) is rough, and reveals a network of collagen fibers (**CF**) like that observed previously on the surface of Howship's lacunae (Fig. 4, p. 31). Furthermore, many of the osteocytes are extensively surrounded by spherules (**Sp** in Figs. 1 and 4), which obscure their surface. It is most probable that the spherules represent a form of inorganic salts or organic matrix undergoing solubilization (compare with the spherules observed in Howship's lacunae). Cells involved in osteocytic osteolysis are also characterized by a low density of both inorganic salts and organic matrix in the area immediately surrounding their lacunae. It is thought that the primary event involves changes in the organic matrix (decrease in its density) that may be caused by an increase in the cell's secretion of acid mucopolysaccharide. The loss of salt most likely accompanies this modification of the organic substrate. Increases in alkaline phosphatase activity have also been observed in hypertrophied osteocytes. Factors that accelerate the maturation of osteocytes or increase their secretion are associated with osteolysis. These include pregnancy, pathological bone resorption, as well as increased levels of parathyroid hormone and vitamin D.

Fig. 1, ×3365; Fig. 2, ×5520;
Fig. 3, ×5965; Fig. 4, ×6015

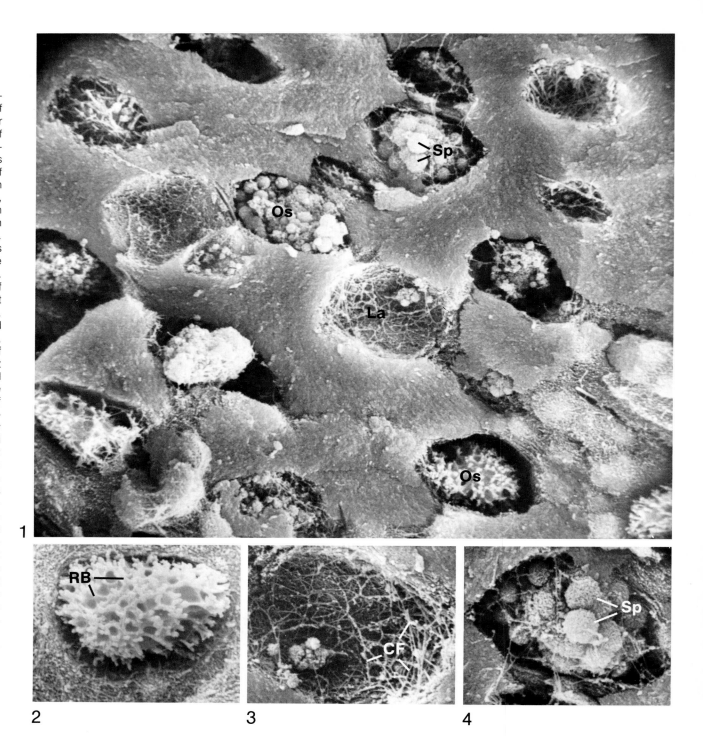

CIRCULATING BLOOD, BLOOD VESSELS, AND BONE MARROW

Introduction

Blood can be considered as a special type of connective tissue in which cells (erythrocytes and leukocytes) and parts of cells (platelets) are suspended in a fluid extracellular substance called plasma. Erythrocytes function in the transport or distribution of oxygen to the cells and tissues of the body. Many of the leukocytes are phagocytic and play a role in the defense of the body against infection. Leukocytes can traverse the walls of capillaries and venules and are frequently located in extracellular compartments containing tissue fluid. Plasma is an aqueous solution containing proteins, inorganic salts, and various organic compounds. Plasma proteins include albumins, complement, gamma globulins (antibodies or immunoglobulins), and fibrinogen (active in coagulation of blood). Although lipids are insoluble in water, they are transported by combining with the hydrophobic portions of albumin. Plasma serves to transport various metabolites from areas of absorption or synthesis to all cells of the organism that require them. The end products of metabolism, as well as various hormones, are also carried in the blood plasma. Carbon dioxide itself or its bicarbonate form can be transported in solution in plasma and in association with the hemoglobin of

erythrocytes. When blood is removed from the body, a clot forms, which contains the formed elements of blood and a clear yellowish liquid called serum. Serum resembles plasma in composition, but lacks several factors active in clotting (e.g., fibrinogen) and contains more serotonin.

Hydrostatic pressure enables water and crystalloids of plasma to pass through the endothelium on the arterial side of capillaries to form tissue fluid. Tissue fluid is resorbed at the venous end of capillary beds because of the colloidal osmotic pressure exerted by the proteins of the plasma. Tissue fluid can also be resorbed in terminal branches of lymphatic vessels. Interruption of the normal formation and absorption of tissue fluid results in swelling, or edema.

Under normal conditions, there are approximately 5000 leukocytes per μl of blood. They may be classified into two groups, granulocytes (neutrophils, eosinophils, basophils) and agranulocytes (lymphocytes and monocytes), on the basis of the presence or absence of specific staining cytoplasmic granules. Leukocytes may also be classified into polymorphonuclear or mononuclear cells on the basis of nuclear appearance. Further, they may be classified as myeloid (bone

marrow) or lymphoid (e.g., spleen, thymus, lymph node) depending upon the site of cell production. Granulocytes frequently possess irregularly shaped, lobed nuclei and cytoplasmic granules that stain neutrophilic, eosinophilic, or basophilic. Agranulocytes (lymphocytes and monocytes) possess a nucleus that is more regularly shaped, but sometimes indented. Nonspecific staining granules, called azurophilic granules, are sometimes present in the cytoplasm of agranulocytes. All leukocytes are capable of migrating between the endothelial cells of capillaries, and are commonly found in connective tissue and tissue spaces.

Neutrophils, the most numerous of the leukocytes, contain both azurophilic and neutrophilic staining granules. The azurophilic granules contain lysosomal enzymes and peroxidase. The smaller specific granules contain alkaline phosphatase and bactericidal substances. Neutrophils are actively phagocytic upon stimulation. Both types of cytoplasmic granules fuse with the membrane of phagocytosed material, such as bacteria, to destroy them. For example, the release of D-amino acids and lysozyme from azurophilic granules can destroy bacterial cell walls.

Eosinophils have a bilobed nucleus and acidophilic staining cytoplasmic granules containing acid phosphatase, cathepsin, and ribonuclease (lysosomes). Eosinophils are motile and can selectively phagocytose such materials as antigen-antibody complexes, which are then destroyed intracellularly by the eosinophilic granules (lysosomes).

The cytoplasmic granules of basophils contain heparin and histamine, as do the granules of mast cells. Basophils are more limited in their ability to move and to undergo phagocytosis. Although their function is not completely understood, it is thought that the cells liberate the contents of their granules in response to certain antigens.

Lymphocytes, which are the second most numerous leukocyte in the circulating blood, are classed as T lymphocytes (thymus dependent) and B lymphocytes (probably derived from solitary lymphatic nodules that populate other areas, such as the spleen). The role of the lymphocyte in cell-mediated and humoral antibody production is considered in chapter 5, which covers lymphatic tissue.

Monocytes typically have a lightly staining, horseshoe- or kidney-shaped nucleus, and frequently there are azurophilic granules (lysosomes) present. Monocytes arise in the bone marrow. After migrating across capillary walls, they transform in the surrounding connective tissue into phagocytic cells called macrophages.

CIRCULATING BLOOD, BLOOD VESSELS, AND BONE MARROW
Blood

This figure illustrates a number of biconcave erythrocytes (**BE**) and rounded leukocytes (**Le**) within an arteriole. Many of the blood cells are in contact with the inner surface of the tunica intima (**TI**) layer of the arteriole. In mammals, mature erythrocytes are biconcave disks, lack nuclei, and are about 7.5 μm in diameter. The biconcave shape is an adaptation for maximal surface area, over which the exchange of oxygen and carbon dioxide can take place. Erythrocytes, which are essentially nonmotile bags of hemoglobin, are very flexible and can accommodate themselves to small capillaries. There are roughly 5 million erythrocytes per microliter of blood. Although they are generally more numerous in men than in women, their life-span is about 120 days. At the end of their life-span, the macrophage system of the body removes the erythrocytes. Production of the many erythrocytes required each day to replace old ones occurs principally in bone marrow. Stem cells capable of differentiating into erythrocytes synthesize numerous polyribosomes, which function in the synthesis of hemoglobin. Near maturity, almost all of the organelles are extruded, including the nucleus, polyribosomes, mitochondria, and Golgi apparatus. A biconcave shape develops after extrusion of the nucleus. The plasma membrane of the erythrocyte is a semipermeable barrier that is responsible for maintaining differences in concentration of such ions as sodium and potassium between the inside and outside of the erythrocyte. Molecules in the erythrocyte membrane function in active transport of cations.

×6130

CIRCULATING BLOOD, BLOOD VESSELS, AND BONE MARROW
Blood

A thrombus, or clot, within a blood vessel of the eye is illustrated in Fig. 1. A meshwork of insoluble threads is a prominent feature of the thrombus. These threads are composed of a substance called fibrin (**Fi**), and in transmission electron micrographs the fibrin threads have an axial periodicity or banding pattern of about 250 Å. Fibrin threads are formed during clotting in a complicated, multistep process. Briefly, the process consists in the localized release of thromboplastin from the platelets and the injured vessel. Thromboplastin in the presence of calcium ions causes the conversion of prothrombin to thrombin. The presence of thrombin results in the transformation of a blood protein fibrinogen into fibrin threads. In addition to the network of fibrin threads shown in Fig. 1, many erythrocytes (**Er**) and several dendritic platelets (**DP**) are present. The biconcave shape of erythrocytes is energy dependent, and is thought to be due to characteristics of the internal stroma and the extrusion of sodium and water ions. Since mature erythrocytes have no mitochondria, they are dependent for energy upon anaerobic glycolysis and the pentose pathway for energy. The shape of the erythrocyte therefore varies with the osmotic pressure of the plasma. In hypotonic solutions, the erythrocyte undergoes hemolysis, in which there is swelling and possible eventual loss of hemoglobin. In hypertonic solutions, erythrocytes may become crenated. A number of crenated erythrocytes (**CE**) are located in the thrombus in Fig. 1, and a single crenated erythrocyte is shown at higher magnification in Fig. 2.

Fig. 1, ×1800; Fig. 2, ×11,480

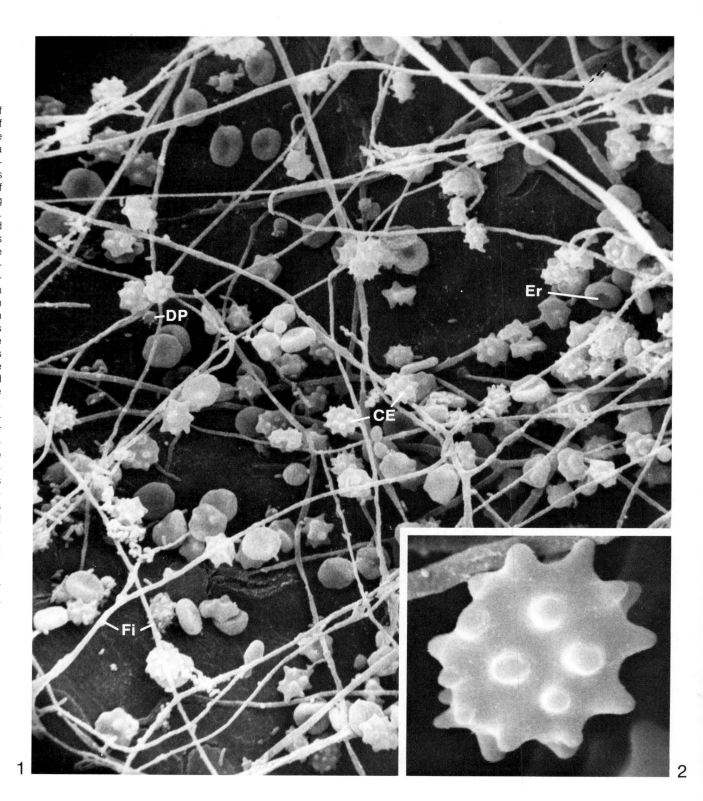

1

2

CIRCULATING BLOOD, BLOOD VESSELS, AND BONE MARROW
Blood

A portion of a blood clot is shown in Fig. 1; fibrin (**Fi**) threads form a meshwork surrounding aggregated platelets and blood cells. Erythrocytes sometimes tend to aggregate into columns—an arrangement that is termed rouleaux formation (**Ro** in Fig. 1) and is probably caused by surface tension. Erythrocytes (**Er**) can gain entrance into the circulatory system by squeezing through the intercellular spaces between the thin cells lining the sinusoids (**Si**) of the spleen (Fig. 2) and marrow. The process, which is termed diapedesis, is illustrated at the **arrows** in Figs. 2 and 3. In the circulating blood, platelets (**Pl**) are anucleate, biconvex disks (Fig. 4) approximately 2–5 μm in diameter. They are present only in mammals, have a life-span of about 6–8 days, and can phagocytose small particles, such as viruses and bacteria. In humans, there are approximately 300,000 platelets per microliter of blood. They are formed as cytoplasmic excrescences from giant cells in the bone marrow called megakaryocytes (illustrated in section on bone marrow). The outer surface of the platelet plasmalemma is coated with a material rich in acid mucopolysaccharides and glycoproteins that forms an adhesive layer. Structures found within platelets include microtubules and microfilaments, and they probably play a role in maintaining the discoid shape as well as in the development of the dendritic appearance after activation. In addition, there are mitochondria, electron-dense granules (containing serotonin), glycogen, and a system of tubules and vesicles. Another type of granule, called an alpha granule, contains lysosomal enzymes, acid mucopolysaccharides, and platelet thromboplastin. Both serotonin and epinephrine are contained within platelets and upon release appear to play a role in the vasoconstriction of blood vessels at the site of injury. When discontinuities occur in the wall of blood vessels, platelets tend to aggregate so as to assist in closing the rupture. [Figs. 1 and 4 courtesy of R. G. Kessel and C. Y. Shih, *Scanning Electron Microscopy in Biology,* Springer-Verlag, 1974.]

Fig. 1, ×4005; Fig. 2 ×8545; Fig. 3, ×12,850; Fig. 4, ×6130

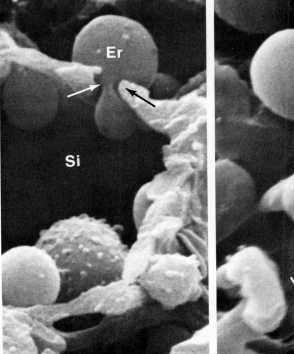
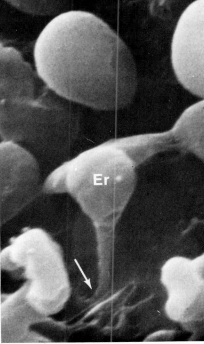

CIRCULATING BLOOD, BLOOD VESSELS, AND BONE MARROW
Blood

The leukocytes of the blood are classified primarily on the basis of their nuclear configuration and on the content and staining reaction of their cytoplasm. Since leukocytes are motile to varying degrees, their overall shape is highly variable. In order to identify leukocytes by scanning electron microscopy, it is necessary to stain a wet-mount preparation, which can then be viewed and photographed in the light microscope. Fig. 1 shows such a preparation, in which it is possible to identify an erythrocyte (**Er**), platelet (**Pl**), monocyte (**Mo**), lymphocytes (**Ly**) and neutrophils (**Ne**). The same preparation is then observed in the scanning electron microscope (Fig. 2), making it possible to observe the surface features of the same cells shown in Fig. 1. An erythrocyte (**Er**), activated dendritic platelet (**Pl**), monocyte (**Mo**), lymphocytes (**Ly**) and neutrophils (**Ne**) are all identified in Fig. 2. The monocyte, which is shown at higher magnification in Fig. 3, is covered by a number of surface folds and projections of variable size. The dendritic form of an activated platelet (**Pl**) is also illustrated in Fig. 3. The two lymphocytes illustrated in Fig. 4 have numerous microvilli projecting from their surfaces, the presence of which has been used as a criterion for distinguishing B lymphocytes from T lymphocytes; the latter have a smoother, less structurally specialized surface. The neutrophils shown in Fig. 5 are covered by a number of shallow surface folds. That the surface morphology of leukocytes may exhibit wide variation has been demonstrated by Wetzel et al. (1973, 1974). Leukocytes may vary in shape from rounded to flattened, depending upon whether they are prepared for scanning electron microscopy from culture or at different times after attachment to a substratum, such as a cover glass. The cells may also vary in the presence or absence and number of such surface specializations as microvilli, microplicae (surface folds), and ruffles (lamellipodia).

Figs. 1 and 2, ×2260; Figs. 3 to 5, ×7120

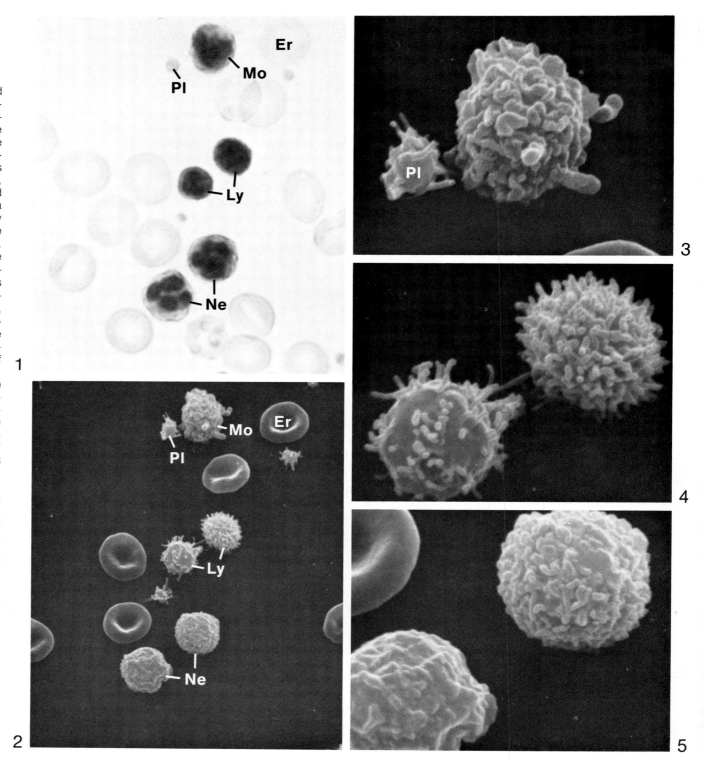

40

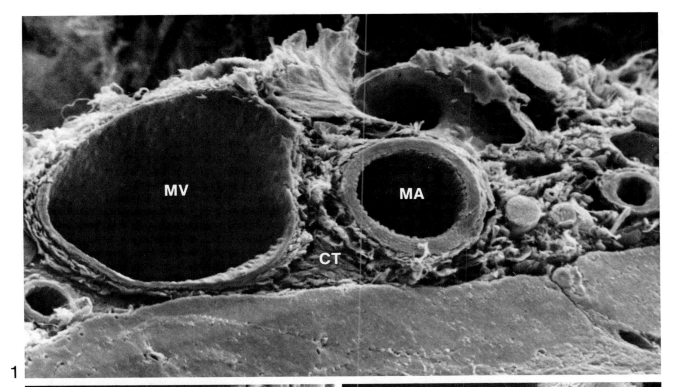

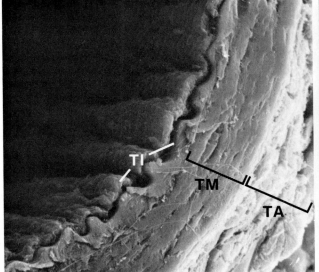

1

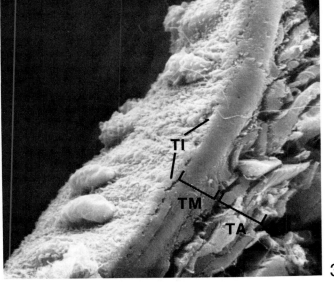

2

3

CIRCULATING BLOOD, BLOOD VESSELS, AND BONE MARROW
Blood Vessels

The blood vascular system consists of an extensive network of vessels that permeate the body. The vessels vary widely in size and composition, but include the following divisions: elastic arteries, muscular (or distributing) arteries, arterioles, capillaries, venules, medium-size veins, and large veins. The walls of the larger branches of both arterial and venous vessels are made up of three distinct layers. From innermost to outermost, these layers are the tunica intima, tunica media, and tunica adventitia. Arterial branches of the circulatory system frequently accompany corresponding divisions of the venous system. A muscular, or distributing, artery (**MA**) and a medium-sized vein (**MV**) are illustrated from the connective tissue (**CT**) surrounding the vas deferens in Fig. 1. Note that the vein has a thinner wall and a larger lumen compared to its companion artery. For comparison, the wall of a medium-sized artery is shown in Fig. 2, and the wall of its companion medium-sized vein is illustrated in Fig. 3. In both vessels, the corresponding width of the tunica intima (**TI**), tunica media (**TM**), and tunica adventitia (**TA**) are denoted. The major cell type found in the wall of blood vessels is unique because as it differentiates from mesenchyme cells, it retains broad functional capacities. This cell can become a smooth muscle cell, and has the capability of secreting elastin, collagen, reticular fibers, and intercellular matrix. The tunica intima of blood vessels consists of a continuous inner layer of simple squamous cells known as endothelium. Endothelium may, in turn, be surrounded by a thick elastic band called the internal elastic lamina. In some cases, a thin layer of subendothelial connective tissue is present between the endothelium and internal elastic lamina. In elastic arteries, such as the aorta, subclavian, common carotid, and pulmonary arteries, the tunica intima is wider and contains a number of circularly arranged elastic fibers.

Fig. 1, ×305; Fig. 2, ×2350; Fig. 3, ×2585

CIRCULATING BLOOD, BLOOD VESSELS, AND BONE MARROW
Blood Vessels

The tunica media of elastic arteries contains numerous, circularly arranged elastic fibers that permit these vessels to expand during systole and to recoil passively during diastole. In contrast, the tunica media of medium-sized or distributing arteries has an abundance of smooth muscle cells arranged in circular layers around the vessel wall. Some collagenic and elastic fibers and some intercellular substance may be present among the smooth muscle. The tunica adventitia of both types of arteries consists of connective tissue. Arterioles serve as pressure-reducing valves in the circulatory system. They can function in this way because they have relatively narrow lumens and thick muscular walls. As a result, blood reaches the capillary bed under greatly reduced pressure so as not to damage the delicate endothelial tubes. Blood pressure is regulated primarily by the degree of tonus in the smooth muscles of the arteriole wall. The figure on this page shows a transverse section through an arteriole. The arteriole lumen (**Lu**) is lined by a continuous layer of endothelial cells (**EC**) whose internal nuclei cause the cells to bulge into the lumen. This sheet of cells resides on a scalloped internal elastic lamina that completely surrounds the vessel. The lamina is actually composed of elastin. Both endothelium and internal elastic lamina constitute the tunica intima (**TI**), but the two layers are separated in this preparation. The tunica media (**TM**) varies in thickness, but usually consists of 1–3 layers of circularly arranged smooth muscle. A few elastic and collagenic fibers and some intercellular substance may also be present in the tunica media. The tunica adventitia (**TA**) consists primarily of collagenous and elastic fibers. This layer varies with the size of the arteriole and may be continuous with surrounding areolar connective tissue.

×3915

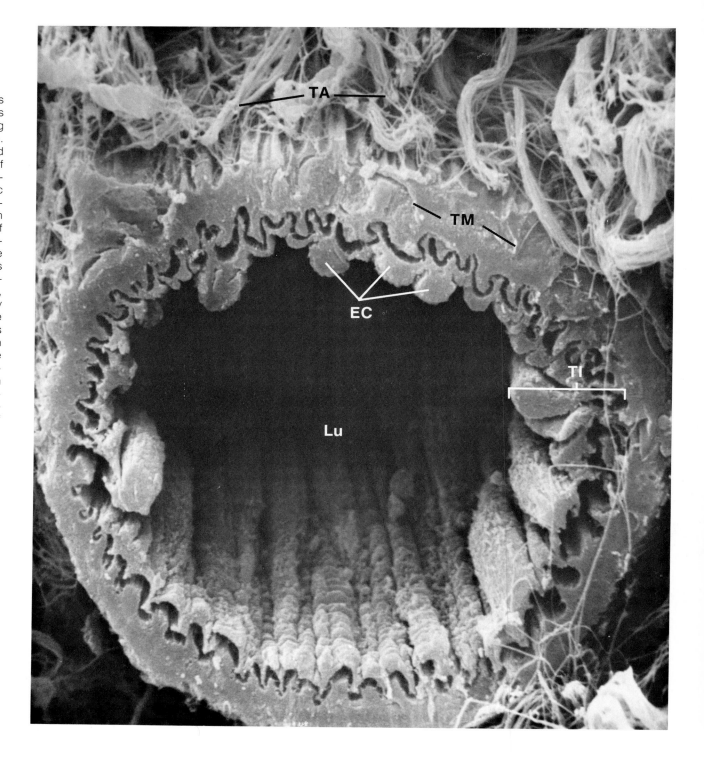

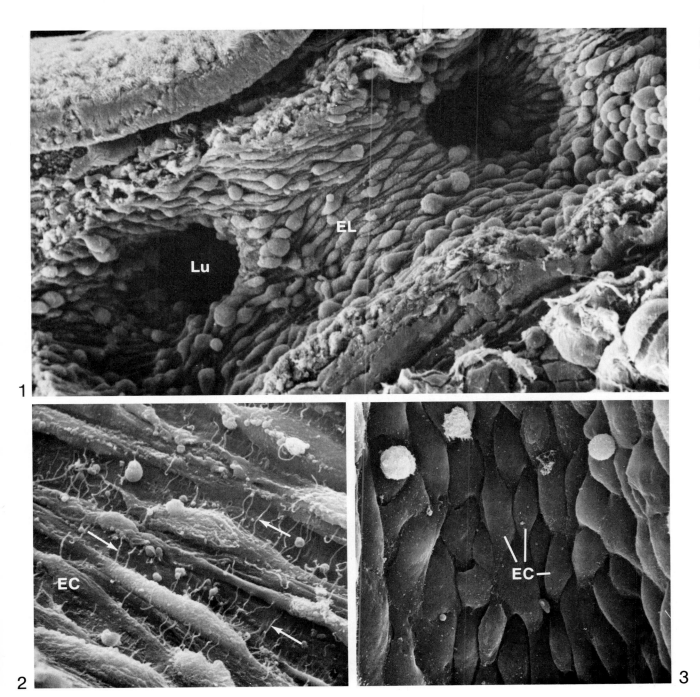

CIRCULATING BLOOD, BLOOD VESSELS, AND BONE MARROW
Blood Vessels

The entire vascular system is lined internally by a single layer of endothelial cells. Capillaries are made up exclusively of endothelial cells. Individual cells are oriented longitudinally with respect to the axis of the vessel. They are exceedingly thin, and their nuclei tend to cause the cells (**EL** in Fig. 1) to bulge into the lumen (**Lu** in Fig. 1). The variation in shape of the individual cells and the extent to which they bulge into the vessel lumen may depend on the amount of contraction of the smooth muscle in the blood vessel wall and on the amount of tension in the internal elastic lamina of arteries and arterioles. Many pathological conditions result from defects in the endothelial lining of blood vessels. Endothelial cells (**EC**) of the tunica intima of several different blood vessels are illustrated in Figs. 2 and 3. The endothelial cells illustrated in Fig. 3 are quite smooth, and the boundaries between adjacent endothelial cells are distinct. Although endothelial cells are remarkably smooth and devoid of surface specializations in some vessels (Fig. 3), this is not always the case. Note that slender cell extensions (**arrows** in Fig. 2) extend at right angles from the long axis of adjacent endothelial cells. Veins consist of the same three layers that make up arteries, but veins, in general, are more loosely constructed. Arteries are better supplied with elastic and muscular tissue, and thus the media is usually the thickest layer. In contrast, the adventitia layer of veins is usually the best developed. Further, collagen is more prevalent in the walls of veins, and elastin is more extensively distributed in arteries. Because of this variation, and in view of the fact that collagen fibers seem to cause platelet aggregation (thrombus formation), blood clots are especially prevalent in veins. Smooth muscle is present in the tunica media of medium-sized veins, especially the superficial veins of the limbs. In large veins, however, such as the vena cava, the smooth muscle is longitudinally oriented in the tunica adventitia.

Fig. 1, ×575; Fig. 2, ×2805; Fig. 3, ×1825

CIRCULATING BLOOD, BLOOD VESSELS, AND BONE MARROW
Blood Vessels

A longitudinal section of a capillary is illustrated in Fig. 1, a transmission electron micrograph. Capillaries range in diameter from 3–10 μm. The nucleus (**Nu**) causes the endothelial cell to bulge slightly into the lumen (**Lu**), which in this case contains two erythrocytes (**Er**). Junctions between endothelial cells include gap junctions, or nexuses, and desmosomes. Tight junctions are also found between capillary endothelial cells of other tissues, such as the central nervous system and thymic cortex. Rough-surfaced endoplasmic reticulum (**ER**) and mitochondria (**M**) are present in the endothelial cell cytoplasm. Protein, fluid, and metabolites may be incorporated into pits (**Pi**) on the luminal side of the cell and transported in membrane-bound pinocytotic vesicles (**PV**) through the cytoplasm to the opposite side of the cell, where it is released (**arrow**) into the perivascular space (**PS**) by membrane fusion. The capillary endothelium is invested with a thin basal lamina (**BL**), and thin wedges of pericyte cytoplasm are closely associated with the basal lamina of the endothelium. Pericytes (**Pe**), or pericapillary cells, are thin, elongate, and have a highly branched cytoplasm. Structurally the cells closely resemble fibroblasts, but contain numerous cytoplasmic filaments. The external surface of the pericyte is coated with basal lamina material (**PBL**). Although the function of pericytes is not completely understood, it has been suggested that the cells are phagocytic, capable of transformation into smooth muscle cells (as in the growth of new venules), and that they produce basement membrane material for endothelial cells. Capillaries have either a continuous endothelium, such as shown in Fig. 1, or a fenestrated endothelium as illustrated in Figs. 2 and 3. The fenestrated endothelium (**FE**) in the renal glomerular capillary contains circular discontinuities (**arrows** in Figs. 2 and 3) that are approximately 500–1000 Å in diameter. Similar fenestrated endothelium may be present in regions where rapid exchange occurs between blood and surrounding tissue (e.g., endocrine glands, intestine, choroid plexus, ciliary body).

Fig. 1, ×29,370; Fig. 2, ×15,485; Fig. 3, ×15,575

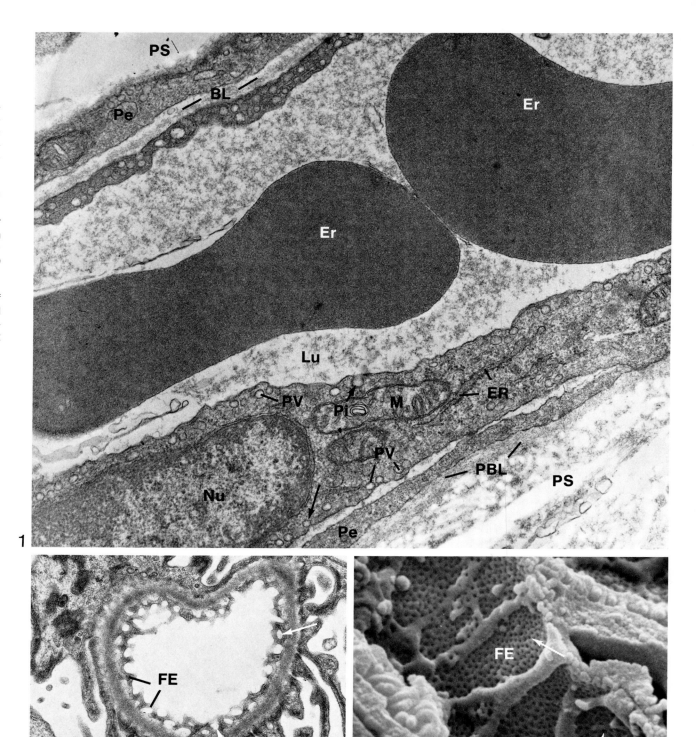

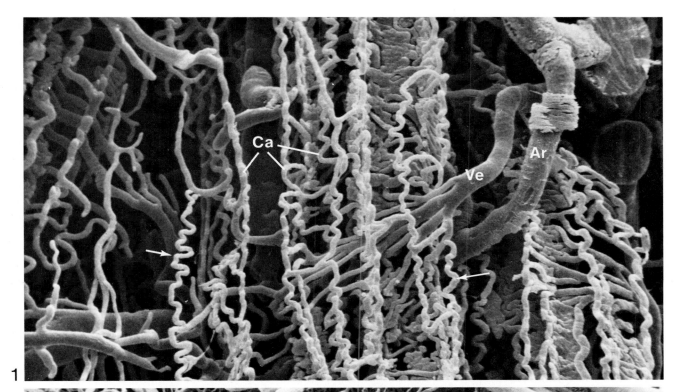

1

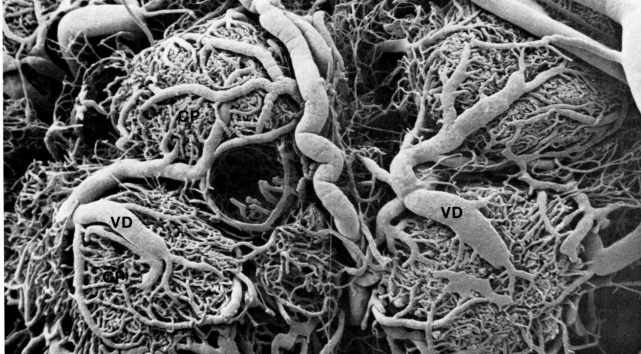

2

CIRCULATING BLOOD, BLOOD VESSELS, AND BONE MARROW
Blood Vessels—Vascular Casts

Many of the important features of the circulation, such as capillary density, variation in branching of arterioles and venules, three-dimensional distribution of the vessels within the tissue, and continuity between the vascular divisions, cannot be fully realized in tissue sections. Casts of the microvasculature of skeletal muscle (Fig. 1) and corpora lutea in the ovary (Fig. 2) reveal many of the above-mentioned features. These casts are prepared by injecting an unpolymerized mixture into the vascular system and allowing it to harden within the vessels. The tissue is subsequently digested using a potassium hydroxide solution, leaving only a replica of the microvasculature. This replica is then viewed with the scanning electron microscope, which permits the three-dimensional distribution of the microvasculature to be studied and photographed. As would be expected, the distribution of the vessels is highly variable, depending upon the organ or tissue studied; often the vessels follow the pattern of tissue organization. For example, in Fig. 1, the capillaries (**Ca**) of the skeletal muscle are oriented longitudinally in the same direction as the muscle fibers they supply. When the fiber shortens during contraction, the capillaries can conform to this change by coiling along their length (**arrows**). The capillary plexus is observed to be supplied by an arteriole (**Ar**) and drained by a venule (**Ve**). There are differences in the terminal branching of both divisions. In Fig. 2, the capillary plexus (**CP**) supplying each of the four corpora lutea is extremely dense and conforms to their spherical shape. Since the function of the corpora lutea depends upon the entrance into the circulation of the hormones they produce, it is not surprising that the vascular supply is well developed here. Each venous division (**VD**) is observed to drain more than one corpus luteum, and the branching of the major vascular divisions is extensive. From these examples, the great complexity in patterns formed by the microvasculature can be more fully appreciated.

Fig. 1, ×355; Fig. 2, ×90

CIRCULATING BLOOD, BLOOD VESSELS, AND BONE MARROW
Bone Marrow

Because most of the cells in the circulating blood have a relatively short lifespan, a constant renewal of these cell populations is required. Erythrocytes, granular leukocytes, and monocytes are produced primarily within marrow cavities of certain bones. T lymphocytes are produced in the thymus, whereas B lymphocytes are formed in lymph nodes, the spleen, tonsils, and in solitary lymph nodules present under moist epithelial membranes. Evidence that a common stem cell can give rise to most of the cell types found in the peripheral blood has become widely accepted. As the common stem cell divides, it gives rise to progeny that can differentiate along particular cell lines, such as erythrocytes or granular leukocytes. The process of blood-cell formation is called hemopoiesis, and bone marrow represents a type of hemopoietic tissue. Thus bone marrow, located within the marrow cavities of bones, consists of differentiating blood cells, reticular cells and reticular fibers, many endothelial-lined sinusoids, and fat cells, or adipocytes. The rat femur was fractured to expose the internal marrow shown here. The developing cells of the marrow (**Ma**) are seen to be closely packed. Networks of channels called venous sinuses (**VS**), or sinusoids, traverse the marrow. In the living state, blood with mature blood cells (**BC**) flows through the sinuses and is segregated from the surrounding immature blood cells of the marrow by endothelial cells of the sinusoidal wall.

×705

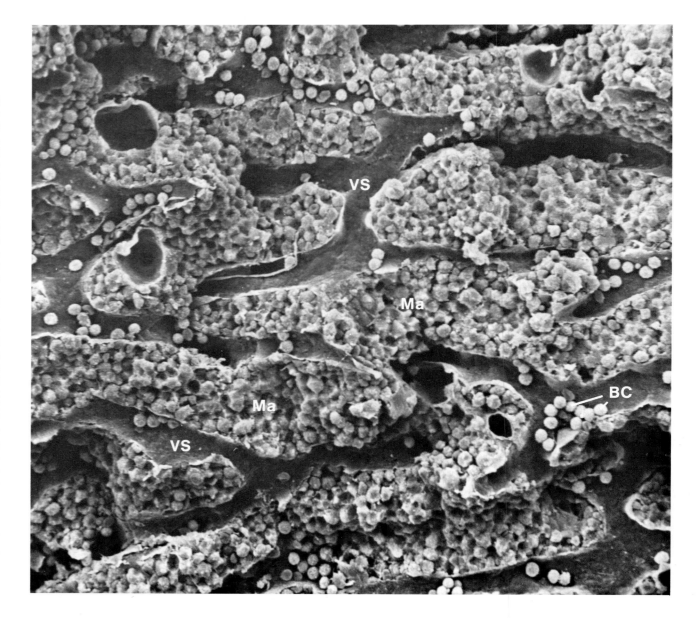

46

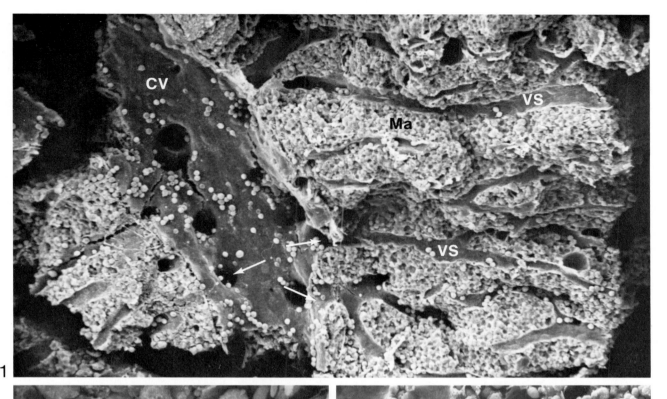

1

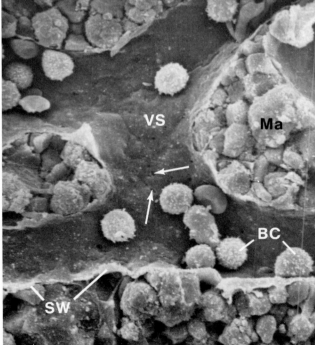

2

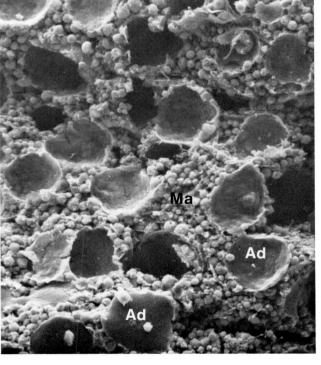

3

CIRCULATING BLOOD, BLOOD VESSELS, AND BONE MARROW
Bone Marrow

The venous sinuses (**VS**) of the bone marrow drain into central veins (**CV**) that converge into large veins carrying blood away from the bone. Areas where venous sinuses open or drain into the central vein are denoted by **arrows** in Fig. 1. Mature blood cells (**BC** in Fig. 2) lying within the venous sinuses are separated from hemopoietic foci of the marrow (**Ma**) by a thin sinusoidal wall (**SW**) consisting of endothelial cells. Recent studies by Becker and DeBruyn, using the scanning electron microscope, indicate that the sinusoidal wall is continuous and has no permanent patent openings into the surrounding marrow. These studies further suggest that blood cells migrate from the marrow into the sinusoids by actually traversing the endothelial cell through transient openings (**arrows** in Fig. 2) rather than passing through the intercellular spaces between endothelial cells. At birth, active marrow is red because of the large number of developing red blood cells and because of the extensive vascularity due to the presence of many venous sinuses. Later in life, portions of the marrow may become less active in blood cell production, in which case there is a decrease in the number of blood-forming cells and an increase in the number of fat-storing cells, called adipocytes. Adipocytes (**Ad**), which are illustrated in the marrow in Fig. 3 may in sufficient quantity give the marrow a yellow color in the living state. Lipid has been dissolved from many of the sectioned adipocytes in Fig. 3.

Fig. 1, ×290; Fig. 2, ×1820; Fig. 3, ×385

CIRCULATING BLOOD, BLOOD VESSELS, AND BONE MARROW
Bone Marrow

Vascular casts shown in these figures demonstrate the general pattern of circulation present in the bone marrow. The tissue was digested away with a strong alkaline solution, leaving only a polymerized replica of the blood vessels. The arteries are easily distinguished from the veins by the presence of small fusiform depressions on the surface of the cast that correspond to the impressions left by the endothelial cells. Moreover, the arteries are more regular in diameter along their length. These distinguishing features have been observed in casts of every organ of the body that has been prepared for study. The nutrient artery enters the marrow after penetrating the shaft of the bone, and branches to form central arteries (**CA**) that follow a longitudinal course through the marrow of long bones. The three central arteries at the left margin of Fig. 1 originated from a larger nutrient artery. The central arteries give rise to long straight arterioles (**Ar**) that are much narrower. The arterioles terminate as slender arteriolar capillaries (**Ca**) that supply the venous sinuses (**VS**). In Fig. 2, the continuity between an arteriolar capillary and a venous sinus can be observed (✱). The presence of thin straight arteriolar capillaries that supply venous sinuses is analogous to that seen in the splenic circulation. At the periphery of the marrow, the venous sinuses are supplied by arteriolar capillaries originating from Haversian arteries that course through the osseous shaft of the bone. The venous sinuses anastomose extensively with one another and are highly variable in diameter. They drain into central veins (**CV** in Figs. 1 and 3), and a constriction is often present at the point of entry into the central vein (see **arrows** in Fig. 3). The central veins converge into a large nutrient vein (**NV** in Fig. 1) that exits the marrow.

Fig. 1, ×60; Fig. 2, ×525; Fig. 3, ×130

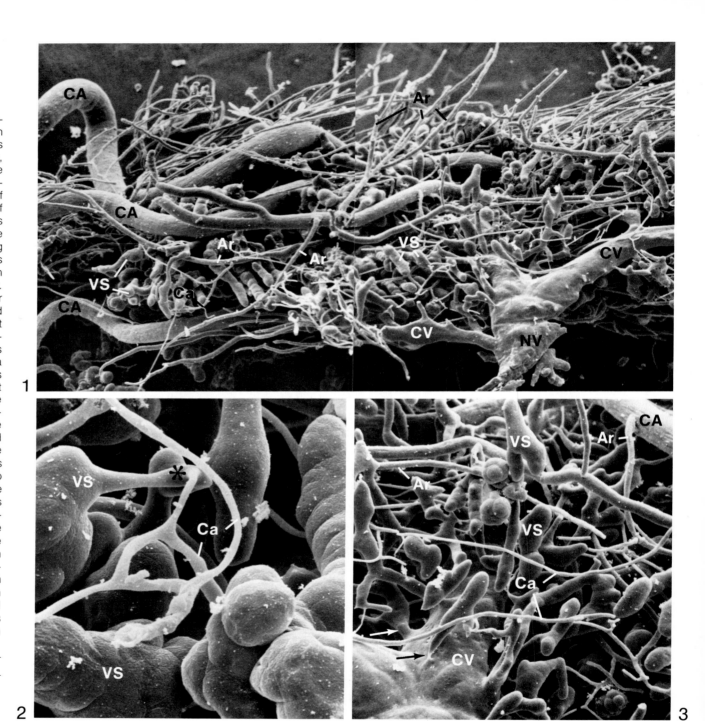

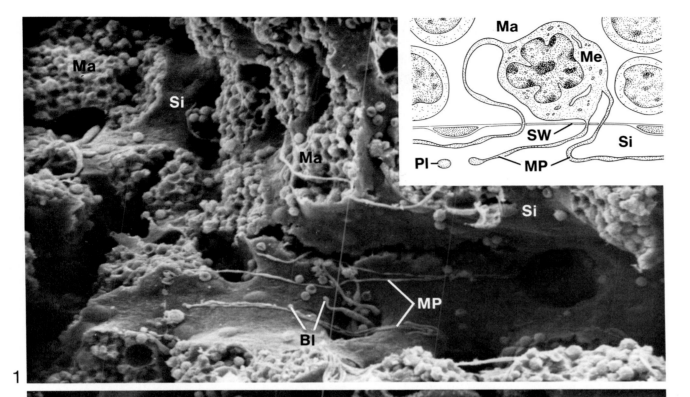

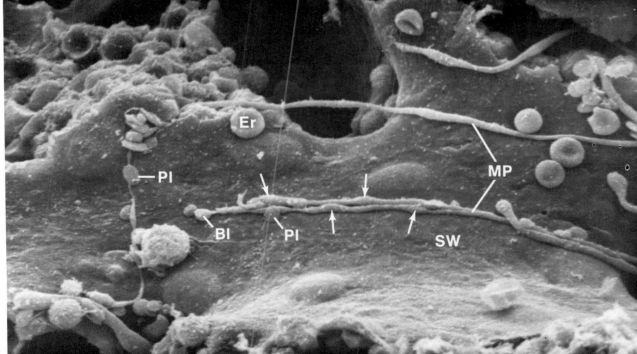

CIRCULATING BLOOD, BLOOD VESSELS, AND BONE MARROW
Bone Marrow

Giant polyploid cells, called megakaryocytes (**Me** in **inset** of Fig. 1), are among the cells found in bone marrow (**Ma**). Platelets of the circulating blood form as excrescences from the cytoplasm of megakaryocytes. Although the cell bodies of megakaryocytes are located in the marrow, long slender extensions of their cell cytoplasm (**MP**) extend into the lumen of the myeloid sinusoids (**Si**). These worm-shaped extensions emanate from the marrow and sinusoidal wall (**SW**) and occupy preferential positions along the longitudinal axis of the sinusoid (Figs. 1 and 2). The megakaryocyte extensions frequently terminate as rounded tips (**Bl** in Fig. 2) that are similar in size and shape to individual platelets (**Pl**) nearby. The isolated platelets (**Pl**) can be compared in size to erythrocytes (**Er**) in Fig. 2. Periodic expansions (**arrows** in Fig. 2) are frequently noted along the length of the megakaryocyte extensions. These relationships and other observations have prompted Becker and DeBruyn (1976) to suggest that individual platelets are released into the circulation by pinching off from intravascular megakaryocyte processes. The periodic constrictions along the megakaryocyte process reflect a stage in the pinching-off process. In summary, it appears that platelet release occurs intravascularly from slender processes originating from cell bodies of megakaryocytes that are located extravascularly.

Fig. 1, ×655; Fig. 2, ×1920

CIRCULATING BLOOD, BLOOD VESSELS, AND BONE MARROW
Bone Marrow

The transmission electron micrograph of bone marrow in Fig. 1 was taken from tissue 10 minutes after an intravenous injection of colloidal carbon. One of the marrow macrophages (**M**) sends a large pseudopod (**large arrow**) into the sinus (**S**) through a large gap in its wall. Erythroid cells (**E**) are present close to the sinus wall, and form erythropoietic islands around macrophages. One normoblast (✱) is undergoing mitosis. Another (**double arrow**) is in the process of extruding its nucleus, before eventual migration into the sinus lumen. One extruded nucleus (**triple arrow**) is phagocytosed by a macrophage. A large megakaryocyte (**Meg**) and a nearby fat cell (**F**) are also present. Cells of the granulocytic series (**My**), are typically produced in nests located away from the venous sinuses. Upon maturation they become motile and can move through the sinus wall into the lumen, thereby gaining access to the bloodstream. In Fig. 2, a granulocyte (**Gr**) is observed in the process of migration between endothelial cells toward the sinus lumen. It is not known what factors cause the endothelial cells that line the sinus to separate at their junctions, thus allowing for the migration of mature marrow cells into the circulation. Erythroblasts (**Er**), basophils (**Ba**), and a megakaryocyte are present in the hematopoietic space. One erythroblast is undergoing nuclear extrusion. Adventitial cells surround the outside of the endothelium in many areas, and their presence varies, depending upon the relative amount of migration taking place (fewer adventitial cells are present during increased states of migration). The adventitial cells also can extend processes into the hemopoietic area of the marrow, thereby incompletely partitioning the marrow into compartments containing specific blood cell lines. [Fig. 1 courtesy of S. C. Luk and G. T. Simon. From A. W. Ham, *Histology* (7th ed.), Lippincott Co., Philadelphia and Toronto, 1974. Fig. 2 by permission of L. T. Chen, L. Weiss, E. Handler, and Grune & Stratton Inc. From "An electron microscopic study of the bone marrow of the rat in an experimental myelogenous leukemia," *Blood* 39:102 (1972).]

Fig. 1, ×1515; Fig. 2, ×11,125

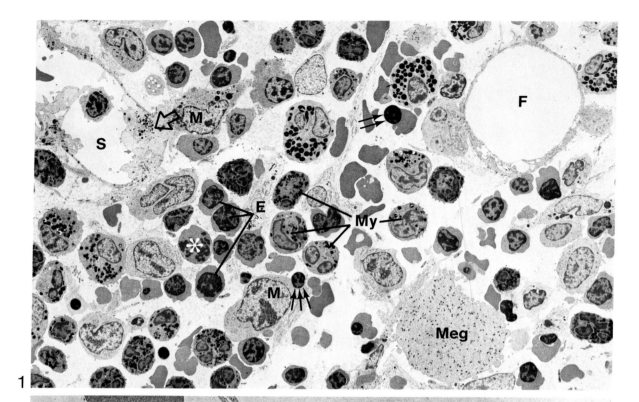

1

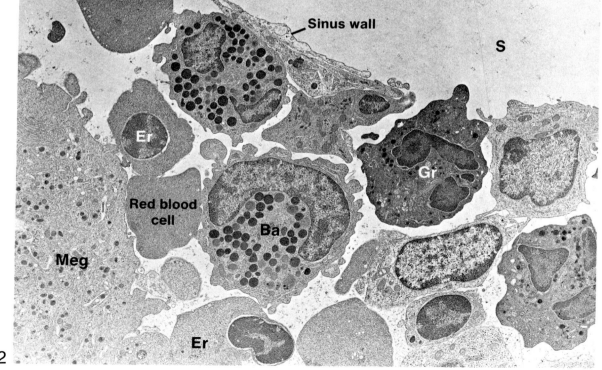

2

LYMPHOID ORGANS

Introduction

THE IMMUNE RESPONSE

The immune system, first appearing in verte-brates, can be considered as an evolutionary decree of independence. Ancestral symbiotic relationships are cast off and replaced with a new ability to distinguish self from nonself. Emphasis is switched from cooperation with the external environment to the environment within, in which internal cell-cell interactions constitute the key to an effective immune system.

The immune response to a foreign substance can be divided into three temporal events, an *initial* or *afferent* phase, a *central* processing phase, and an *effector* response. The initial phase comprises the events taking place between the entry of antigen or foreign substance and its first encounter with cells (primarily T lymphocytes) of the immune system. In this initial encounter, "self" must be distinguished from "nonself." This is accomplished only by the population of lymphocytes (in most instances, T lymphocytes) that possess specific surface receptors having a high binding affinity for the particular antigen. It is quite remarkable that of the enormous variety of antigens encountered, there appears to be a circulating T lymphocyte within the entire population of lymphocytes that is capable of recognizing and binding to a specific antigen. In the initial phase, it seems that the antigen must first be presented in the appropriate form and location before the specific lymphocyte can evoke a response. The macrophages of the immune system are thought to serve this function. Populating lymphoid organs, such as the spleen and lymph nodes, these phag-

ocytic cells are capable of trapping and engulfing antigens as they flow past. In a manner that is not clearly defined, the macrophage then "processes" the antigenic molecule and subsequently presents it upon its surface in a form that is readily recognized by specific lymphocytes circulating in the vicinity. Antigens borne by the blood first encounter macrophages in the spleen, whereas those in the lymph first encounter macrophages in lymph nodes.

The central phase of the immune response is largely concerned with cooperation between T and B lymphocytes that are specific for the antigen, followed by their subsequent proliferation. Overall, the cell proliferation in the central phase gives rise both to "memory" lymphocytes and to lymphocytes that are involved in the effector (antigen destruction) phase. The memory cells are long-lived and may include representatives of both T and B lymphocyte populations. The memory cells continue to circulate as "primed policemen" of the immune system until the specific antigen is reintroduced at a later date, at which time they immediately become stimulated to mount an attack. Alternatively, the effector lymphocytes (T or B) produced in the central phase are involved in the immediate response to and destruction of the antigen. Clonal expansion of B lymphocytes with specific antigen receptors gives rise to progeny that differentiate into plasma cells. The plasma cells subsequently synthesize and secrete into the blood a specific antibody to the antigen. The effects that follow lead to a "humoral" response. Proliferation of T lymphocytes may give rise to "sensitized" T cells involved in a

"cell-mediated" response to antigens. Overall, the central phase is characterized by the selective proliferation of lymphocytes that are specific for the particular antigen. This clonal expansion has been considered by many to be an amplifying step of the immune response.

The humoral and cell-mediated responses are events that characterize the effector (antigen destruction) phase of the immune response. In both types, further amplification of the response is achieved by the recruitment of nonspecific destructor cells into the immediate vicinity of the antigen. The nonspecific destructor cells of the humoral response include neutrophils (polymorphonuclear leukocytes), whereas the macrophage and cytotoxic ("killer") T cell are the major effector cells of the cell-mediated response. In the humoral response, antibodies bind to the antigenic site and, in concert with complement, trigger the movement (by chemotaxis) of neutrophils into the area where they may phagocytose and destroy the antigen. In addition, the combination of antibody and complement factors alone may cause the demise of a foreign cell, a process involving damage to the target-cell membrane, resulting in cell lysis. In the cell-mediated response, nonspecific effector cells, such as macrophages or cytotoxic killer cells (differentiated form of the T lymphocyte), are activated and directed to the location of the antigen by specific factors called lymphokines. Lymphokines are thought to be released into the area primarily by sensitized T lymphocytes. Sensitized T lymphocytes are those lymphocytes that have proliferated during the central phase from T lymphocytes possessing specific receptors to the antigen. Some lymphokines are chemotactic, resulting in the migration of effector cells into the area, whereas others are activating factors that cause the transformation of effector cells into a cytotoxic form (i.e., transformation of macrophages into "angry" macrophages). Furthermore, a number of lymphokines released are directly cytotoxic (lymphotoxins) to the target cell. One of the net results is the migration of nonspecific effector cells into the area, at which time they are "activated" to destroy the target cells. Innocent bystander cells may also be destroyed in the process as a result of the "nonspecific" action of activated effector cells. In response to a given antigen, both the humoral and cell-mediated responses may occur concommittantly, as superimposed events. The cell-mediated response, however, is particularly dominant in the rejection of transplanted tissue, delayed hypersensitivity reactions, autoimmune diseases, and tumor destruction.

ORGANIZATION OF LYMPHATIC TISSUE

Lymphatic tissue is a special type of connective tissue consisting of reticular cells that produce many reticular fibers arranged in a meshwork. The supportive meshwork formed by reticular cells constitutes the "fixed," or stable, component of lymphoid organs, such as the spleen and lymph nodes. Free cells of the immune system (i.e., macrophages and lymphocytes) are located within the spaces of the reticular meshwork, and it is here that they may interact with foreign antigens that are brought into the area via the blood or lymph. Stimulated lymphocytes that result from such interactions may proliferate within the interstices of the reticular meshwork and form compact, densely populated areas called lymphatic nodules. Blood vessels and sinuses are widely distributed within lymphatic tissue and provide a

pathway for the continual recirculation of lymphocytes between blood and lymph. Lymphatic tissue is also considered as a type of hemopoietic tissue, since it is concerned with the production of the lymphocytes found in the circulating blood. Lymphatic tissues, such as the thymus, lymph nodes, and spleen, are encapsulated, hence they can also be considered as organs. Other types of lymphatic tissue, such as tonsils and solitary lymphatic nodules, may be distributed under moist epithelial membranes of the oral cavity, pharynx, and digestive tract.

The lymphatic circulatory system consists of numerous vessels of variable size. The system begins as many closed-ended lymph capillaries that are widely distributed in most tissues and organs of the body. The capillaries are continuous with progressively larger vessels that ultimately end in two large lymph vessels that empty into the venous circulatory system. Lymph nodes are inserted at various intervals into an extensive lymphatic circulatory system. Therefore, the position of the nodes is particularly well suited to filter lymph. Lymph nodes serve to produce B lymphocytes as well as plasma cells that can synthesize and secrete humoral antibody. They also constitute a region for the interaction between antigens, macrophages, and lymphocytes. Lymph nodes also contain the so-called thymus-dependent areas that consist of cell lines of T lymphocytes that originate in the thymus. Macrophages within the tissue are capable of phagocytosing particulate material of various kinds, and play an important intermediary role in the interaction between antigens and lymphocytes.

The spleen is a lymphatic organ designed to filter blood. It produces both T and B lymphocytes, which enter the circulation by migrating through the walls of splenic blood vessels. In response to blood-borne antigens, the spleen can form plasma cells (via suitable B lymphocytes that transform initially into immunoblasts, then into plasmablasts, and ultimately into plasma cells) that are capable of producing specific humoral

antibodies. Because of the large numbers of phagocytic macrophages, the spleen is involved in the removal of microorganisms from the circulating blood and in the removal of large numbers of worn-out blood cells. The spleen also serves as a storage depot for blood.

The thymus is the primary site for the production and maturation of T lymphocytes. A variety of genetic codes is expressed in the population of T lymphocytes as reflected in the many different immunoglobulin-like receptors on the surface of different cells in the population. After leaving the thymus, mature T lymphocytes circulate in the blood vascular system and colonize specific regions of the spleen, lymph nodes, and other lymphatic tissue. A large population of T lymphocytes is produced by the thymus, and even though the cells have a similar structure, there is a vast array of differences between cells in the nature of their surface receptors. Therefore, specific populations of T lymphocytes are capable of responding to specific antigenic molecules. A few hundred homologous receptor molecules are built into the plasma membrane of one T lymphocyte. In contrast, the number of immunoglobulin-like receptors on B lymphocytes varies from approximately 50,000 to 150,000.

Bone marrow is the primary site for the formation of erythrocytes, granular leukocytes, monocytes, and megakaryocytes. Some B lymphocytes, however, can be produced by the bone marrow, and probably serve to populate lymphatic tissues. For example, if an animal is irradiated so as to destroy its lymphoid organs and is subsequently inoculated with various bone marrow cells from an unirradiated animal, it has been determined that cells of the marrow, and only such cells, are capable of repopulating those irradiated lymphoid organs. There is also evidence that very early in embryonic development, undifferentiated lymphoid stem cells originate in the bone marrow, to be carried in the blood vascular system to the thymus, spleen, and lymph nodes, where they constitute the precursors of T and B lymphocytes.

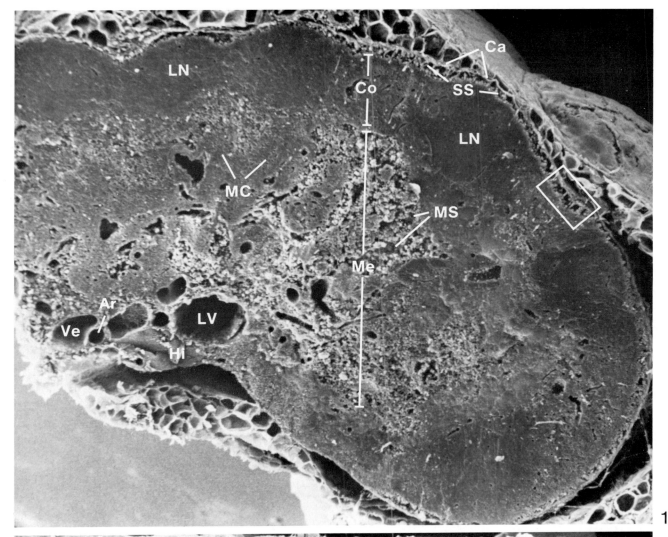

LYMPHOID ORGANS
Lymph Node

Lymph nodes are encapsulated, oval, or kidney-shaped structures interposed at intervals along lymphatic vessels of the body. They serve to filter lymph and to add new lymphocytes to the circulation. The interior of the node is rather indistinctly divided into an outer cortex (**Co** in Fig. 1) and an inner medulla (**Me** in Fig. 1). The cortex contains lymphatic nodules (**LN**) that consist primarily of closely packed aggregations of developing B lymphocytes; these appear as smooth, solid areas in this fractured preparation. The nodules continue into the medulla as solid cords of cells referred to as medullary cords (**MC**). A connective tissue capsule (**Ca** in Figs. 1 and 2) covering the lymph node surface is commonly associated with adipose tissue. Connective tissue divisions of the capsule, called trabeculae (**Tr** in Fig. 2), extend into the interior of the node. Lymph enters via branches of afferent lymphatic vessels, which are equipped with one-way valves that prevent backflow of lymph. The afferent lymphatic vessels pierce the capsule on the convex surface of the lymph node and open into an underlying subcapsular sinus (**SS**). The organization of the subcapsular sinus is illustrated in greater detail in Fig. 2, which is representative of the area enclosed by the rectangle in Fig. 1. Lymphocytes (**Ly**) and macrophages are abundant in this region. The subcapsular sinus is continuous with loosely organized tunnels, called cortical sinuses, which extend into the cortex and continue into the medulla as medullary sinuses (**MS**). Lymph flowing through these sinuses exits the node via an efferent lymph vessel (**LV**) at the hilus (**Hi**) region. Arteries (**Ar**) and veins (**Ve**) also enter and leave the node at the hilus.

Fig. 1, ×120; Fig. 2, ×2080

1

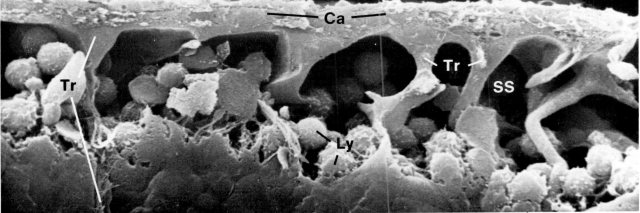

2

LYMPHOID ORGANS
Lymph Node

The lymph node preparation shown in Fig. 2 was cut open rather than fractured. As a result, it is possible to observe more cellular detail in the cortically placed lymphatic nodules (**LN**) and in the medullary cords (**MC**). Lymph nodules are the site of proliferation and differentiation of stimulated B lymphocytes, and consist primarily of lymphoblasts, lymphocytes, macrophages, and a supportive framework of reticular cells and fibers. Upon stimulation, circulating B lymphocytes within blood may enter and "seed" cortical nodules of lymph nodes by migrating through the walls of postcapillary venules located in the immediate vicinity of the nodules. Once in the nodular region, these stimulated cells may give rise to a specific cell line of antibody-producing cells. Intermediate forms of differentiating B lymphocytes are commonly observed in the nodule and include the highly proliferative lymphoblast (immunoblast). Progeny of the lymphoblast migrate to the periphery of the nodule and into the medullary cord region as they transform into plasma cells. Plasma cells synthesize and release specific humoral antibodies against those antigens that initially stimulated their formation from B lymphocytes. The preparation is also useful for indicating the flow of lymph through the intercommunicating sinuses. The pathway of flow, as denoted by arrows, originates from afferent lymphatic vessels piercing the capsule (**Ca**) and proceeds into the subcapsular sinus (**SS**), through cortical or peritrabecular sinuses (**CS**), and into the medullary sinuses (**MS**). In the afferent phase of the immune response, antigens filtering through the sinuses are often trapped and "processed" by macrophages that reside there and which subsequently interact with specific lymphocytes. The enclosed rectangle within the sinus is enlarged (Fig. 1) to reveal the morphological correlate of the functional macrophage-lymphocyte interaction. A group of lymphocytes (**Ly**) adhere to a single macrophage (**Ma**), which has long, thin tendrils projecting from its surface.

Fig. 1, ×3700; Fig. 2, ×430

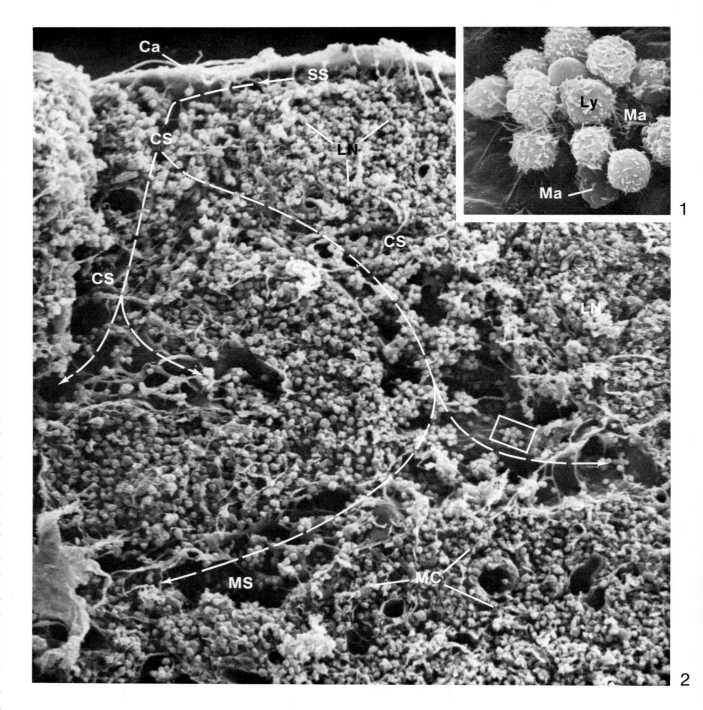

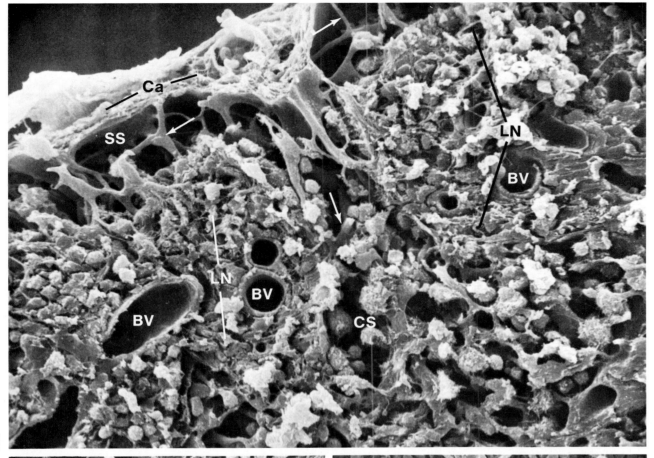

1

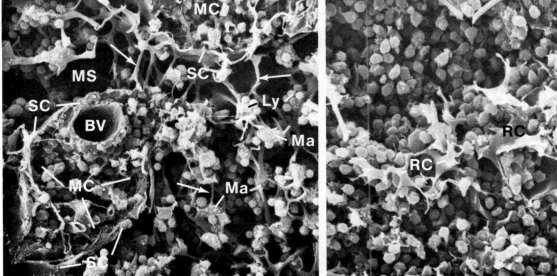

2

3

LYMPHOID ORGANS
Lymph Node

Details of the organization of the cortex are depicted in Figs. 1 and 3 and of the medulla in Fig. 2. A portion of the connective tissue capsule (**Ca**) surrounding the node is present in Fig. 1. Processes (**arrows**) of the highly branched reticular cells are evident in the subcapsular sinus (**SS** in Fig. 1), the cortical sinus (**CS** in Fig. 1), and the medullary sinus (**MS** in Fig. 2). Reticular cells (**RC**) are also demonstrated in a cortical lymphatic nodule shown in Fig. 3. Other cell types that might be encountered in such a region include macrophages, T lymphocytes, lymphoblasts, and differentiating B lymphocytes. Portions of the two highly cellular lymphatic nodules (**LN**) in Fig. 1 are traversed by blood vessels (**BV**) of variable diameter. Postcapillary venules in paranodular areas can be traversed by both T and B small lymphocytes. Most of the lymph has been washed free of the medullary sinus in Fig. 2, but groups of lymphocytes (**Ly**) remain attached to fixed macrophages (**Ma**) that reside on the reticular cell network (**arrows**). The medullary cords (**MC**) in Fig. 2 are separated from the sinuses by a thin lining composed of squamous cells (**SC**), sometimes called littoral cells. Blood vessels (**BV** in Fig. 2) are also present in the medullary cords. Cellular elements of the medullary cords include lymphoblasts, plasma cells, reticular cells, and macrophages. Some of the cellular elements of the medullary cord, such as plasma cells, most likely can traverse the endothelial lining of the sinus and thereby enter the lymph as it flows through the node.

Fig. 1, ×1145; Fig. 2, ×585; Fig. 3, ×825

LYMPHOID ORGANS
Lymph Node

These figures illustrate the medullary cords (**MC**) and medullary sinuses (**MS**) of a lymph node exposed by cryofracture. The area contained within the rectangle in Fig. 1 is enlarged in Fig. 2 to illustrate the structural details and relationships of these two parts of a node. Closely packed cells are present in the medullary cord (**MC** in Fig. 2). Lymphoblasts, lymphocytes, plasma cells, macrophages, reticular cells, and reticular fibers can all be found within medullary cords. The medullary cord is separated from the medullary sinus (**MS** in Fig. 2) by a single layer of squamous cells constituting the sinus lining (**SL**). Within the sinus, the highly branched form and smooth surface of the reticular cell (**RC**) are characteristic features. Several macrophages (**Ma**) adhere to the reticular cells, and are distinguished by numerous and long slender extensions of the macrophage plasma membrane. Large numbers of lymphocytes (**Ly**) are also present in the sinus, and many of them are in contact with the macrophages.

Fig. 1, ×735; Fig. 2, ×1570

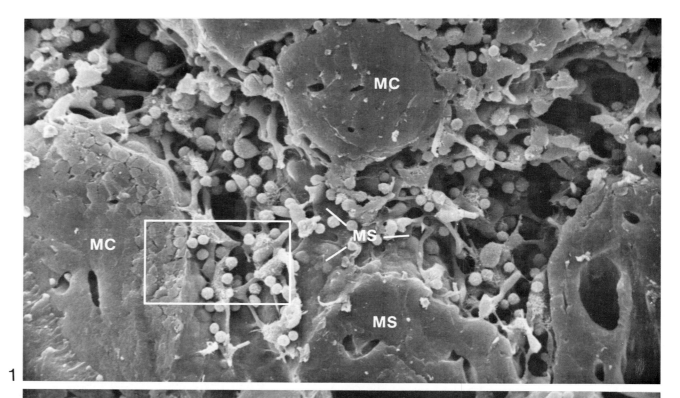

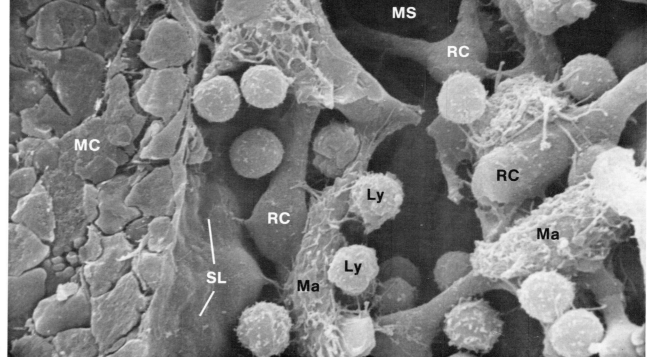

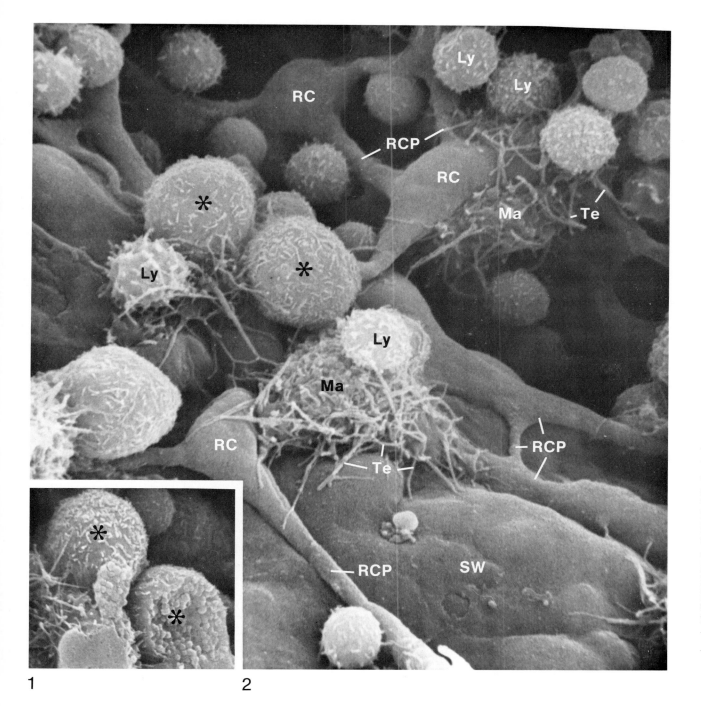

LYMPHOID ORGANS
Lymph Node

The trabecular meshwork of reticular cells (**RC**) and reticular-cell processes (**RCP**) traversing the lymph node sinus in Fig. 2 probably serves to create turbulence in the lymph flow, thus slowing its movement through the node. The fixed macrophages (**Ma**), which possess a number of long, slender tendrils (**Te**), are attached to the reticular meshwork and are ideally positioned to interact with antigens and lymphocytes in the lymph as it "percolates" through the sinus. The endothelial lining of the sinus wall (**SW**) is also depicted, as are a number of lymphocytes (**Ly**) adhering to the macrophage surface. The large rounded cells (✱) present in the sinus appear to contain cytoplasmic granules when fractured open (Fig. 1). Once the macrophages have phagocytosed antigens, they are capable of "processing" them, with the result that more antigenic sites become exposed. In addition to trapping and processing antigens, macrophages are now believed to play a key role in mediating the T and B lymphocyte interaction required for a humoral immune response. In order for a B lymphocyte to become stimulated, it is usually necessary for a high concentration of antigen to be localized around the surface receptors of the B cell. In this respect, instead of bringing the antigen to the B cell, it is much more efficient to hold a high concentration of antigen stationary in a position accessible to a wide variety of B cells. In fact, this seems to be the function of the macrophage. The T lymphocyte apparently deposits an antigen receptor complex from its cell membrane onto the macrophage surface, where it may be exposed to the congested traffic of B lymphocytes filtering through the sinus. In such a position, the antigen may interact with those specific B lymphocytes that have an affinity for the antigen.

Fig. 1, ×3455; Fig. 2, ×3915

1

2

LYMPHOID ORGANS
Lymph Node

A small area of a medullary sinus is illustrated in Figs. 1 and 2. The reticular cells (**RC**) have a smooth surface, as do the reticular-cell processes (**RCP**) that extend from the cell body. In contrast, the macrophages (**Ma**) are unbranched and are characterized by many rounded surface blebs (**Bl**) and finger-like extensions, or tendrils (**Te**), of the surface membrane. Lymphocytes (**Ly**) are numerous and usually closely adhere to a portion of the macrophage. During formation of the lymph node, the reticular cells produce reticular connective-tissue fibers (**RF**), which make up a part of the stroma of the node. After synthesis by the reticular cells, the fibers become extracellular, but those in the sinuses of the node are ensheathed by the cytoplasmic extensions of the reticular cells. Consequently, in the sinus portion of the node, the reticular fibers are not freely exposed to the circulating lymph. In the preparation illustrated in Fig. 2, a portion of a reticular-cell process (**RCP**) was fractured away during tissue preparation. As a result, it is possible to observe a bundle of reticular fibers (**RF**) that normally would be surrounded by a branch of the reticular cell. These reticular fibers probably provide support to the highly branched reticular-cell processes.

Fig. 1, ×5015; Fig. 2, ×7320

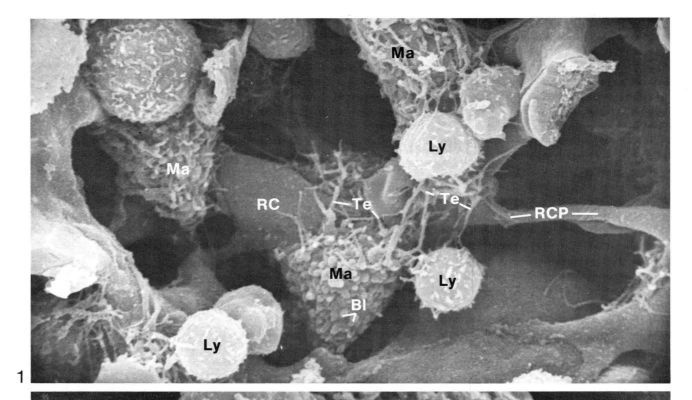

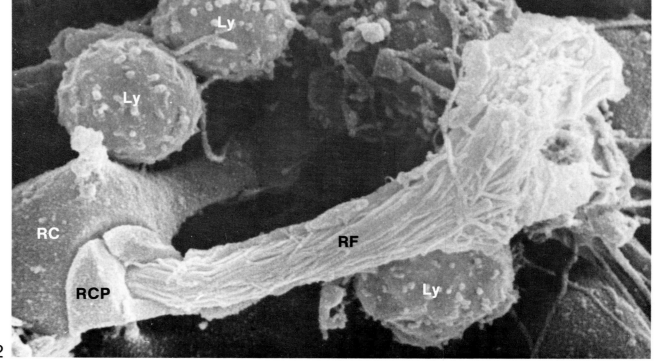

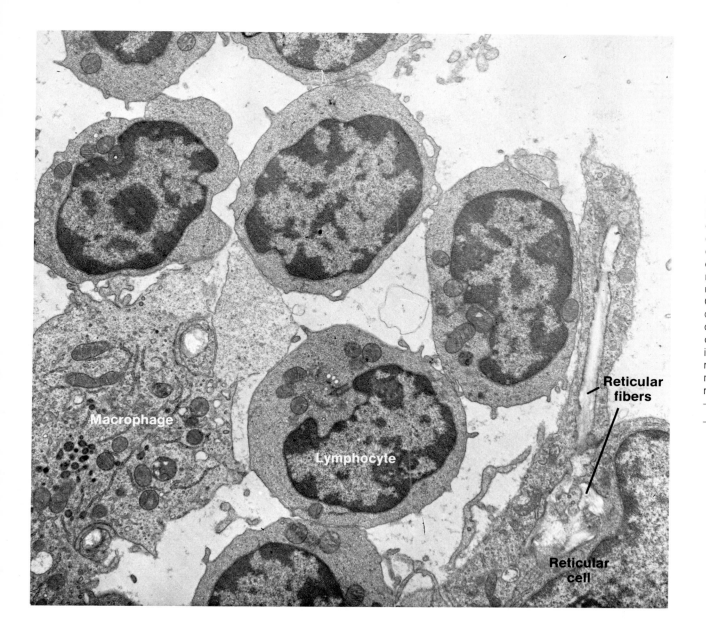

Macrophage

Lymphocyte

Reticular fibers

Reticular cell

LYMPHOID ORGANS
Lymph Node

This transmission electron micrograph of a small region of a medullary sinus of a rat lymph node shows a portion of the cytoplasm of the cells that make up the sinus. A number of lymphocytes are closely associated with a macrophage. The small lymphocytes illustrated in the electron micrograph are about 6 μm in diameter, and their nuclei contain heterochromatin, which forms a peripheral rim adjacent to the nuclear envelope. An occasional short, narrow microvillus is associated with the cell surface. The cytoplasm contains numerous free ribosomes and several small spherical mitochondria. The rough-surfaced endoplasmic reticulum is not usually present in the cytoplasm, and the Golgi apparatus is small. In some sections, an occasional small dense granule, a lysosome, can be observed in the cytoplasm. A portion of a reticular cell and its process also appears in the illustration. Reticular fibers are surrounded by the reticular cell, so that they are not directly exposed to the lymph in the surrounding space of the sinus.

×8545

LYMPHOID ORGANS
Lymph Node

These transmission electron micrographs illustrate cellular interrelationships between macrophages, reticular cells, and lymphocytes within a medullary sinus of a rat lymph node. Lymphocytes are closely associated (**arrows**) with the macrophage cell surface in Fig. 1. A section through a macrophage is illustrated in Fig. 2. It is characterized by the extensions (tendrils) projecting from the cell surface and by the presence of numerous lysosomes and digestive vacuoles within its cytoplasm. The macrophage is wedged between two reticular-cell processes. The junction between the surfaces of reticular-cell processes and the macrophage is denoted by the broken lines. The reticular-cell processes are easily distinguished in section, since bundles of electronlucent reticular fibers are commonly ensheathed by the reticular-cell processes.

Fig. 1, ×9680; Fig. 2, ×11,390

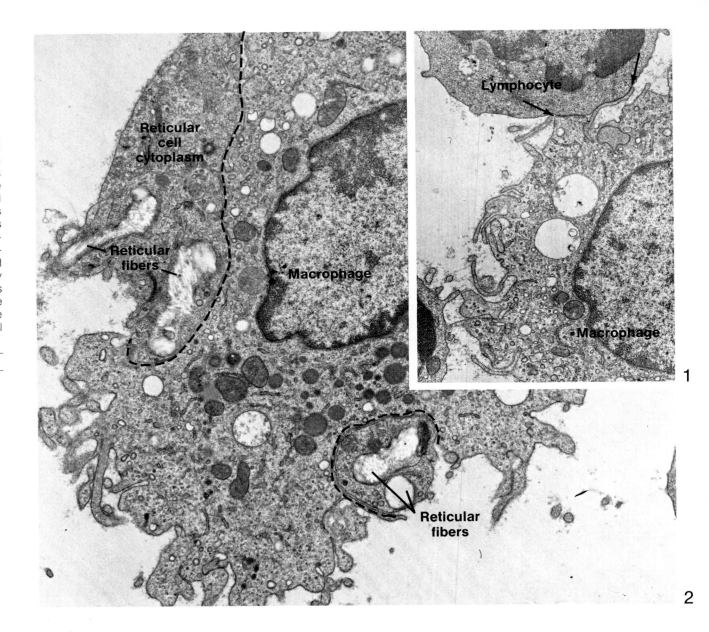

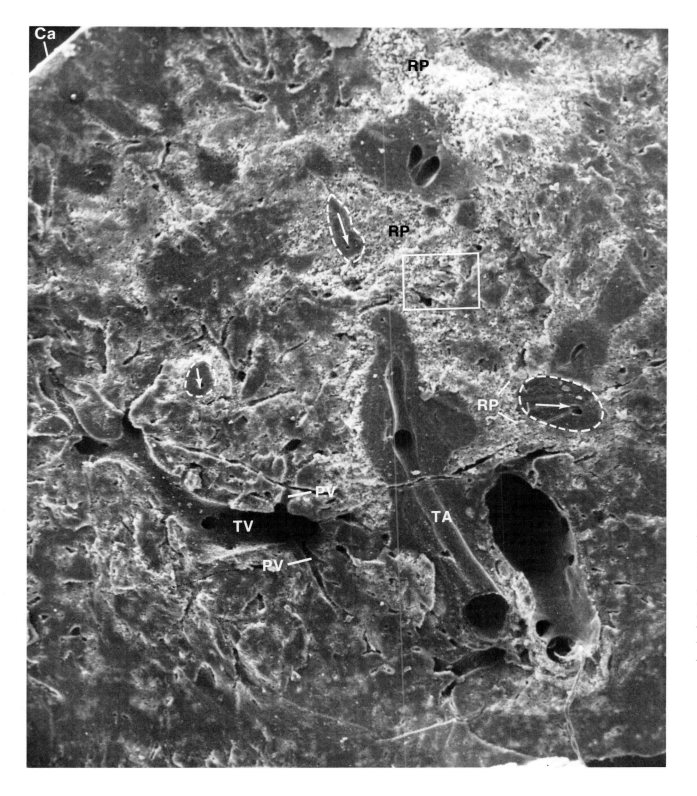

LYMPHOID ORGANS
Spleen

The spleen is enclosed by a capsule (**Ca**) of dense connective tissue consisting of both collagen and reticular fibers, fibroblasts, and smooth-muscle cells. At intervals, the capsule extends as trabeculae into the interior of the spleen. Together, the capsule, trabeculae, and reticular connective tissue form a framework for the cellular elements (parenchyma) of the spleen. The parenchyma of the spleen is also known as splenic pulp, and is divided into red pulp and white pulp. The red pulp (**RP**) appears red in the living state due to heavy infiltration of red blood in this region. It consists of many large, thin-walled vessels, called splenic sinuses, and intervening thin plates of cells called the splenic cords (cords of Billroth). Both sinuses and cords are illustrated at very low magnification within the rectangle in this figure. The small spaces represent the sinuses, and the adjacent, apparently solid structures are the cords. Removal of effete red blood cells occurs within the red pulp, whereas production of new lymphocytes occurs primarily within the white pulp. The white pulp, for the most part, takes the form of cylindrical sheaths, containing closely packed aggregations of lymphocytes, immunoblasts (lymphoblasts), reticular cells, and macrophages. Because each sheath of lymphoid tissue characteristically invests a centrally located artery (**arrow**), they have been referred to as periarterial lymphatic sheaths. Several periarterial lymphatic sheaths are outlined by **broken lines,** and appear as solid structures in this cryofractured preparation. The majority of the lymphoid cells that populate the periarterial lymphatic sheaths are T lymphocytes that have been derived from the thymus. Details of the circulatory system identified at this magnification include a branch of a trabecular artery (**TA**), a trabecular vein (**TV**), and red pulp veins (**PV**), which drain into trabecular veins.

×75

LYMPHOID ORGANS
Spleen

A single periarterial lymphatic sheath (**LS**) of the white pulp of the rat spleen is illustrated in transverse section in Fig. 1. The densely packed lymphoid cells within the sheath appear smooth, and their borders are indistinguishable because the preparation was cryofractured. A central artery (**CA**) is present within the sheath, which is surrounded by a marginal zone (**MZ**) of red pulp. The marginal zone is highly vascular and contains a tight meshwork of reticular cells and fibers. The marginal zone is thought to be a major site in which antigens passing from the blood into the spleen are first concentrated. Initial interactions between antigens, macrophages, and lymphocytes may occur in this region. A cast of a portion of the splenic circulatory system comparable to the region shown in Fig. 1 is included in Fig. 2. All of the cellular constituents have been removed by digestion with a strong alkaline solution. Blood enters the spleen via the splenic artery, which then divides into a number of trabecular arteries that are distributed within the connective tissue septa (trabeculae). After leaving the trabeculae, the arterial branches become invested by the lymphatic sheaths of the white pulp. In this position, these branches are called central arteries, white pulp arteries, or follicular arteries. The empty area surrounding the central artery (**CA**) normally contains white pulp (**WP**) of the lymphatic sheath. As the central artery passes through the white pulp, it may give rise to two types of vessels. In one case, the central artery may branch into a number of slender follicular arterioles (**FA**) (arteriole capillaries) that traverse the white pulp and terminate (**arrows** in Fig. 2) in perinodular or marginal sinuses (**MS**) of the red pulp. The replicated marginal sinuses are surrounded by spaces that would be filled with red pulp cords in the living state. In the other case, the central artery may branch perpendicularly or terminate distally as penicillar arteries that pass into the red pulp, where they may branch further to communicate with sinuses of the red pulp (closed splenic circulation) or open directly into the nonvascular spaces between the red pulp cords (open splenic circulation).

Fig. 1, ×460; Fig. 2, ×170

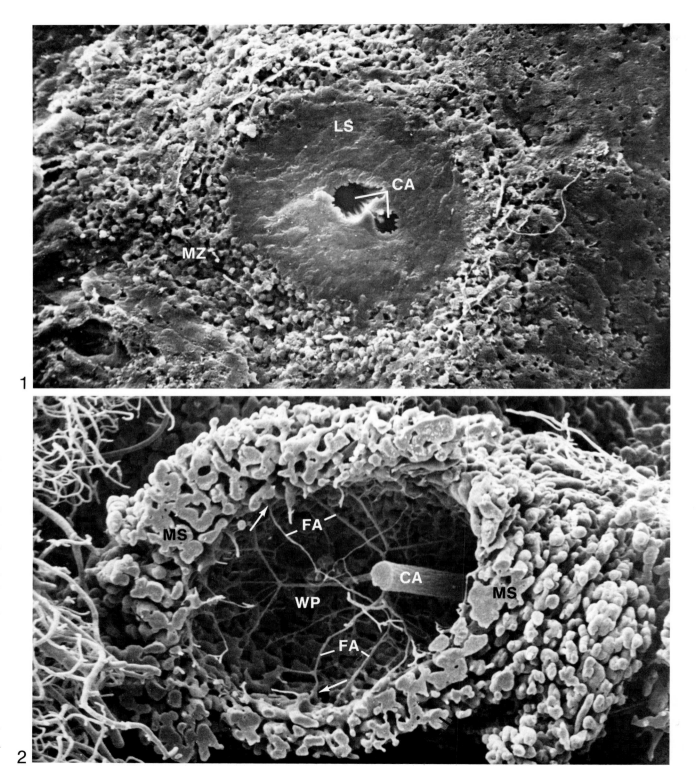

LYMPHOID ORGANS
Spleen

The photomicrograph in Fig. 1 illustrates a transverse section through a portion of a periarterial lymphatic sheath (**LS**). In addition to the T lymphocytes and macrophages that constitute the majority of the lymphatic sheath, spherical foci of B lymphocytes are often disposed along the longitudinal axes of the lymphatic sheaths, and are referred to as splenic nodules, or primary follicles. Active proliferation of cells in the center of a splenic nodule frequently leads to the formation of a germinal center (**GC** in Fig. 1) that stains lighter because the cells are larger, their nuclei are more euchromatic, and they possess more cytoplasm than the surrounding cells. Splenic nodules with germinal centers are termed secondary follicles and may displace the central artery of the lymphatic sheath or they may directly surround the central artery (**CA**), as indicated in Fig. 1. The progeny of dividing cells in the germinal center come to surround it, and form the outer shell, or mantle layer (**ML**), of the splenic nodule. There is no sharp demarcation between the mantle layer of the splenic nodule and the periarterial lymphatic sheath. The regional distribution of T and B lymphocytes within the white pulp is determined by using fluorescent-labeled antibodies that bind to specific cell surface markers characterizing the T and B cells. A small portion of the red pulp (**RP**) within the marginal zone adjacent to the lymphatic sheath is included in Fig. 1. Details of the red pulp are illustrated in the photomicrograph in Fig. 2. Many splenic sinuses (**SS**), lined with a thin layer of endothelium, riddle the surrounding splenic cords of cells (**SC**). The splenic cords, as will be illustrated in subsequent figures, consist of a meshwork of reticular cells and reticular fibers. Macrophages may be either fixed or attached to the reticular meshwork or they may exist free in the red pulp. Other cell types that can be found in the splenic cords include plasma cells and all formed elements of the circulating blood (i.e., erythrocytes, granular leukocytes, agranular leukocytes, and platelets).

Fig. 1, ×490; Fig. 2, ×785

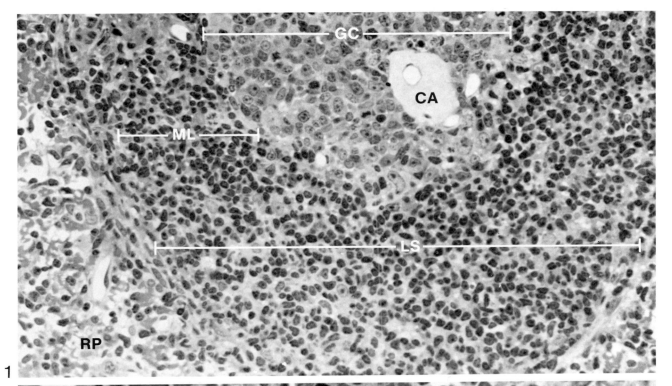

1

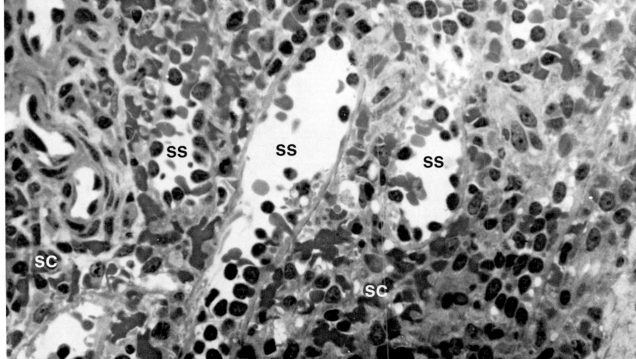

2

LYMPHOID ORGANS
Spleen

A casting medium that has filled and replicated only a portion of the splenic microvasculature is shown in Fig. 1. The vessels of the lymphatic sheaths and immediately surrounding marginal sinuses of the red pulp are replicated here, but the extensive network of venous sinuses between marginal sinuses is not. Such preparations are obtained by limiting the volume of casting medium injected; they are useful in displaying the sheath-like arrangement of marginal sinuses that invests the white pulp. In preparations more extensively replicated (as will be shown subsequently), the spaces evident between marginal sinuses in Fig. 1 would be occupied by a vast network of venous sinuses. The space between the central artery (**CA**) and marginal sinuses (**MS**) of each lymphatic sheath represents the location occupied by white pulp (**WP**) before tissue digestion. Trabecular arteries (**TA**) pass in a direction perpendicular to the lymphatic sheath to pierce the marginal sinus (**arrows** in Fig. 1). The details of the circulation shown within the rectangle in Fig. 1 are enlarged in Fig. 2. The central artery (**CA**) arises from a branch of the trabecular artery (**TA**) within the substance of the white pulp. The course of this artery parallels that of the lymphatic sheath. As previously described, two kinds of vessels may arise from the central artery. One vessel type, called the follicular arteriole (**FA**), extensively ramifies within the location of the white pulp (**WP**). As these arteriole capillaries reach the periphery of the sheath, they terminate and become continuous with the marginal sinuses (**MS** in Fig. 2). The continuity between central artery (**CA**), follicular arterioles (**FA**), and marginal sinuses (**MS**) is illustrated in Fig. 3. Another vessel type, the penicillar artery (**PA** in Figs. 1 and 2), extends from the central artery through the marginal sinus to the surrounding red pulp. In some cases, the penicillar arteries are thought to communicate with, and hence to supply, the anastomosing network of venous sinuses (excluding the marginal sinuses) that make up the red pulp circulation.

Fig. 1, ×45; Fig. 2, ×115; Fig. 3, ×220

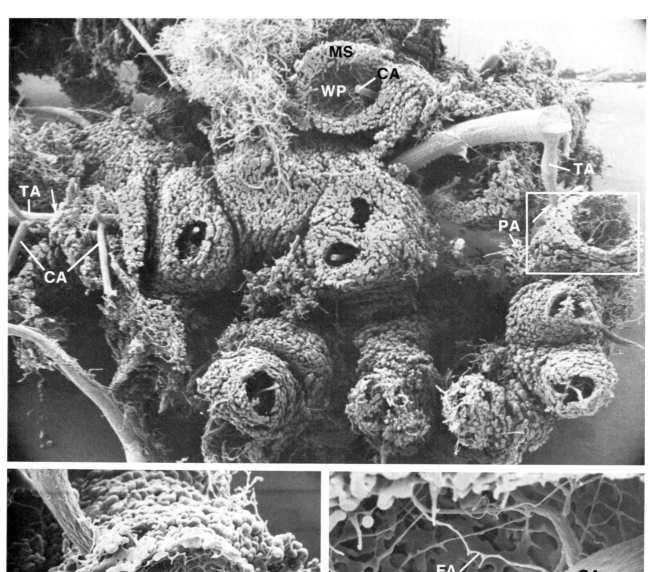

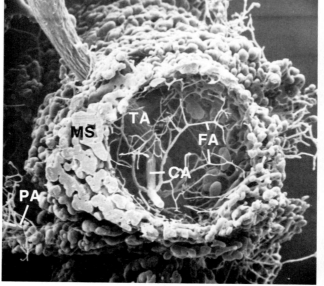

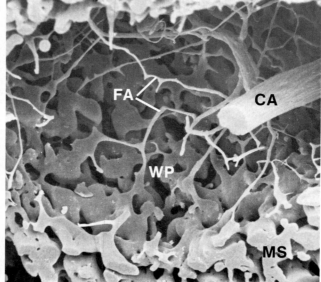

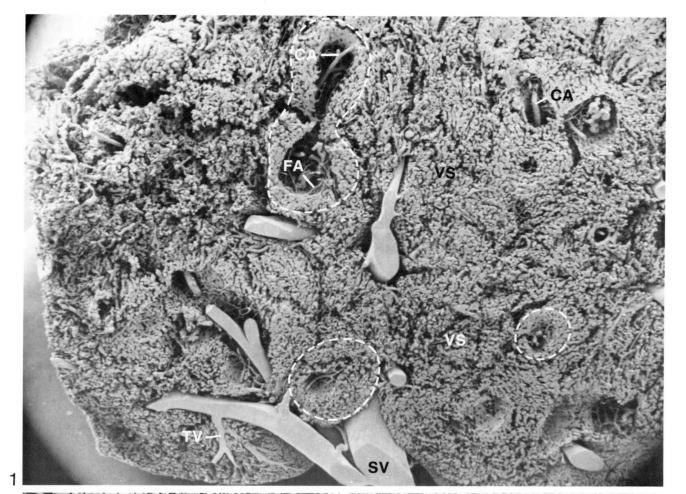

1

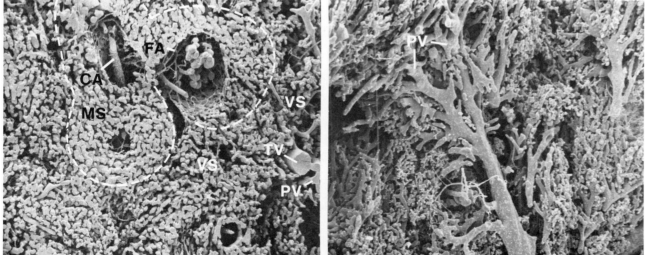

2   3

LYMPHOID ORGANS
Spleen

In contrast to previous preparations, those shown on this page represent a more complete replication of the splenic microvasculature. Although the sheath-like arrangement of marginal sinuses (**broken lines**) can still be distinguished, the surrounding vicinity is occupied by replicated venous sinuses (**VS**). Central arteries (**CA**) and follicular arterioles (**FA**) are identified. That portion of the replica in the upper right of Fig. 1 is illustrated at higher magnification in Fig. 2, where the relationships between the central artery (**CA**), follicular arterioles (**FA**), marginal sinuses (**MS**), and venous sinuses (**VS**) of the red pulp can be seen. According to the theory of closed splenic circulation, the penicillar arteries communicate directly with venous sinuses. In contrast, proponents of the theory of open splenic circulation believe that there is a discontinuity between the end of the penicillar arteries and the venous sinuses so that blood opens directly into the red pulp cords and associated spaces. Although an extensive literature exists on the nature of the splenic circulation, it is now held that both vascular patterns exist. The large number of venous sinuses in the splenic red pulp drain into numerous pulp veins (**PV** in Figs. 2 and 3), which ultimately join to become trabecular veins (**TV** in Figs. 1 to 3.) Trabecular veins coalesce to form the larger splenic vein (**SV** in Fig. 1), which exits at the hilus.

Fig. 1, ×30; Fig. 2, ×45; Fig. 3, ×75

LYMPHOID ORGANS
Spleen

Portions of two longitudinally sectioned splenic sinuses (**SS**) are illustrated in Fig. 1. In addition, cellular elements of the splenic cords (**SC**) are visible between the sinuses. A small region of red pulp containing splenic sinuses (**SS**) and splenic cords (**SC**) is also illustrated in Fig. 2 from a cryofractured specimen of rat spleen. Blood cells (**BC**) are present in the sinus lumen, but they can migrate (✱) between the cords and sinuses via small fenestrations (**arrows**) located between the endothelial cells (**EL**) lining the sinuses. Effete erythrocytes entering the interstices of the red pulp via the open pathway of splenic circulation are likely to become phagocytized by numerous macrophages present in the area. In addition, macrophages located in the bone marrow or liver may serve this function. Macrophages contain lysosomes whose enzymes are used in the intracellular digestion of phagocytosed erythrocytes. When red blood cells are destroyed, such by-products as the pigment bilirubin and iron (in the form of ferritin or hemosiderin) are released into the blood stream to be reutilized in the bone marrow for erythropoiesis or excreted from the body. In the fetus, the spleen produces erythrocytes and granular leukocytes. Such activity does not normally occur after birth, but under certain pathological conditions, the splenic red pulp can once again function to produce erythrocytes and granular leukocytes. Although the spleen performs a number of important functions, it can be removed without serious effect. Upon removal of the spleen, other hematopoietic tissues are capable of performing many of its functions.

Fig. 1, ×2475; Fig. 2, ×3895

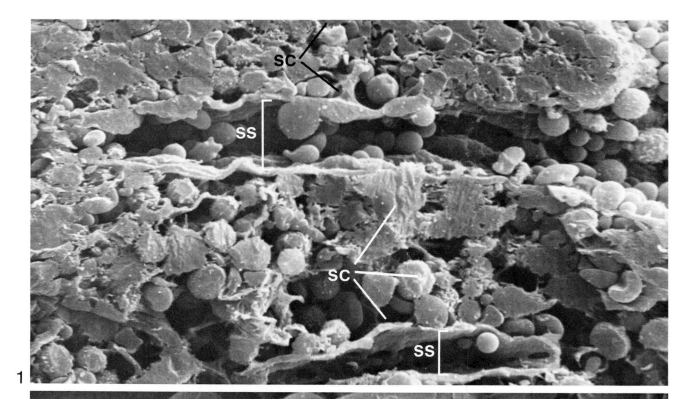

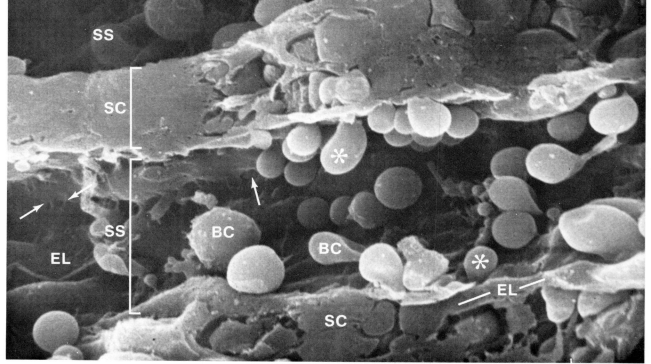

66

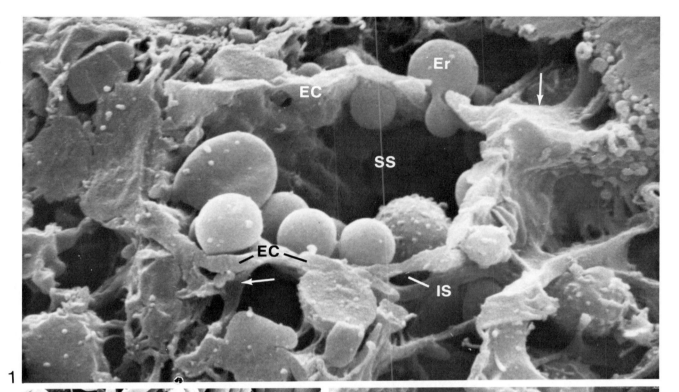

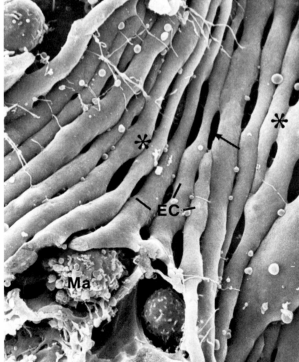

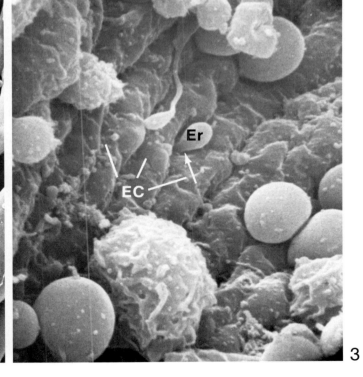

LYMPHOID ORGANS
Spleen

The splenic sinus (**SS**) illustrated in transverse section in Fig. 1 is surrounded by cellular elements of the red pulp cords. The endothelial cells (**EC**) that make up the wall of the sinus are identified. The processes of many reticular cells in the surrounding splenic cords terminate (**arrows**) or attach to the outer surface of the sinus endothelium. The external surface of the sinus endothelium has an incomplete basement membrane that is usually present in the form of bands that surround the sinus at right angles to its longitudinal axis. Adhering junctions between endothelial cells of the sinus are lacking in most areas; consequently, blood cells can easily migrate between red pulp cords and sinuses by squeezing through the intercellular spaces (**IS**) between endothelial cells. One erythrocyte (**Er** in Fig. 1) is illustrated in the process of such a migration, and is partially in the sinus and partially in the pulp cord region. A lymphocyte and several red blood cells are present in the splenic sinus. The diameter of the sinuses varies from about 10–40 μm, depending upon the amount of blood in the spleen. The interior of the splenic sinus, as illustrated in Figs. 2 (human) and 3 (rat), is lined with endothelial cells (**EC**), which are fusiform in shape. Their long axis parallels that of the sinus, and junctional complexes are only present where lateral processes of adjacent endothelial cells are in contact (✲) with each other. Thread-like microprojections often extend from the endothelial cells, and the majority of the cell surface is smooth. Macrophages (**Ma**) are commonly observed in the spaces of the red pulp, directly underlying the splenic sinuses. They possess long tendrils and irregular surface projections. Neutrophils also predominate within the meshwork of the red pulp. As noted above, the migration of erythrocytes (**Er**) is frequently observed at sites of separation (**arrows**) between endothelial cells. [Fig. 2 courtesy of T. Fujita. From "A scanning electron microscope study of the human spleen," *Arch. Histol. Jap.* 37(3):187 (1974).]

Fig. 1, ×6835; Fig. 2, ×2405; Fig. 3, ×7735

LYMPHOID ORGANS
Spleen

A small portion of the splenic red pulp is illustrated in Fig. 2, a transmission electron micrograph. The venous sinus contains red blood cells (**RBC**) and the endothelial cells (**EC**) lining the sinusoid are identified. Splenic sinuses are somewhat larger in size compared to common capillaries. In Fig. 1, an erythrocyte (**RBC**) is illustrated in the process of migrating between the endothelial cells (**EC**) of the sinus. The sinus lumen (**Lu**) is also indicated. As the erythrocyte moves through the intercellular space (**IS**) between cells of the sinus wall, it becomes greatly attenuated (**arrows**). It is currently believed that only a certain percentage of the erythrocytes circulating through the spleen enter the red pulp spaces via the open circulation, whereas others are shunted directly into the venous sinuses via the closed circulation. Physiological factors, such as changes in arterial resistance, blood flow, or blood pressure, may influence the relative degree to which either pathway is favored. Those erythrocytes that do enter the red pulp can gain access to the venous outflow only by migrating through the sinus wall into its lumen. This process, as illustrated in Fig. 1, requires great cellular plasticity, which seems to be characteristically lacking in aged red blood cells. The inability of aged erythrocytes to replace lost constituents of its membrane (i.e., proteins and lipids) or repair its membrane may be responsible for the increase in rigidity exhibited by aging cells. Changes in membrane structure may also increase the cells' susceptibility to phagocytosis by macrophages. Consequently, it appears that only red blood cells that enter the red pulp spaces are subjected to the "filtering" process of the spleen. Although some erythrocytes may at first bypass filtration through the red pulp by traveling via the closed circulatory route, subsequent recirculation of these cells through the open splenic vasculature will inevitably test their degree of plasticity.

Fig. 1, ×21,000; Fig. 2, ×5875

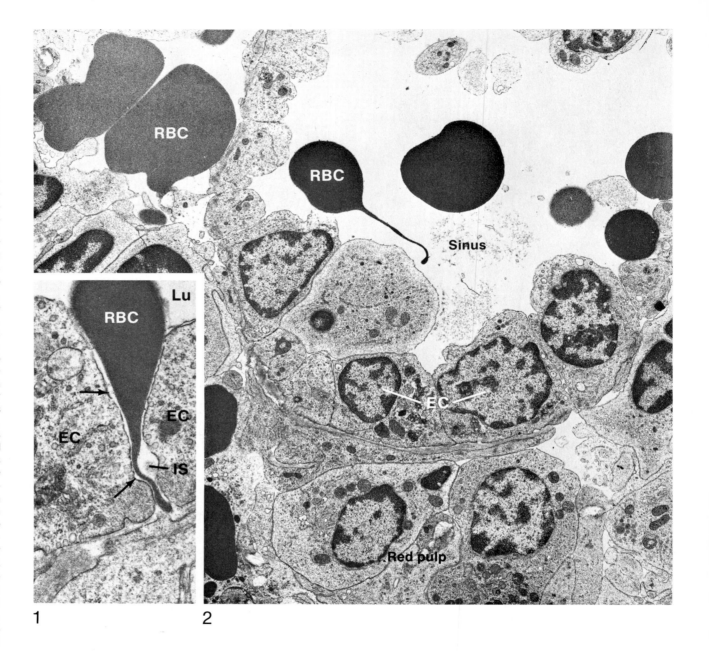

1   2

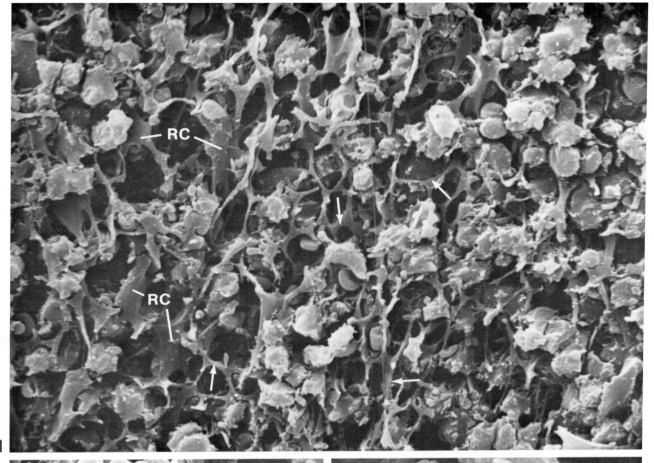

LYMPHOID ORGANS
Spleen

The red pulp, in addition to containing venous sinuses, consists of a three-dimensional framework (Fig. 1) of reticular cells (**RC**) and reticular fibers upon which resides numerous macrophages. The fiber bundles are surrounded (and thus hidden from view) by branching processes (**arrows**) of reticular cells. Effete red blood cells entering the red pulp spaces may become entrapped within the reticular framework, at which time they are likely to become phagocytosed by fixed macrophages located on the surface of reticular cells. In addition, the presence of aged erythrocytes in the red pulp apparently serves as a stimulus for the differentiation of circulating monocytes into free and fixed macrophages. Plasma cells also make up a portion of the red pulp, and, together with reticular cells, macrophages, and blood cells, they collectively form the cords of Billroth, as observed with the light microscope. In a cryofractured preparation of the red pulp (Fig. 2), it is possible to observe the cell bodies of reticular cells (**RC**) and a number of their extensions (**arrows**). The fixed macrophages (**Ma**) can be distinguished by the thin projections (**Pr**) of their surface. In Fig. 3, two free macrophages (**Ma**) can be observed on the endothelial surface of a trabecular vein.

Fig. 1, ×820; Fig. 2, ×2410; Fig. 3, ×3270

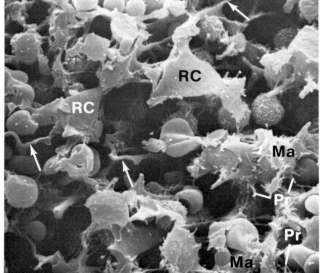

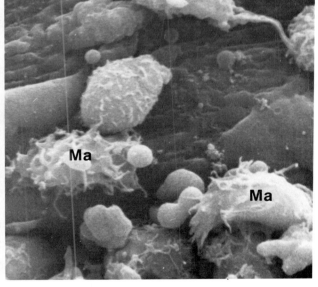

LYMPHOID TISSUE
Spleen

The transmission electron micrographs shown here illustrate macrophages, plasma cells, and red blood cells (**RBC**) within the red pulp. Macrophages contain a large number of membrane-bound inclusions called digestive vacuoles (or secondary lysosomes), which result from the fusion of phagocytosed material (e.g., erythrocytes) with primary lysosomes. The macrophage shown in Fig. 2 contains an entire nucleus derived from an ingested cell. Plasma cells can be recognized by a characteristic arrangement of heterochromatin in the nucleus, which adheres to the nuclear membrane. In addition, the plasma cell typically possesses a prominent system of rough-surfaced endoplasmic reticulum (**RER**) whose cisternae are frequently dilated with synthesized product. This product is a specific humoral antibody that is released from the cell.

Fig. 1, ×11,390; Fig. 2, ×5875

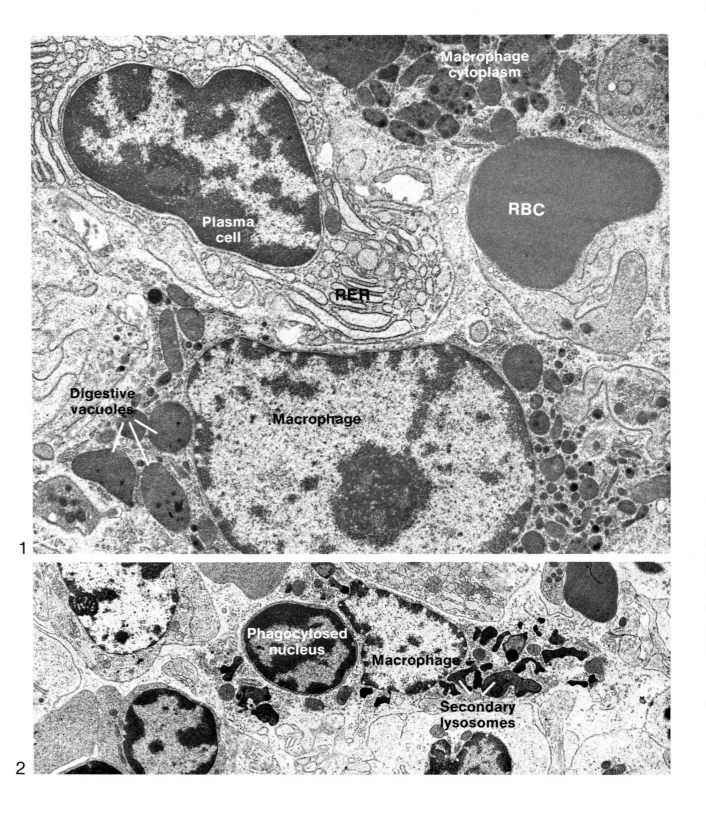

1

LYMPHOID ORGANS
Thymus

The thymus is a bilobed structure located in the mediastinum anterior to the major vessels of the heart. It is largest at birth or shortly thereafter and then gradually decreases in size. The thymus (Fig. 1) is surrounded by a thin capsule of connective tissue. At intervals the capsule extends into the parenchyma of the thymus to form septa (**Se**), which partially subdivide the lobes into lobules (**Lo**). Each lobule contains an inner or central medulla region (**Me**) and an outer cortex (**Co**). The approximate division between the cortex and medulla of a lobule is denoted by the **broken line** in Fig. 1. As a result of incomplete lobulation, the cortical and medullary zones of one lobule may be continuous with similar regions of adjacent lobules. A portion of the cortex (**Co**) of two thymic lobules divided by a connective tissue septum (**Se**) is illustrated at higher magnification in Fig. 2. Large numbers of closely packed cells can be observed in the cortex, where they are more concentrated than in the medulla of the lobule. Progenitor T cells are thought to arise in the bone marrow and seed the outer cortex of the thymus. Here the cells proliferate, giving rise to immature T cells. As the T cells, or thymocytes, migrate from the cortex into the medulla, they undergo further development and maturation, so that upon reaching the medulla they become immunocompetent cells. In their mature form, each T cell possesses its own homologous population of surface receptors capable of binding to a particular antigen. Mature T lymphocytes leave the thymus by entering vessels of the circulatory system in the medulla. T lymphocytes have a long life-span, and in addition to being found in the circulating blood, populate specific regions of lymph nodes (thymus-dependent areas), spleen, and solitary lymphatic nodules.

Fig. 1, ×35; Fig. 2, ×935

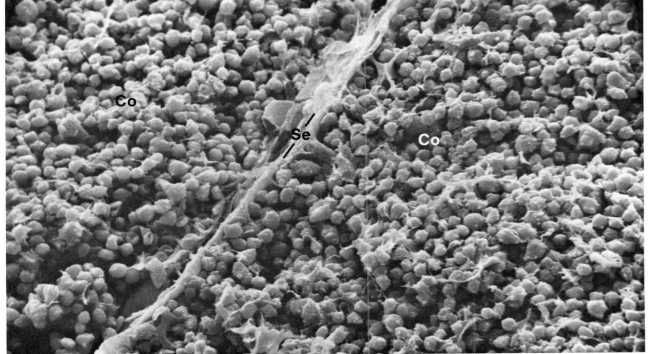

2

LYMPHOID ORGANS
Thymus

The cortex of the incomplete thymic lobules also consists of T lymphoblasts, which are capable of extensive division. The T lymphoblasts, which are larger than the T lymphocytes, are capable of producing the enormous population of T lymphocytes found in the thymus. The rate of lymphoblast cell division is greatest at birth, and it has been estimated that a milligram of mouse thymus may produce one million lymphocytes a day. The thymus develops in the embryo as an outgrowth of the endodermal epithelium of the third branchial pouches. Epithelial reticular cells derived from endoderm are thus characteristic of the thymus. During embryonic development, the thymus is colonized by stem cells derived from blood islands of the yolk sac, and perhaps from hematopoietic cells of the liver. Progenitor T cells from the bone marrow are thought to be capable of colonizing the thymus after birth. The highly branched epithelial-reticular cells are present in both the cortex and the medulla. The branches of one epithelial reticular cell may be anchored by desmosomes at their ends to branches of nearby epithelial reticular cells. Unlike other lymphoid tissues of the body, reticular fibers are not associated with the reticular cells, and the highly branched epithelial reticular cells form a *cytoreticulum,* with many T lymphocytes and T lymphoblasts packed within the interstices of the meshwork formed by the cytoreticulum. Both the T lymphocytes (**Ly**) and the epithelial reticular cells (**ER**) are illustrated in Figs. 1 and 2. In addition to providing a supporting framework for the thymic lobule, it has been suggested that the epithelial reticular cells may be the source of a variety of hormones produced by the thymus that are thought to play a role in the proliferation, differentiation, and maturation of lymphocytes derived from the thymus. In the medulla, the epithelial reticular cells may also be arranged into concentric spherical layers of cells surrounding a keratin or a calcified product. Such bodies are called thymic corpuscles, or Hassall's corpuscles. Macrophages may also be present in the medulla, but usually in reduced frequency compared to the cortex. Plasma cells are not normally found in the thymus.

Fig. 1, ×2180; Fig. 2, ×3240

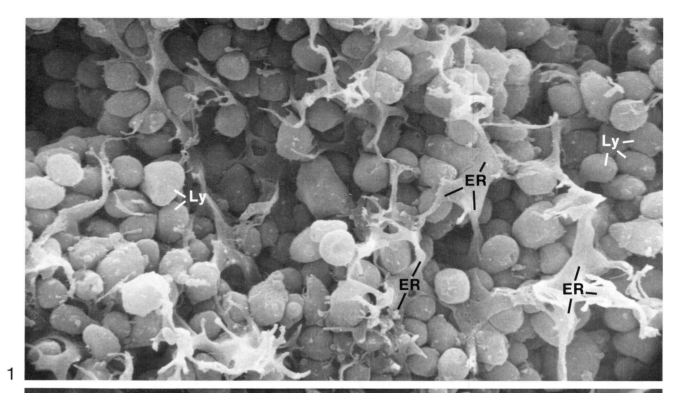

1

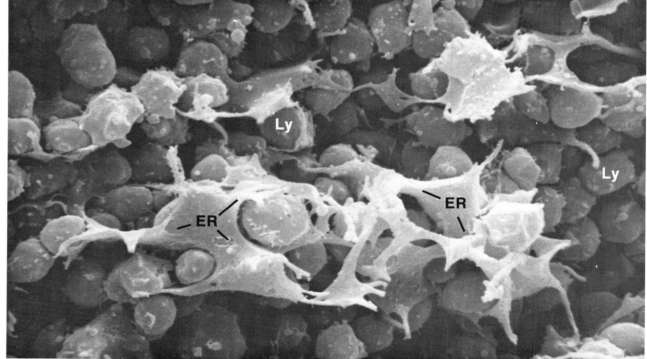

2

LYMPHOID ORGANS
Thymus

It is believed that the population of thymic lymphocytes is extremely heterogeneous in terms of their genetic code for the synthesis of diverse immunoglobulin-like receptors. As a result, a large number of T lymphocytes are produced that are capable of recognizing a wide variety of foreign antigens. Under normal conditions, however, T lymphocytes are not produced to react with the proteins of the individual or host. One theory proposes that developing T lymphocytes that might recognize a normal body protein as foreign are destroyed in the thymic cortex by a premature encounter with the proteins of the host before entering the general circulation. Thus there is enormous cell death in the thymus. The nature of the circulatory system is important in these events. Macrophages are widely distributed in both the thymic cortex and the medulla, and may possibly function in the removal of wrongly programmed lymphocytes. Alternatively, they may play a role in controlling the proliferation and differentiation of the immature thymocytes. This role of the macrophage may be similar to that seen in the bone marrow, where erythroblastic islands are formed around macrophages. Large macrophages (**Ma**) in the thymic cortex are illustrated in Figs. 1 to 3. The macrophages are surrounded by many lymphocytes (**Ly**), which appear to be attached to the macrophage surface. One of the rosette formations shown in Fig. 1 is illustrated at higher magnification in Fig. 2. A cryofracture through a similar region (Fig. 3) reveals the vacuoles (**Va**) within the macrophage cytoplasm. The vacuoles may represent a form of lysosome (digestive vacuoles), the contents of which were removed during specimen preparation. The close association between these cells and surrounding lymphocytes is still apparent. The **arrows** in Fig. 3 denote regions in which lymphocytes appear to be in an early stage of being phagocytosed by the macrophage.

Fig. 1, ×1715; Figs. 2 and 3, ×3425

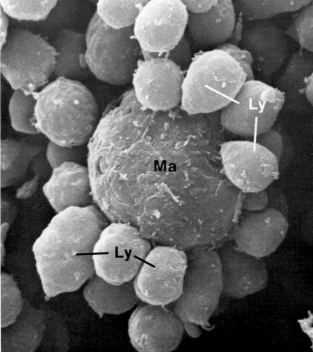

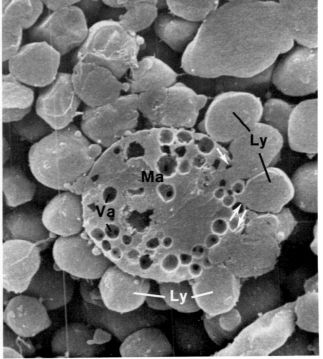

LYMPHOID ORGANS
Thymus

In Fig. 1, the cryofracture through a macrophage (**Ma**) and associated lymphocytes (**Ly**) reveals an area in which lymphocytes may be in the process of being engulfed (**arrow**) by the macrophage. In addition, a small lymphocyte (✳) appears to be sequestered within the vacuolated (**Va**) cytoplasm of the macrophage. Phagocytosis of thymocytes by macrophages has been demonstrated by transmission electron microscopy. As shown in Fig. 2, lymphocytes appear to be in contact with the plasma membrane of the macrophage, and one lymphocyte can be seen within the macrophage cytoplasm. The thymic cortex is supplied only by capillaries (Fig. 3) that originate from arterioles situated at the junction between the cortex and medulla. The cortical capillaries (**Ca**) are surrounded by a continuous basement membrane (**BM**), and pericytes (**Pe**) are present in the subendothelial space (**SS**). Macrophages may also be present at intervals in this subendothelial space. Epithelial reticular cells (**ER**) form a sheath that invests the capillary and subendothelial space. The epithelial reticular cells also have basement membrane (**EBM**) material associated with their inner surface. The arrangement of the structures just described, together with the development of tight junctions between endothelial cells, constitutes the thymic blood barrier, a device thought to prevent antigens from entering the thymic cortex after birth, at a time when lymphocytes are differentiating. The blood thymic barrier is thought to be leaky during embryonic development, allowing the body proteins to enter the cortex, with the result that forming lymphocytes that might react to normal body proteins are destroyed. The barrier becomes tight just before birth to protect the forming lymphocytes from a premature encounter with a foreign protein that might enter the circulation. Medullary capillaries (**Ca** in Fig. 4) do not possess the organization characteristic of those present in the cortex, and do not form a thymic blood barrier. Lymphocytes (**Ly**) and erythrocytes (**Er**) are identified in Fig. 4. [Fig. 2 courtesy of P. E. Lipsky and A. S. Rosenthal. From "Macrophage-lymphocyte interaction," *Jour. Exper. Med.* 138:900–924 (1973). The Rockefeller University Press.]

Fig. 1, 3745; Figs. 2 and 3, ×4050; Fig. 4. ×5140

74

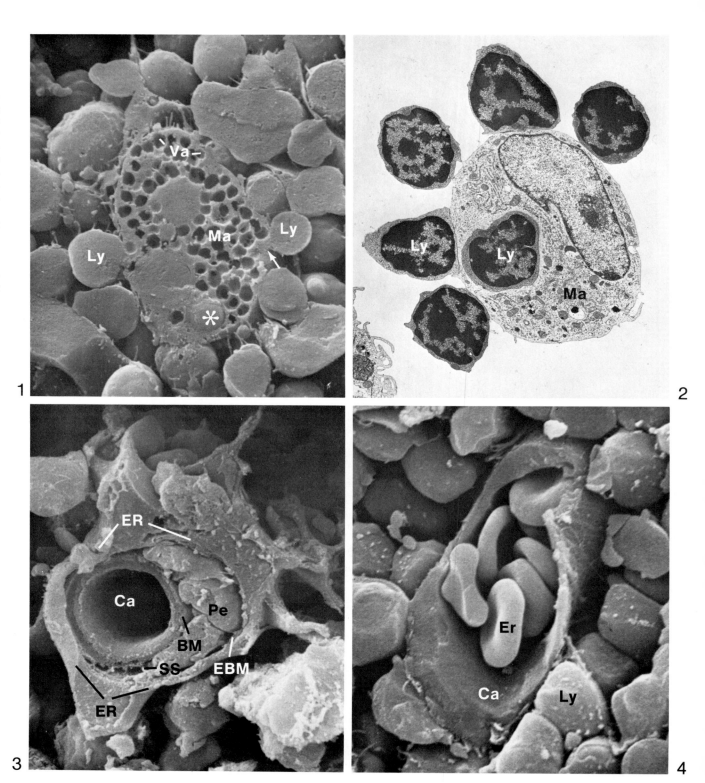

NERVOUS TISSUE

Introduction

The neuron—the basic constituent of all nervous tissue—is unique in that it possesses the specialized properties of irritability and conductibility. For the most part, these properties reside in the neuron's plasma membrane, since that is where electrical excitation is initiated and conduction of the resulting impulse occurs. In an attempt to understand those properties that permit the neuron's plasma membrane to perform such specialized functions, investigators have synthesized prototype membranes that exhibit similar features. A number of molecules have recently been introduced (e.g., alamethacin), that can assemble to form "ion channels" of low resistance within a synthesized membrane (lipid bilayer). The ability of such molecules to form these channels is a function of electrical potential (voltage) applied to the membrane. Such ion selection and controlled membrane resistance closely resemble the events leading to the initiation and propagation of an action potential as observed in living neurons.

Although the initiation and conduction of an electrical impulse are important requirements for a functional nervous system, the ability to transmit these impulses from one neuron to the next is equally important. Historically, neurons were first thought to form a connecting network, but it is now accepted doctrine that processes of neurons are closely associated, but do not touch. They are separated by a specialized intercellular synaptic gap (measuring approximately 200 Å in width) over which impulses are transmitted from one neuron to the next. The transmission of the impulses occurs by the release of specific chemicals (neurotransmitters) contained within membrane-bound packets located within the cytoplasm of the nerve terminus. An electrical impulse propagated to the end of a neuronal process triggers a series of events that leads to fusion of the intracellular membrane-bound vesicles with the plasma membrane, thus causing the release of the chemical contents into the synaptic gap by exocytosis. The neurotransmitter then diffuses across the intercellular space and, upon reaching the membrane of the adjacent neuronal process, elicits an electrical potential. In some cases, the chemical transmitter may increase the probability that an action potential may result in the postsynaptic neuron; in other cases, the net effect may be inhibitory. Since a number of neuronal processes may converge upon one postsynaptic neuron, the net effect depends on the summation of synaptic activity of all the neurons converging at the synapse. Furthermore, the chemical synapse is polarized, so that impulses may be transmitted in only one direction.

The chemical synapse has been widely studied because, as a special interneuronal space, it is accessible to therapeutic influence through the administration of specific drugs, which, upon gaining access to the synaptic gap region, may act as an agonist or antagonist of specific neurotransmitters. Once in the synaptic gap, such chemicals may mimic the effect of a neurotransmitter, inhibit the metabolic inactivation of the neurotransmitter, or compete for the transmitter receptor site on the postsynaptic neuronal membrane. Moreover, since certain neuronal pathways are characterized by specific neurotransmitters (e.g., acetylcholine, norepinephrine, dopamine, gamma-amino-butyric acid, serotonin), such pathways may be selectively stimulated or inhibited by specific drugs. Such pathway specificity with respect to neurotransmitters has provided the basis for therapeutics in the field of neuropharmacology. But since many pathways are mediated by the same neurotransmitter, a therapeutic administration of a drug is often accompanied by side effects (e.g., the use of anticholinergic agents as decongestants may also produce a dry throat).

Although it is beyond the scope of this chapter to include details of specific neuronal pathways, it is useful to understand the cellular organization and the constituents of the nervous system. Anatomically, nervous tissue can be divided into the central nervous system, which consists of the brain and spinal cord, and the peripheral nervous system, which consists primarily of neuronal processes (dendrites and axons) and small aggregations of nerve cells referred to as ganglia. The central nervous system is further distinguished by areas of gray and white matter. The gray matter contains nerve cell bodies, their respective processes, and nonneuronal cells (neuroglial cells). The white matter, however, is devoid of nerve cell bodies and consists primarily of axons enwrapped in a myelin sheath, which is formed by successive wrappings of the plasma membrane of specialized neuroglial cells. The myelin sheath, acting as an insulator, allows electrical impulses to propagate much faster down the length of an axon by the process known as saltatory conduction. In saltatory conduction, gaps (nodes of Ranvier) present along the lengths of the myelin sheath enable the electrical impulse to "jump" from one nonmyelinated gap to the next, thus increasing the speed of conduction. Myelinated axons are also present in the peripheral nervous system, but the myelin sheath here is formed by cells embryologically derived from the neural crest (Schwann cells).

Since the central nervous system develops embryologically as a hollow neural tube, it is not surprising that fluid-filled cavities persist after its formation. The four communicating cavities in the brain are called ventricles, and consist of two lateral ventricles in the cerebral hemispheres, a third ventricle in the diencephalon, and a fourth ventricle in the medulla oblongata. The ventricles are filled with cerebrospinal fluid, partially derived from specialized areas of tissue vasculature (choroid plexuses) that project into the ventricles at localized areas.

From a morphological standpoint, neurons may be classified by the number of processes they possess. In general, the neuron may take the form of a pseudounipolar, bipolar, or multipolar cell. Pseudounipolar neurons are primarily sensory, and are found at such sites in the peripheral nervous system as the dorsal root ganglia. The pseudounipolar neuron is characterized by a single process that projects from the cell body (perikar-

yon) and divides at a short distance from the cell into two branches traveling in opposite directions. Electrical impulses are received by one of the branches and are conducted toward the perikaryon, whereas the other branch conducts impulses away from the cell body. The bipolar neuron, as its name implies, possesses two processes, one axon and one dendrite. Each process extends from opposite ends of a fusiform cell body. Examples of bipolar neurons are found in the inner nuclear layer of the retina, the vestibular and cochlear ganglia, and make up the sensory portion of the olfactory epithelium. The bulk of the neurons found in the central nervous system are multipolar, and possess several poles at which processes emerge. There are usually numerous dendrites that possess secondary branches, and the axon arises from a small conical elevation of the cell body called the axon hillock.

Neurons have also been classified on the basis of axonal length (the Golgi classification). The term Golgi type I neuron refers to those nerve cells that possess long axons, particularly those whose axons leave the central nervous system via the ventral roots of spinal nerves or the main trunks of cranial nerves. The term Golgi type II neuron refers to nerve cells with short axons, most of which remain inside the central nervous system (association or interneuron type).

Nerve cells and neuroglial cells together constitute a substantial component of the central nervous system. The neuroglial cells ("nerve glue") include three characteristic types: the astrocyte (fibrous and protoplasmic), the oligodendrocyte, and the microglia. Since the central nervous system contains little or no connective tissue for support, it has generally been accepted that neuroglial cells perform this function, but it is now held that these cells provide more than just support.

Astrocytes, the largest of the glial cells, possess

numerous long, narrow processes. The cytoplasmic extensions of the astrocyte are thought to provide substantial support for neuronal elements. Many of the processes terminate on the surface of blood vessels and have been referred to as "vascular feet." In this respect, astrocyte processes may mediate the transport of nutrients and metabolites between capillaries and neurons. By means of an active ion pump in their cell membrane, astrocytes also function to maintain the appropriate electrolyte concentration in the vicinity of neurons. Oligodendrocytes appear smaller than astrocytes and possess fewer processes. In the gray matter of the central nervous system, they are often closely associated with the surface of the nerve cell body. Recent evidence has indicated that a type of symbiotic metabolism may exist between the neuron and the oligodendrocyte. In the white matter, the oligodendrocyte provides the myelin sheath that surrounds the numerous axonal processes. In contrast to astrocytes and oligodendrocytes, which originate from the ectoderm of the neural tube, microglia are derived from mesoderm. It is currently believed that microglia have phagocytic capabilities, particularly during injury, and may represent the central nervous system counterpart of the macrophage.

The first portion of this chapter will focus primarily on the basic ultrastructure and surface properties of the multipolar neuron and neuroglial cells. The ventricular cavities of the brain, their lining cells (ependyma), and the choroid plexuses will also be illustrated. The major portion of the chapter, however, will concentrate on the specialized neurosensory organs (i.e., taste buds, olfactory epithelium, the eye, and the ear). With the proper techniques and fine dissection, these organs are particularly well suited for study by scanning electron microscopy.

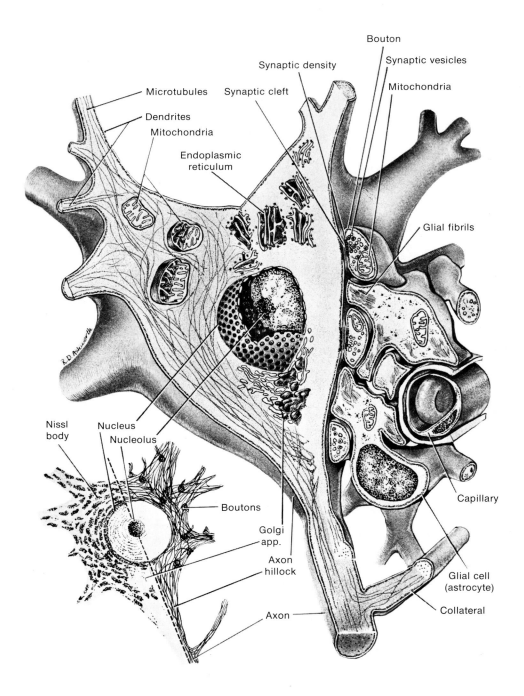

NERVOUS TISSUE
Neurons and Glial Cells

The neuron and associated structures diagrammed here illustrate features typically observed at the electron microscopic level. Neurons characteristically consist of a cell body (perikaryon) with numerous cell processes (axon and dendrites). The nucleus, nucleolus, chromatin, and nuclear pores are represented, as well as the intracellular location of the Golgi apparatus, mitochondria, and endoplasmic reticulum. The bundles of neurofilaments and microtubules present in both the axons and dendrites are thought to play a role in the transport of molecules along the length of the processes. Synaptic endings, glial cell (astrocyte) processes, and a nearby capillary containing an erythrocyte are also illustrated. The drawing at the lower left shows the organization of the perikaryon as it would be viewed with the light microscope. The portion of the drawing to the left of the broken line represents structures observed with the Nissl stain. These include the nucleus and nucleolus as well as Nissl bodies (rough endoplasmic reticulum), which are observed in both the cytoplasm of the perikaryon and in proximal dendrites. Note the absence of Nissl bodies in the axon and axon hillock. With the Nissl stain the Golgi apparatus appears as a negative image. To the right of the broken line are structures commonly observed when metal stains are used, and these include neurotubules and neurofilaments. [From William D. Willis, Jr., and Robert G. Grossman, *Medical Neurobiology* (2nd ed.). St. Louis: C. V. Mosby, 1977.]

NERVOUS TISSUE
Neurons and Glial Cells

On the surface of the multipolar neurons shown in Fig. 1 are the remains of numerous synaptic boutons. The boutons appear as small bulbous terminations of long slender axons, the majority of which have been broken along their length during preparation. Each bouton represents the specialized junctional area between axon and perikaryon that forms the chemical synapse. The neuron present in the center of Fig. 2 is characterized by a cylindrical axon (**Ax**) that projects from its surface and functions to conduct impulses away from the perikaryon. The length and diameter of axons vary according to the type of neuron. Surface features characteristic of the multipolar neuron illustrated in Fig. 3 correspond to those diagrammed on the previous page. The **arrow** in Fig. 3 identifies a synapse between a neuronal process and perikaryon surface. The two small, irregularly shaped cells observed in close association with the surface of the perikaryon (**Pe**) are called satellite cells (**SC**). The term "satellite cell" refers primarily to the astrocyte and oligodendroglia. Oligodendroglia are smaller than astrocytes; their processes are shorter and less numerous. In grey matter, oligodendroglia remain closely associated with the neuronal surface, resembling the satellite cells illustrated in Fig. 3. There appears to be a metabolic dependency between neurons and oligodendroglia, such that factors affecting the metabolism of the neuron will be reflected by changes in metabolism of the oligodendroglia (and vice versa). Using cinematography, the oligodendroglia have been observed to undergo rhythmic pulsatory movements as observed in tissue culture. The survival of cultured neurons is relatively low if the satellite cells are removed, but the significance of this symbiotic association has yet to be appreciated. [Figs. 1, 2, and 3 courtesy of Nabil Azzam.]

Fig. 1, ×1740; Fig. 2, ×1960; Fig. 3, ×9970

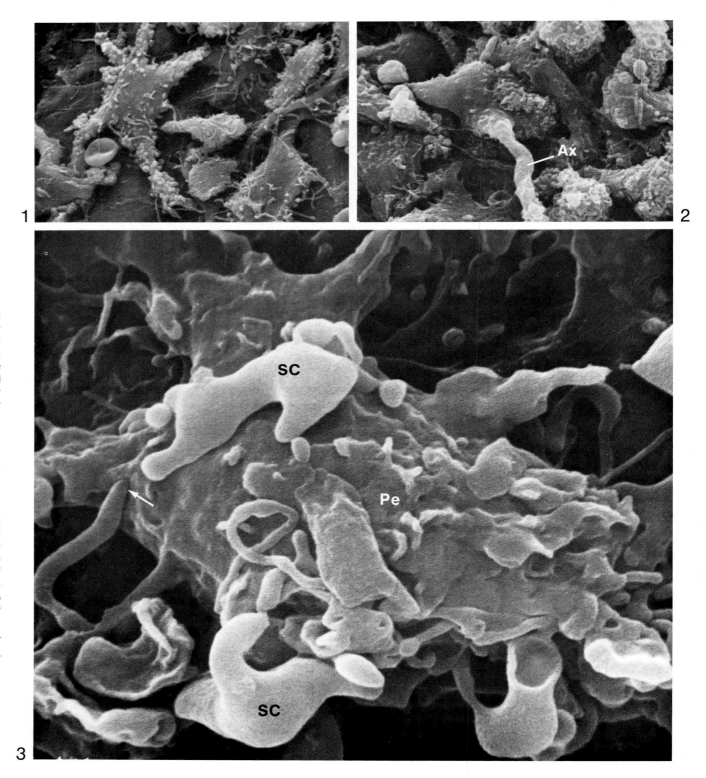

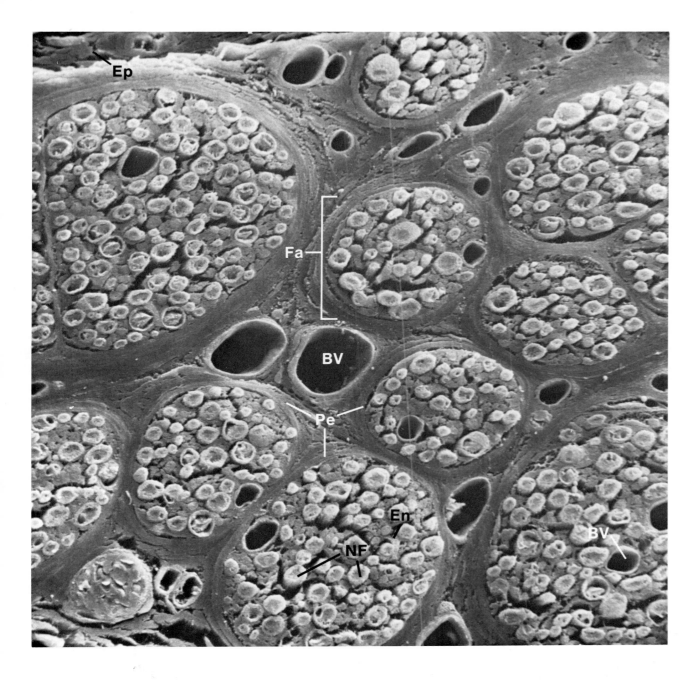

NERVOUS TISSUE
Nerve Fibers

The nerve axons and dendrites that course outside the central nervous system are organized into parallel nerve fiber bundles referred to as fascicles (**Fa**). Individual fascicles do not run as isolated cables but commonly branch at acute angles and connect with adjacent fascicles to interchange nerve fibers. A band of loose connective tissue called epineurium (**Ep**) encircles groups of fascicles binding them together so that they collectively form an anatomical nerve, a portion of which is shown in this micrograph. When the epineurium divides into distinct sheaths of dense connective tissue enclosing individual fascicles, it is then called perineurium (**Pe**). A further division of the perineurium, the endoneurium (**En**), forms thin layers of loose connective tissue, which surround single nerve fibers (**NF**). Numerous blood vessels (**BV**) within the endoneurium and perineurium supply the nerve fibers with nutrients. The nerve fibers within each fascicle are of both myelinated and unmyelinated types, and, in addition, there is considerable variability in their diameter.

×900

79

NERVOUS TISSUE
Nerve Fibers

When a portion of the connective tissue sheath surrounding nerve fascicles is removed (Fig. 1), underlying myelinated nerve fibers (**NF**) are observed. The external surface of the Schwann cells that surround myelinated axons is illustrated in greater detail in Fig. 2. Note the variation in axon diameter. The endoneurium surrounding individual axons is made up of numerous strands of collagenic fibers (**CF**), and capillaries are often observed in close association with the connective tissue. The deep transverse folds in some regions along the myelinated nerves may represent nodes of Ranvier, although positive identification is not possible without correlative transmission electron microscopy. Shallow folds are present on the Schwann cell surface. A cryofracture preparation of the sciatic nerve (Figs. 3 and 4) illustrates the myelin sheath (**MS**) formed from the successive wrapping of the Schwann cell plasma membrane about the axon. Fig. 3 shows the relationship between a Schwann cell nucleus (**SN**) and the myelin sheath. In transverse sections of fixed and fractured myelinated nerve fibers, the axoplasm commonly has a fibrillar appearance (Fig. 3). This appearance is probably related to coagulation of the axoplasm and to the numerous neurofilaments and neurotubules that extend the length of the processes. The connective tissue endoneurium (**En**) is particularly evident in Fig. 4.

Fig. 1, ×1620; Fig. 2, ×1550;
Fig. 3, ×2775; Fig. 4, ×6940

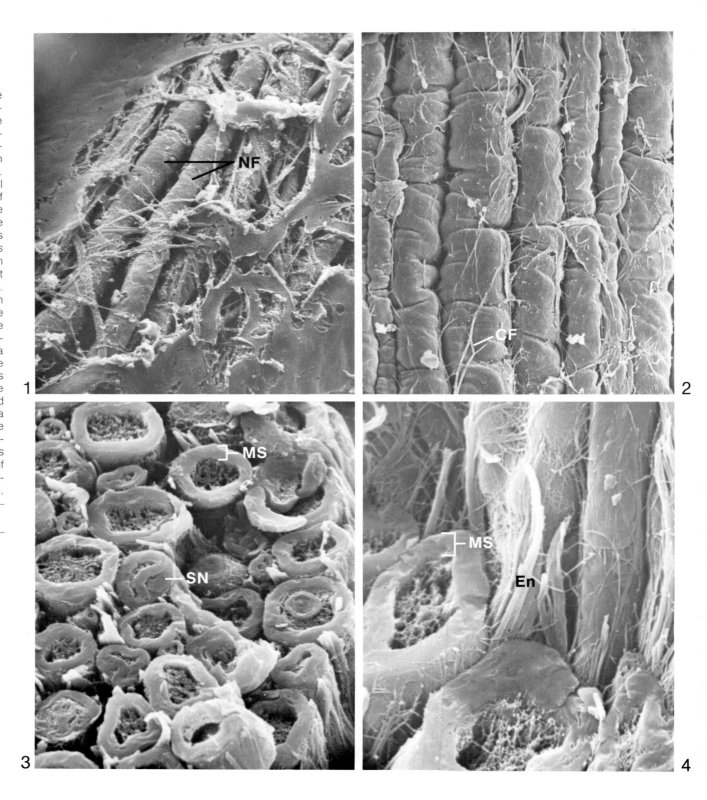

NERVOUS TISSUE
Ventricles

Much of the ventricle surface (as illustrated from the rabbit in Fig. 1) is lined by a layer of ependyma cells, which range in shape from cuboidal to squamous. The majority of the ependyma cells possess cilia (Ci) which project from their apical surfaces into the ventricular cavity. A higher-magnification view of the ciliated cells is shown in Fig. 2. The rhythmic beat of the cilia may assist in the movement of cerebrospinal fluid throughout the four communicating ventricular cavities of the brain. In some areas of the ventricles, the ciliated cells are less concentrated and appear to be organized into linear rows of cells. Adjacent to these areas there is often a transition (Fig. 3) into an area devoid of cilia: the nonciliated cells of the ventricle in this area have been termed tanycytes (Ta). A network of multipolar cells resembling neurons is commonly observed on the tanycyte surface; these cells have been referred to as supraependyma cells (SC). Both tanycytes and supraependyma cells have received considerable attention because they have been observed in areas of the ventricle that directly overlie the circumventricular organs. For example, on the floor of the third ventricle, the basal processes of tanycytes terminate on the underlying perivascular space surrounding the portal capillaries of the hypophyseal circulation. Evidence accumulated so far suggests that the supraependyma cells may secrete bioactive peptides (hormones and neurotransmitters) into the cerebrospinal fluid in addition to those liberated from the free ends of axons terminating in the ventricle. Both supraependyma cells and tanycytes are thought to constitute a network that may also function as receptors, sensing local changes in the concentration of molecules within cerebrospinal fluid. They may even transport them into the portal hypophyseal vascular bed, which might ultimately affect and regulate endocrine secretion. In this manner, these specialized cells may play the role of transducers, functionally integrating the blood, brain, and cerebrospinal fluid into one neuroendocrine unit.

Fig. 1; ×1825; Fig. 2, ×3010; Fig. 3, ×1780

NERVOUS TISSUE
Choroid Plexus

The well-vascularized areas derived from pia mater that evaginate from the roof of the ventricles are known as the choroid plexuses. Choroid plexuses are thought to be the major sites for the formation of cerebrospinal fluid. They are not the only site of production, however, as evidenced by the fact that formation of cerebrospinal fluid still remains 40% of normal upon their removal. Vascularized areas within the subarachnoid space are other possible production sites. In Fig. 1, a portion of a choroid plexus (**CP**) projects into a ventricle (**Ve**). The choroid plexus is highly folded, and its surface is covered by a specialized cuboidal layer of epithelium, which differs somewhat from the ciliated ependyma cells (**EC**) that line most of the ventricular surface. Because a portion of the choroid plexus has been sectioned, a number of blood vessels (**BV**) can be visualized internal to the epithelial covering. The cuboidal epithelial cells (Fig. 2) possess numerous irregular microvilli (**Mv**) whose free ends are frequently dilated. The dilation may be artifactual, but it is more likely that it is a manifestation of changes in functional state, such as increased secretory activity. A vascular cast of the choroid plexus, from which all tissue constituents have been removed, is illustrated in Fig. 3. A large capillary network (**CN**) is interposed between an arterial supply and venous drainage. Although most of the cerebrospinal fluid is derived from the extensive capillary bed, it is important to realize that cerebrospinal fluid is not merely a blood filtrate. The choroid epithelium is capable of transporting, both actively and passively, ions and large molecules into the cerebrospinal fluid. The high concentration of the enzyme carbonic anhydrase within these cells may be related to the transport of hydrogen, bicarbonate, sodium, and chloride ions and water. [Fig. 2 courtesy of Nabil Azzam.]

Fig. 1, ×325; Fig. 2, ×3410; Fig. 3, ×150

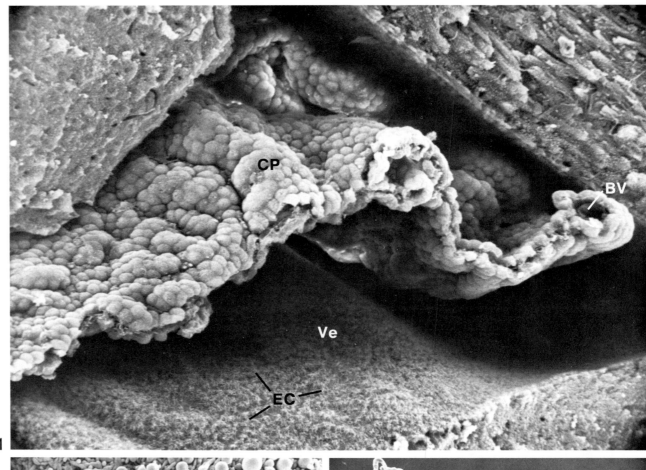

1

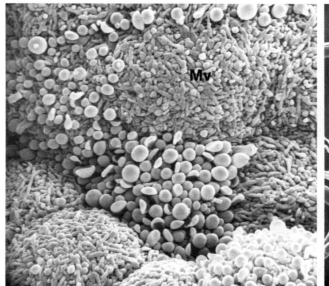

2

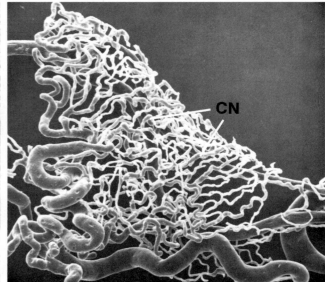

3

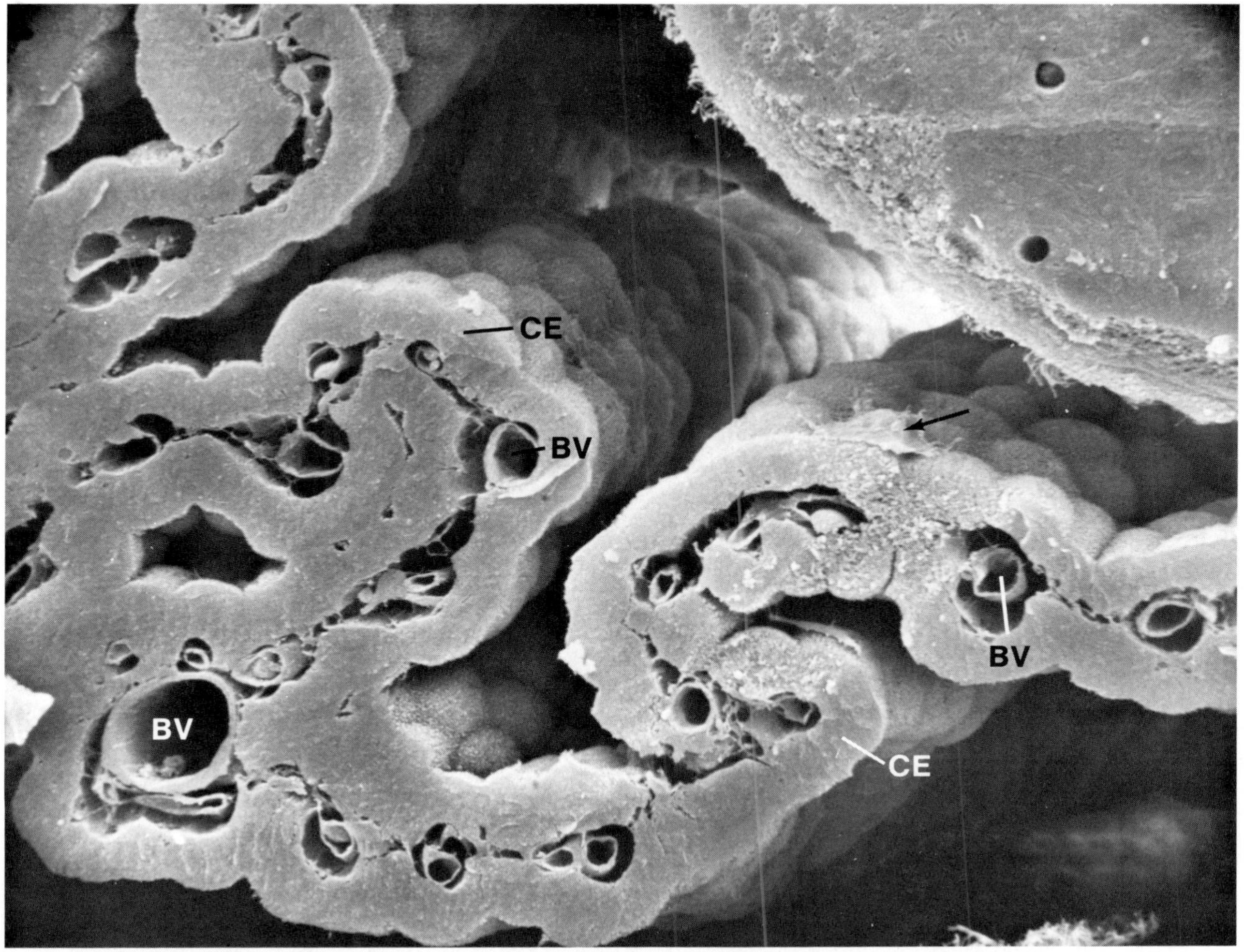

NERVOUS TISSUE
Choroid Plexus

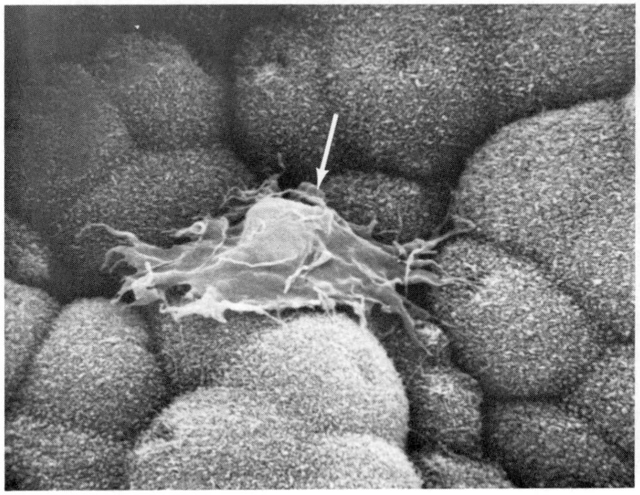

The cryofractured preparation of the choroid plexus shown in Fig. 1 is useful for demonstrating the relationship of the cuboidal epithelium (**CE**) to the internal plexus of blood vessels (**BV**). Ions that diffuse through the fenestrated endothelial lining of the capillaries are actively transported by the cuboidal epithelium into the ventricles. Water osmotically accompanies these ions, contributing to the formation of cerebrospinal fluid. Tight junctions between epithelial cells prevent the extracellular passage of large molecules into the ventricle. Within the connective tissue stroma surrounding the blood vessels, numerous macrophages are frequently observed. Macrophages (**arrows**) may also be present on the epithelial surface of the choroid plexus. As observed at higher magnification in Figs. 2 and 3, the macrophages often possess a ruffled cell border exhibiting thin cytoplasmic extensions that spread over the microvillous surface of the choroid epithelial cells. [Fig. 3 courtesy of Nabil Azzam.]

Fig. 1, ×1010; Fig. 2, ×1930; Fig. 3, ×2780

NERVOUS TISSUE
Eye—Organization

This diagram illustrates the overall organization of the eye as it would be viewed in a median section. Specific parts will be illustrated in succeeding diagrams, as well as in scanning and transmission electron micrographs. The external coat of the eye consists of a sclera that becomes continuous anteriorly at the limbus with the transparent cornea. The dense connective tissue of the sclera serves to protect the eye, to maintain turgor pressure and shape of the eye, and is the attachment site for extraocular muscles used in moving the eye. A middle layer, called the uvea, consists of a choroid coat that is continuous anteriorly with the ciliary body and iris. An inner photosensitive region, the retina, terminates anteriorly at the ora serrata. There are three internal chambers; the two smaller anterior and posterior chambers (with respect to the iris) are filled with aqueous humor, and the large vitreous body is located posterior to the lens. Cloquet's (hyaloid) canal represents the remains of the embryonic hyaloid artery. The lens is suspended by zonule filaments that extend from the ciliary part of the retina and ciliary body. The anterior surface of the lens is closely associated with an aperture in the iris, called the pupil, the size of which can be varied by action of the circular and radial smooth muscle in the iris. Many optic nerve fibers, which are axons of neurons in the retina, pass via the optic nerve to the brain. The eye has a multiple embryonic origin. The retina and optic nerve result from an outgrowth of the embryonic forebrain toward the superficial ectoderm of the head region. In contrast, the lens and anterior corneal epithelium are induced to form as a thickening of superficial ectoderm by the outgrowth of the embryonic brain. The choroid, ciliary body, sclera, and remainder of the cornea differentiate from mesoderm. [Diagram © Copyright 1962 CIBA Pharmaceutical Company, Division of CIBA-GEIGY Corporation. Reproduced with permission from *Clinical Symposia,* illustrated by Frank H. Netter, M.D. All rights reserved.]

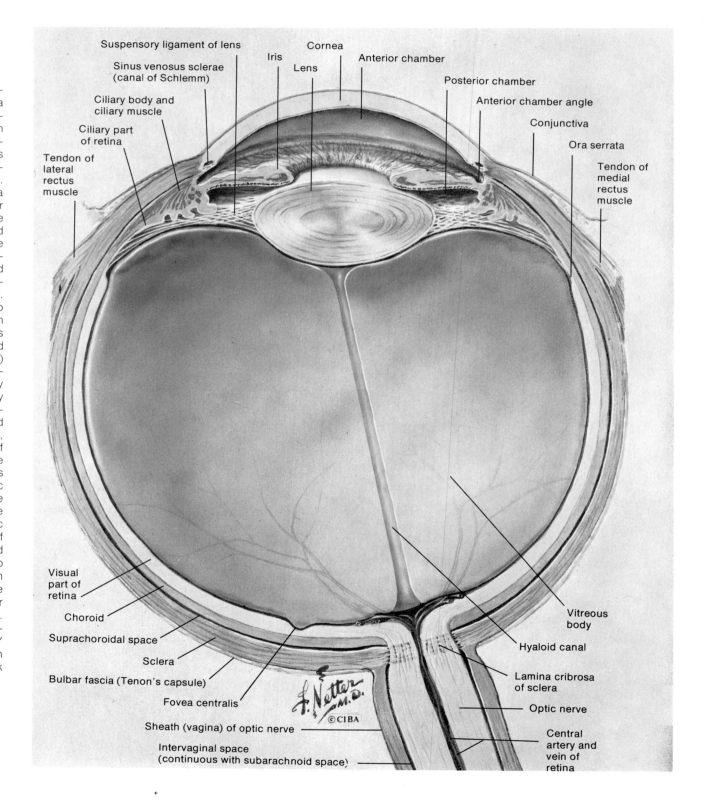

84

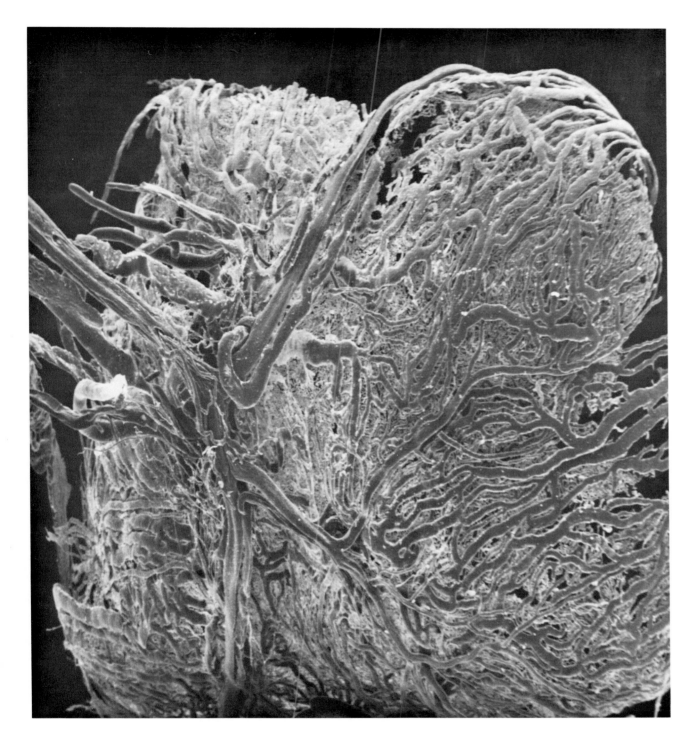

NERVOUS TISSUE
Eye—Circulation

The eye is extensively vascularized, but most of the vessels are restricted to the uvea. Capillaries in the inner layers of the choroid are unique in their arrangement into a layer that is capable of supplying nutrients to the retina. Branches of the ophthalmic artery supply blood to the eye. Shown in this figure is a cast of the blood vascular system of the guinea pig eye, as viewed from the posterior surface. From such a cast, the extensive supply of blood vessels in the uvea can be better appreciated than from examination of sections. The position formerly occupied by the optic nerve would be in the middle left of the figure. Most of the blood supply is derived from arterial branches in the uvea, but the central retinal artery is distributed within the inner portion of the retina. The posterior ciliary arteries are of different lengths. Short branches of the posterior ciliary artery enter the choroid around the optic nerve and supply the posterior part of the retina to the equator via branches called the posterior choriocapillaris. Long branches of the posterior ciliary artery supply, via branches called the anterior choriocapillaris, the retina from the ora serrata posteriorly to the equator. The choriocapillaris is an extensive network of highly branched capillaries. The anterior ciliary arteries pass through the sclera to supply the ciliary body. Some of the branches of this vessel in the ciliary body pass posteriorly to become continuous with the anterior choriocapillaris. Venous drainage is through a system of vortex veins. The vortex veins, after joining to form several internal sclera veins, may form ampullae, which then pierce the posterior sclera in several places and are then termed posterior scleral veins. Most of the venous drainage of the eye leaves through the posterior scleral veins, but a few venous vessels in the iris and ciliary body join the episcleral veins near the limbus region.

×50

NERVOUS TISSUE
Eye—Choroid

The choroid (**Ch**) layer illustrated in Fig. 1 extends from the edge of the optic nerve anteriorly to the ora serrata, where it is continuous with the ciliary body. It is subdivided into three layers, including a vessel layer (**VL**), the capillary layer (**CL**), also called the choriocapillaris, and a thin glassy membrane, Bruch's membrane (**BM**), adjacent to a single layer of pigmented epithelium (**PE**) of the retina. A portion of the sclera with closely packed collagenic fiber (**CF**) bundles and the broken tips of rod outer segments (**RC**) are included in Fig. 1. The vascular casts in Figs. 2 and 3 show the extensive capillary network that makes up the posterior choriocapillaris (**CC**), which is supplied by short branches of the posterior ciliary artery (**CA** in Fig. 3). The choriocapillaris network, viewed from the vitreous side in Fig. 2, is arranged into a sheet within the inner layers of the choroid, close to the outer surface of the retina. The choriocapillaris network, viewed from the exterior surface of the eye, drains into venules (**Ve** in Fig. 3) within the choroid and then into a vortex system of veins. The vessels of the choriocapillaris are larger than normal capillaries, and the endothelium is fenestrated. The fenestrations in the capillary endothelium probably serve to enhance the transfer of nutrients from the vessels in the choroid to the adjacent retina. Pressure in both the arterial and venous blood vessels of the eye must be maintained higher than intraocular pressure for normal blood circulation and proper nutrition of the retina and other portions of the eye.

Fig. 1, ×2025; Fig. 2, ×220; Fig. 3, ×195

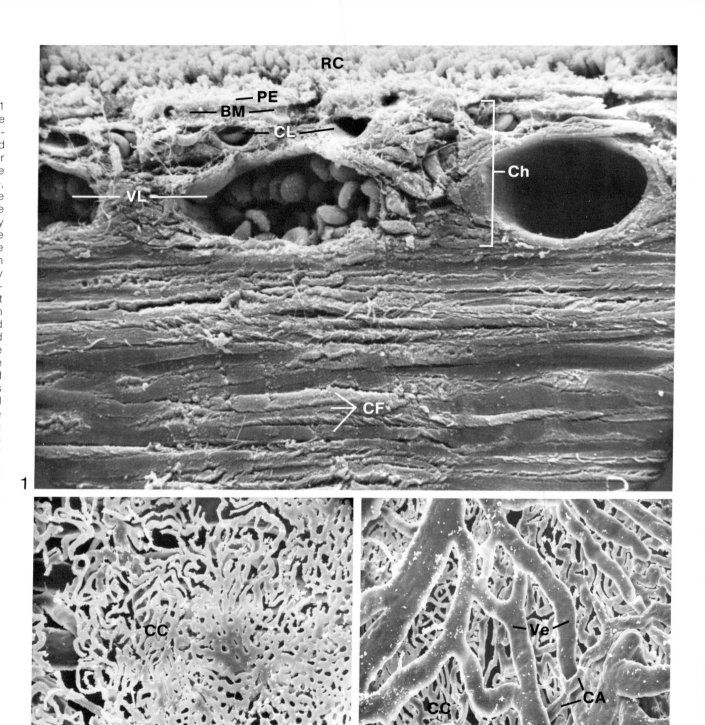

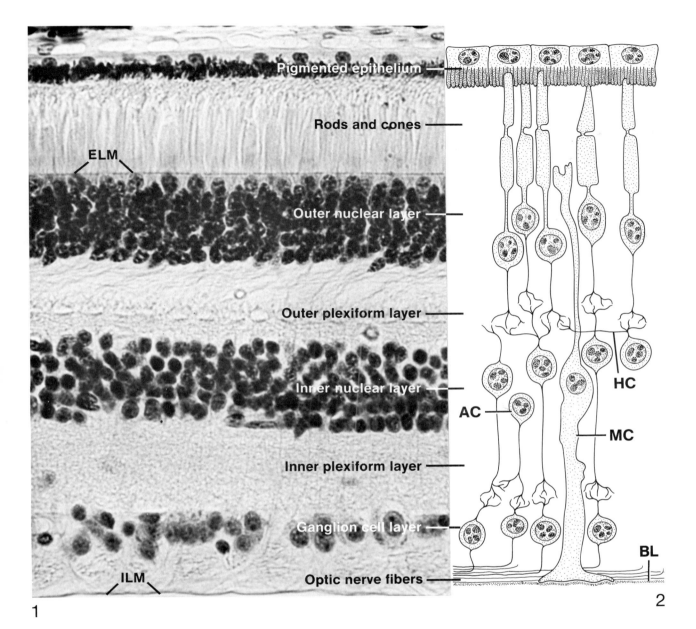

Pigmented epithelium

Rods and cones

ELM

Outer nuclear layer

Outer plexiform layer

Inner nuclear layer

Inner plexiform layer

Ganglion cell layer

ILM

Optic nerve fibers

1

AC

HC

MC

BL

2

NERVOUS TISSUE
Eye—Retina

Before light impinges on the photosensitive rod and cone cells, it must pass through several layers of nerve cells and their processes, which also reside within the retina. The layers of the retina are identified in the light photomicrograph in Fig. 1. A corresponding diagram in Fig. 2 illustrates the synaptic relationships between the cells that form each layer. A layer of pigmented epithelial cells partially invests the tips of photosensitive outer segments of rod and cone cells. Nuclei of rod and cone cells are located in the outer nuclear layer. Axons of these cells extend into the outer plexiform layer and synapse with the dendrites of bipolar neurons. Nuclei of the bipolar neurons make up the inner nuclear layer. Axons of the bipolar neurons extend into and partially form the inner plexiform layer. In this layer, the axons of the bipolar neurons synapse with the dendrites of neurons that form the ganglion cell layer. Axons of those neurons located in the ganglion cell layer extend as a layer of optic nerve fibers to and through the optic nerve to the brain. Supportive cells of the retina, called Müller cells, span much of the retina. Their nuclei are located in the inner nuclear layer. The internal limiting membrane (**ILM**) represents the basal lamina (**BL**) of the Müller cells. The external limiting membrane (**ELM**) is actually a linear arrangement of junctional complexes between localized regions of rods and cones and adjacent Müller cells. Horizontal cells (**HC**), shown in Fig. 2, synapse with axons of a number of rod cells, and in so doing, integrate impulses between rod cells. The amacrine cell (**AC** in Fig. 2) is also important in the integration of electrical transmission between retinal cells. The amacrine cell synapses with axons of the bipolar neurons and dendrites of ganglion cells, but the nature of its influence has yet to be understood.

×1070

NERVOUS TISSUE
Eye—Retina

For comparison with the light photomicrograph and diagram on the previous page, a scanning electron micrograph of the rabbit retina is shown here. The pigmented epithelium (**PE**), or layer 1, is closely associated with the ends of the rod outer segments (**OS**). Junctions between the rod outer and inner segments (**IS**) can be easily distinguished because of their different diameters. Inner and outer segments of the rods and cones constitute the bacillary layer, or layer 2. The nuclei of the extremely elongate rod cells are located in the outer nuclear layer (**ON**), or layer 4, of the retina. An extremely thin external limiting membrane (**ELM**), or layer 3, consists of junctional complexes between the inner segments and the Müller cells. The outer plexiform layer (**OP**), or layer 5, consists of terminations of rod and cone cells that synapse with dendrites of the bipolar neurons. Cell bodies of the bipolar neurons form the inner nuclear (**IN**) layer, or layer 6. Axons of the bipolar neurons extend into the inner plexiform (**IP**) layer (layer 7) and synapse with a layer of ganglion cells (**GC**) designated as layer 8. Axons from the ganglion cells extend for some distance toward the optic nerve as a layer of optic nerve fibers (**ONF**), or layer 9, and eventually form the optic nerve. The basal lamina of Müller cells constitutes layer 10. The complex shape of Müller cells (**MC**) varies in different layers of the retina. In the nuclear layers, including the ganglion cell layer, the lateral cell processes form a continuum that surrounds the neuronal elements. In the inner nuclear layer, impressions (✱) in the Müller cells correspond to the shape of the neuronal somata that were removed during preparation. In the plexiform layers, the Müller cells are not readily identified because this portion of the cell consists of a multitude of slender processes that insinuate between the dense arrangement of dendrites and axons. Within the innermost retinal layers, the Müller cells possess a broad base. Müller cells can synthesize and store glycogen and also possess lactic acid dehydrogenase activity. In addition to their supportive function, they are thought to supply glucose to the nerve cells.

X2645

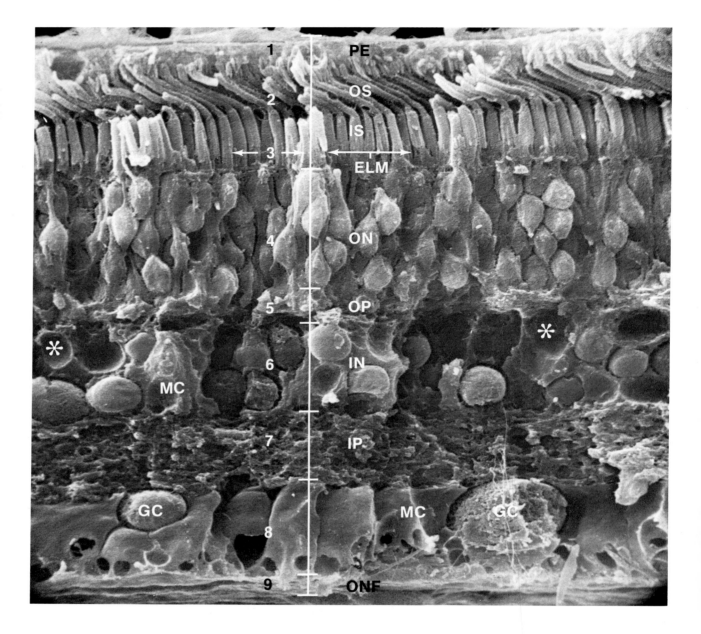

NERVOUS TISSUE
Eye—Retina

Müller cells (**MC**), spanning almost the entire thickness of the retina, form stout radial processes (**RP** in Fig. 2) that originate from each end of its cell body in the inner nuclear layer (**INL**) and radiate through the retinal layers. In the outer and inner plexiform layers (**PL** in Figs. 1 to 3), fine horizontal fibers extend laterally to insinuate themselves between the multitude of axonal and dendritic processes. Conical expansions at the base of the Müller cells form the internal limiting membrane (**ILM** in Fig. 1). In the nerve fiber layer, round spaces (✱ in Fig. 2) exist between the basal processes of Müller cells, which accomodate the nerve fibers traversing this area. In the nuclear layers, including the ganglion cell layer, Müller cells form cytoplasmic lamellae that occupy the interstices of the somata, and hence enclose them (see Fig. 6). The multipolar ganglion cells (**GC** in Fig. 1) vary considerably in size and have been classified into different populations. Their surface texture is often rough, perhaps because of the shearing of many processes during tissue exposure. The bipolar neurons (**Bp**) in the inner nuclear layer are spherical and, for the most part, possess a smooth surface. However, a fibrillar-appearing material (**Fi** in Fig. 6) is sometimes observed to "cap" a small region on the surface of the bipolar neuron. In addition, many of the perikarya in the inner nuclear layer possess an elongate cleft (**Cl** in Fig. 6) that contains a finger-like process. The fibrillar material or the cleft may possibly be an area of synapse on the perikaryon. At the junction between the outer plexiform and inner nuclear layers, there are rough-surfaced cells (✱ in Fig. 1, enlarged in Fig. 5) that appear different in shape and texture from the bipolar cells. These differences, and the location of these cells, suggest that they may be horizontal cells. In Figs. 3 and 4, details are shown of the outer nuclear layer, which contains the cell bodies (**CB**) of rods, the outer rod fibers (**ORF**), and the inner rod fibers, or axons (**Ax**), of the rod cells. The rod axons terminate in the outer plexiform layer and form dilated, oval synaptic endings called spherules (**arrows** in Figs. 3 to 5; Fig. 4 is an enlargement of Fig. 3).

Fig. 1, ×1040; Fig. 2, ×1600; Fig. 3, ×1760; Fig. 4, ×4005; Fig. 5, ×4375; Fig. 6, ×4005

NERVOUS TISSUE
Eye—Retina

The rod photoreceptor is very long, slender, and highly specialized. Rods of the rat retina are illustrated in the scanning electron micrograph (Fig. 1). The internal organization of the rod cell is depicted in the diagram (Fig. 2). Outer segments (**OS**) and inner segments (**IS**) of the rat rod cell are approximately equal in diameter. The junction (**Ju**) between these two areas is clearly demarcated in the scanning electron micrograph and is characterized internally by a modified cilium. The outer segment consists of stacked membranous disks formed by folds of the external cell membrane. Rod outer segments contain numerous membranous disks stacked perpendicular to the long axis, and it is here that the protein rhodopsin (visual purple) is concentrated. In the human and primate, the disks are unconnected to the covering plasma membrane. Inner segments contain mitochondria, elements of the rough and smooth endoplasmic reticulum, microfilaments and tubules, as well as glycogen. The inner segment narrows slightly before expanding into the widest portion of the cell, which contains the nucleus (**Nu**). These nuclei collectively make up the outer nuclear layer (**ONL**) of the retina. The cell then narrows abruptly before synapsing with the dendrites of bipolar neurons. This region of the rod cell contains small presynaptic vesicles. The long, thin filamentous structures (**arrows**) associated with the inner rod segments, oriented parallel to their long axis, are processes of the Müller cells that extend beyond the external limiting membrane to form the "fiber baskets" that embrace the photoreceptors. Portions of Müller cells (**MC**) can also be observed in the outer nuclear layer between the cell bodies of the photoreceptors. [Fig. 2 redrawn by permission from T. L. Lentz, *Cell Fine Structure*, W. B. Saunders Company, 1971.]

Fig. 1, ×2305

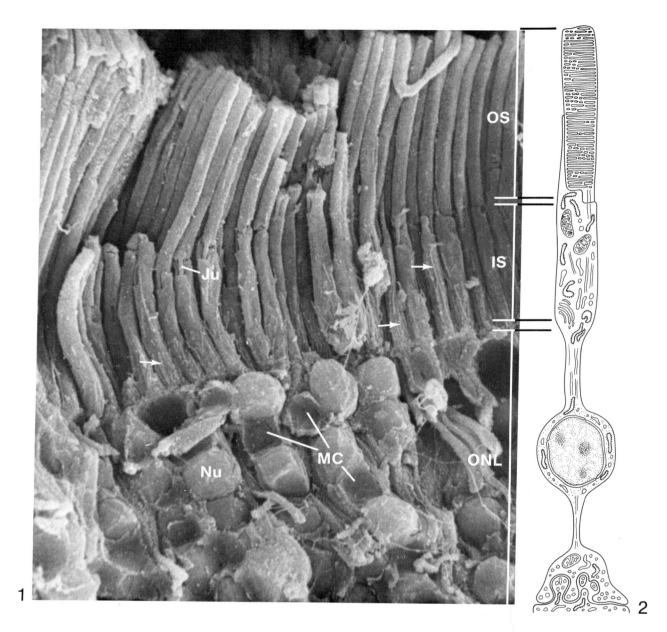

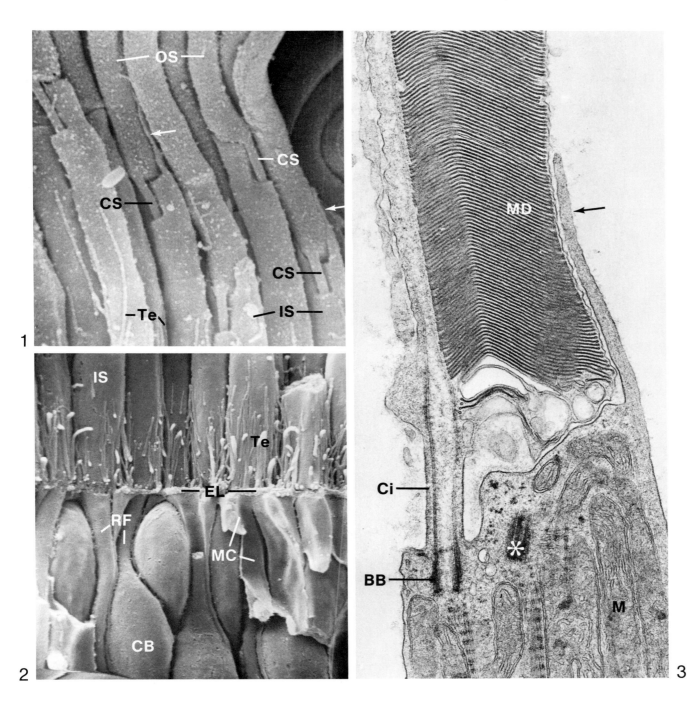

1

2

3

NERVOUS TISSUE
Eye—Retina

Junctions between the inner and outer segments of several rod cells in the rat retina are shown in Fig. 1. A connecting stalk (**CS**) joins the inner segment (**IS**) and the outer segment (**OS**). The connecting stalk has been shown by transmission electron microscopy to contain a modified connecting cilium (**Ci**) that consists of nine peripheral doublet microtubules, but lacks the central pair of microtubules (Fig. 3). The modified cilium is connected to a basal body (**BB**). Another member of the paired basal bodies is located nearby (✽). In Fig. 3, the stacked membranous disks (**MD**) can be seen in the outer segment, and mitochondria (**M**) are visible in the inner segment. The plasma membrane of the outer and inner segments are continuous only in the region of the connecting stalk. Thin, tongue-shaped extensions of the inner rod segment project for variable distances around the proximal one-third of the outer segment, and are identified by **arrows** in Figs. 1 and 3. Several thread-like tendrils (**Te**) extend from the supportive Müller cells and are closely associated with the surface of the inner segments in Figs. 1 and 2. The cell bodies (**CB**) of several rod cells are shown in Fig. 2. The cell body tapers into a thin, short outer rod fiber (**RF**), which then expands as it becomes the inner segment (**IS**) of the rod cell. The cell body also tapers in the opposite direction into an inner rod fiber that forms a synaptic ending (rod spherule) with dendrites of bipolar ganglion cells. The position of the external limiting membrane (**EL**) is denoted in Fig. 2, and portions of Müller cells (**MC**) are present between the photoreceptor cell bodies and outer rod fibers. [Fig. 3 reproduced by permission from M. J. Hogan, J. A. Alvarado, and J. E. Weddell, *Histology of the Human Eye,* W. B. Saunders Company, 1971.]

Fig. 1, ×11,260; Fig. 2, ×4895; Fig. 3, ×32,040

NERVOUS TISSUE
Eye—Retina

Diagrams illustrating the organization of the rod (left) and cone (right) outer segments are shown in Fig. 1. The photosensitive outer segments of rods and cones consist of many stacked membranous disks. The rod disks contain visual purple, which consists of vitamin A aldehyde (retinal) complexed with a class of proteins called opsins. When this pigment absorbs light, a change in the molecular configuration of retinal and its relationship to opsins ensues, and an electrical discharge is produced. Visual purple is rapidly reconstituted. There are three types of cones, each sensitive to a particular spectral color: red, green, and blue. Rods, on the other hand, are sensitive to all wavelengths of visible light and, therefore, cannot distinguish color. The rod inner and outer segments of the guinea pig retina are illustrated in their entire length in Fig. 2. The inner segments (**IS**) and outer segments (**OS**) are joined by a connecting stalk (**arrows**). Finger-like extensions of the rod inner segment invest a portion of the adjacent outer segments at their junction. The tendrils of Müller cells are also visible around the base of the inner segments in the lower portion of the figure. The external portions of the outer segments bear a close and important relationship to the pigment epithelial layer (**PE**). The pigmented epithelial cells (**PE**) also possess cytoplasmic processes (**CP** in Fig. 3) that invest tips of the outer segments (**OS** in Fig. 3). A nucleus (**Nu**) of one of the pigmented epithelial cells is exposed. The close relationship between the outer segment tips and the pigmented epithelium reflects one function of the epithelium—namely, to phagocytose and digest old disks that are shed from the outer segments. Renewal of rod disks begins in the inner segments, where new proteins are synthesized. They are subsequently conducted through the connecting stalk to the base of the outer segment, where new disks are formed by infolding of the plasma membrane. Older disks are displaced progressively toward the end of the outer segment and eventually are shed in the region of the phagocytic pigmented epithelial cells. [Fig. 1 reproduced from R. W. Young, "Visual Cells." Copyright © 1970 by Scientific American, Inc. All rights reserved.]

Fig. 2, ×2805; Fig. 3, ×7050

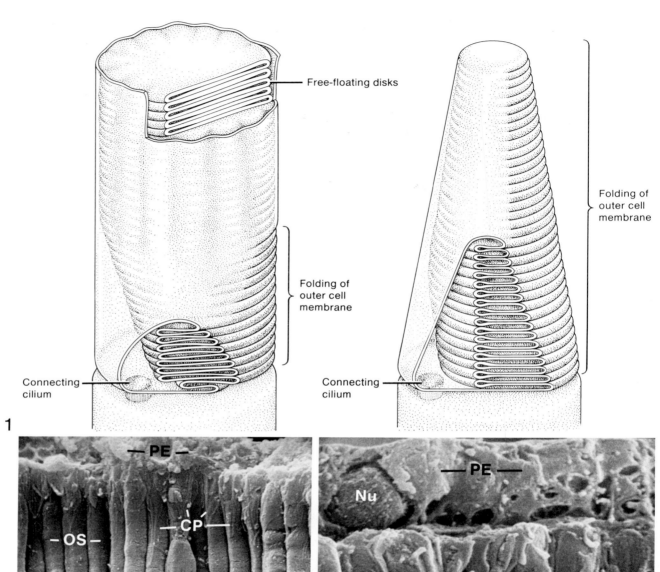

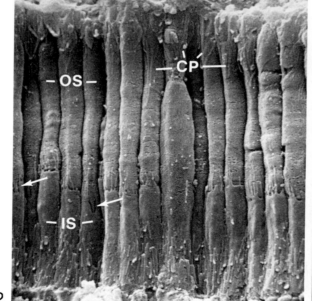

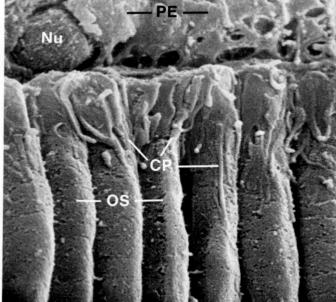

92

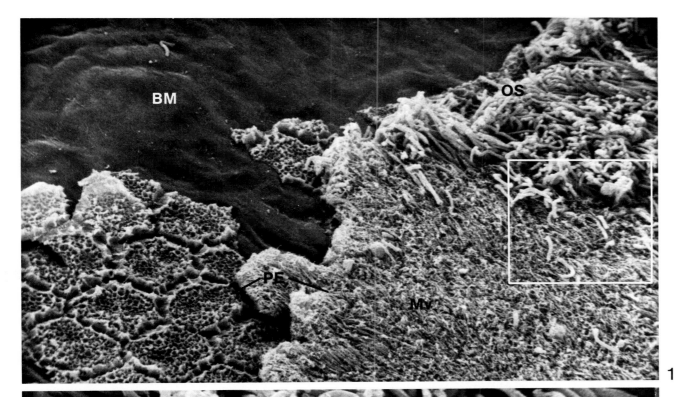

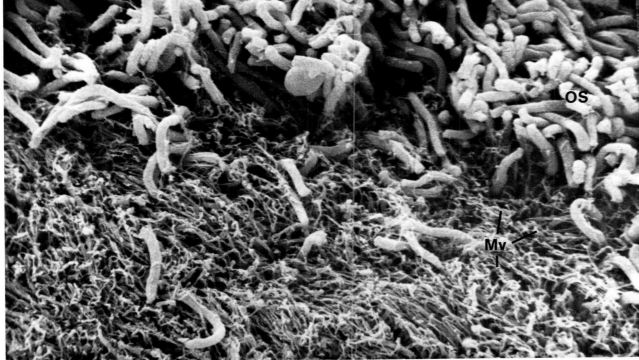

NERVOUS TISSUE
Eye—Retina

In this view from within the rat eye, much of the retina has been stripped from the pigmented epithelial layer (**PE** in Fig. 1). The outer segments (**OS**) of some rods can be seen in parts of both figures. The pigmented epithelium is adjacent to Bruch's membrane (**BM**) of the choroid, and consists of basal lamina material as well as collagenic and elastic fibers. The retinal pigment epithelium consists of hexagonally packed cuboidal cells that strongly adhere to Bruch's membrane and to the ends of rod outer segments. As a result, when the layers are separated by shearing, the cuboidal pigmented epithelial cells are often exposed internally. The vacuolated appearance of the cytoplasm is probably the result of the removal of melanin pigment granules during tissue processing. Microvilli (**Mv**) are present on the inner surface of the pigmented epithelial cells and surround the tips of the rod outer segments. The area within the rectangle in Fig. 1 is enlarged and illustrated in Fig. 2, where it is possible to observe numerous microvilli (**Mv**) extending from the pigmented epithelial cells as well as their close relationship to the outer segments (**OS**).

Fig. 1, ×1210; Fig. 2, ×1795

NERVOUS TISSUE
Eye—Retina

The retina (**Re**) has been removed from the pigmented epithelium (**PE**) in a portion of Fig. 2. In the area enclosed by the rectangle, the tips of the rod outer segments are still attached to an epithelial cell of the pigment layer. This region is enlarged in Fig. 1 and illustrates the close relationship between the rod outer segments (**OS**) and the inner surface of the pigmented epithelium (**PE**). The **arrows** in Fig. 1 denote the presence of many microvillous extensions of the pigmented epithelium, which closely invest the tips of the rod outer segments. In Fig. 3 several pigmented epithelial cells (**PE**) have been exposed internally to reveal a vacuolated cytoplasm normally occupied by many melanin pigment granules in the living state. The outer surface of the pigmented epithelial cell layer is closely associated with Bruch's membrane (**BM**) of the choroid. In Fig. 3 it is possible to observe many fine collagenic fibrils, which make up a portion of this layer.

Fig. 1, ×4300; Fig. 2, ×1075; Fig. 3, ×4840

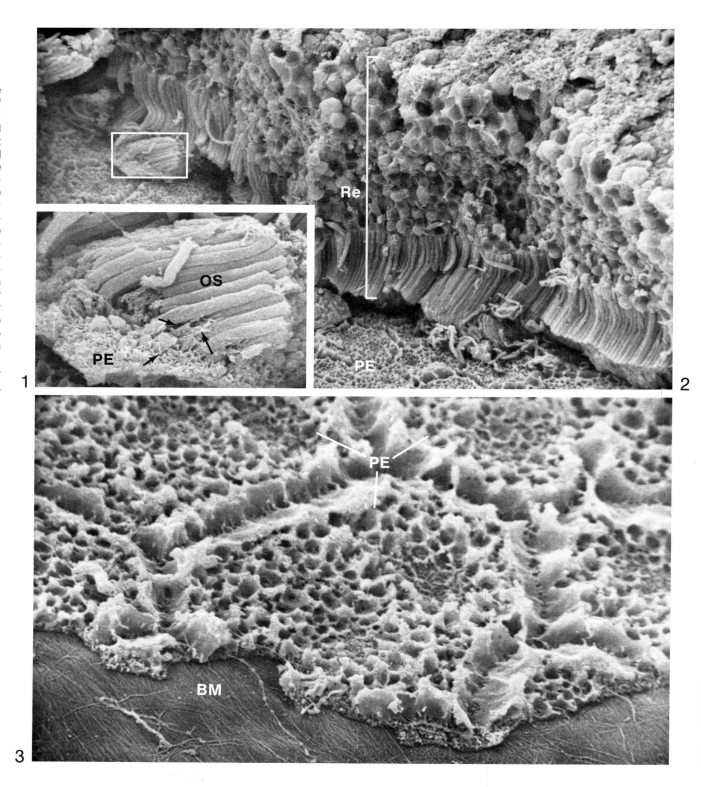

94

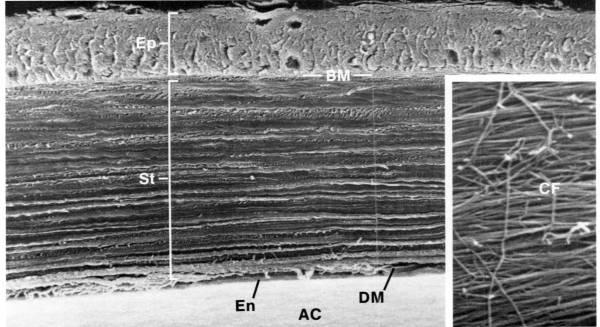

1

Descemet's membrane — Endothelium — Cornea
Schwalbe's line
Trabeculae and spaces of Fontana
Canal of Schlemm
Scleral spur
Anterior chamber angle
Pectinate ligament
Conjunctiva
Anterior ciliary vein
Sclera

Anterior chamber

Iris rolls

Lens

Meridional fibers | Radial fibers | Circular fibers

Suprachoroidal space

Ciliary muscle

Ciliary process

Suspensory ligament of lens (zonula)

Posterior chamber

Dilator muscle of iris

Pigment epithelium

Sphincter muscle of iris

Ciliary body

f. Netter M.D.
©CIBA

NERVOUS TISSUE
Eye—Anterior Organization

The interrelationships of the constituent tissues in the anterior portion of the eye are diagrammed in Fig. 1. The sclera continues anteriorly as a transparent cornea. The trabeculae and spaces of Fontana as well as the canal of Schlemm are located in the limbus at the junction of the sclera and cornea. The ciliary body and ciliary processes represent the anterior extension of the choroid. Meridional, radial, and circular smooth muscles are present in the ciliary body. The iris extends from the ciliary body toward the anterior lens surface and contains an aperture (the pupil). Suspensory ligaments or zonule filaments serve to suspend the lens by attaching to the equator of the lens and the ciliary body. The anterior chamber contains aqueous humor and is located between the posterior corneal surface and the anterior surfaces of the iris and lens. The smaller posterior chamber, also containing aqueous humor, is located posterior to the iris. The cornea (Fig. 2), as illustrated in transverse section, is covered externally by an outer stratified epithelial (**Ep**) layer. The connective tissue stroma (**St**) makes up most of the cornea and consists of layers of collagen fibers oriented parallel to the surface epithelium, but in layers at various angles to each other. Bowman's membrane (**BM**) is located between the epithelium and stroma. When the corneal stroma is teased apart, the size and arrangement of individual collagenic fibrils (**CF** in Fig. 3) can be observed. The inner layer of the cornea consists of a thin endothelial layer (**En**) adjacent to the aqueous humor of the anterior chamber (**AC**). Decemet's membrane (**DM**) is a thin layer of ground substance (basement membrane) between the endothelium and stroma. [Fig. 1 © Copyright 1962 CIBA Pharmaceutical Company, Division of CIBA-GEIGY Corporation. Reproduced with permission from *Clinical Symposia*, illustrated by Frank H. Netter, M. D. All rights reserved.]

Fig. 2, ×890; Fig. 3, ×10,125

Ep
BM
St
En AC DM
CF

2

3

NERVOUS TISSUE
Eye—Cornea

The cornea is transparent, avascular, richly innervated, and resistant to deformation and infection. It also has a limited capacity to regenerate following injury. The diagram on this page illustrates the organization of the corneal epithelium, which consists of 5–6 layers of cells with a turnover period of about one week. The surface of the epithelium is smooth and covered by a precorneal tear film about 7 μm thick. The presence of microplicae (**a**) and microvilli on the outer surface of the external layer of cells may serve to hold the tear film against the cells. A single row of basal cells is anchored by hemidesmosomes to an underlying thin basal lamina. Two or three layers of polygonal cells form the intermediate region of the epithelium. The two layers of flattened surface cells have junctional complexes (zonulae occludentes), which seal off the intercellular spaces. A corneal nerve (**b**) extends across Bowman's layer (**c**) and pierces (**d**) the basement membrane (**f**) of the epithelium, where the Schwann sheath is lost. The unmyelinated nerve then extends toward the surface in the intercellular spaces of the epithelium. A lymphocyte (**e**) is located between two basal cells in the diagram. Bowman's membrane consists entirely of collagen and ground substance. In contrast to the regular arrangement of collagen fiber lamellae in the corneal stroma, the collagen in Bowman's layer is more random in disposition. Some collagen fibers in the superficial layer of the corneal stroma appear to merge with Bowman's membrane (**g**). The transparency of the cornea is related to the smoothness of the epithelium, the lack of blood vessels, and the arrangement of the collagen, cells, and ground substance. The chemical composition, state of metabolism, and hydration of the corneal constituents are also important in maintaining transparency. The collagenic fibrils in the stroma are uniformly separated, and stromal cells are thin, flattened, and elongate. The cornea has a great affinity for water, a characteristic probably due to the ground substance rather than the cells of the stroma. [Reproduced by permission from M. J. Hogan, J. A. Alvarado, and J. E. Weddell, *Histology of the Human Eye,* W. B. Saunders Company, 1971.]

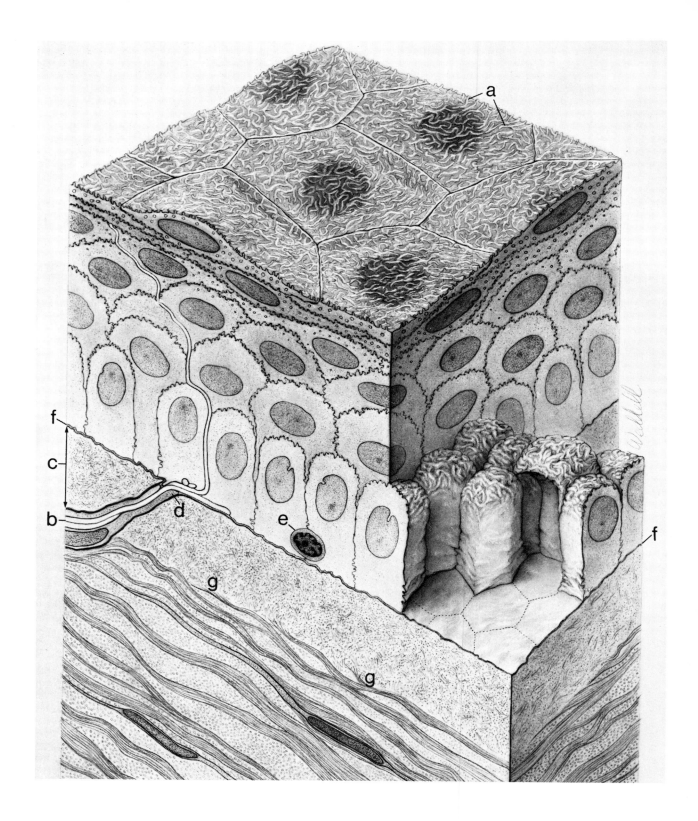

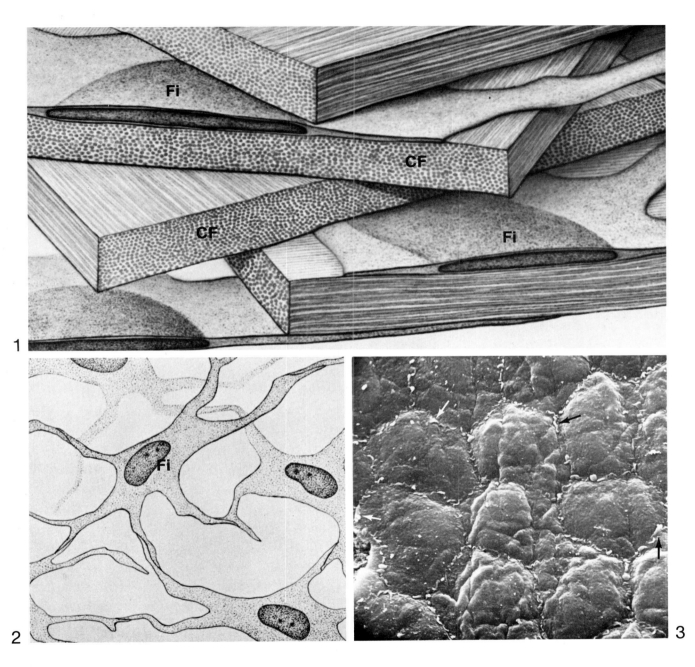

NERVOUS TISSUE
Eye—Cornea

Fibroblasts (**Fi** in Figs: 1 and 2) are present between the lamellae of collagen fibers in the corneal stroma. The cells have long branching processes that frequently contact processes of nearby fibroblasts (Fig. 2). The collagen is arranged into planar lamellae composed of parallel collagen fibrils (**CF**) that extend over the entire length of the cornea. Adjacent lamellae are oriented at an angle to one another (Fig. 1). The distance between neighboring collagen fibrils is uniform. The ground substance of the corneal stroma, consisting of mucoproteins and glycoproteins, occupies the space between fibroblasts and collagenic lamellae, and is thought to play a role in maintaining the appropriate orientation of the fibrils. The surface of the corneal endothelium is illustrated in Fig. 3 and consists of a single layer of thin, polygonal cells. The cell boundaries of the endothelial cells are apparent in Fig. 3, and marginal folds or ruffles of the plasma membrane are identified by the **arrows**. A few short microvilli and a single cilium are frequently present on the surface of the endothelial cells exposed to the aqueous humor. The corneal endothelial cells ultrastructurally resemble cells engaged in active transport. They possess numerous mitochondria in addition to other organelles, and are thought to be involved in the movement of osmotically active substances and may play a role in synthesizing some of the constituents in Decemet's membrane. Using thorotrast as a marker molecule, it has been possible to trace the transport of particulate material, via micropinocytosis, across the endothelium into Decemet's membrane. [Figs. 1 and 2 reproduced by permission from M. J. Hogan, J. A. Alvarado, and J. E. Weddell, *Histology of the Human Eye,* W. B. Saunders Company, 1971.]

×1840

NERVOUS TISSUE
Eye—Ciliary Body, Iris, Lens

A low-magnification view of the inner surface of the iris (**Ir**) and ciliary processes (**CP**) of the ciliary body of the guinea pig eye is illustrated in Fig. 1. The lens was removed in Fig. 1 to reveal the posterior surface of these structures. The inner surface of the ciliary body is subdivided into two regions: the folded pars plicata and the flattened pars plana. The radially arranged smooth muscle of the iris (**Ir**) is contracted in this preparation, so that the pupil (**Pu**) is dilated. A portion of the preparation illustrated in Fig. 1 is enlarged in Fig. 3, in which the ciliary processes (**CP**), the inner surface of the iris (**Ir**), and the zonule filaments (**ZF** in Fig. 3) that constitute the suspensory filaments of the lens can all be observed. The zonule filaments (**ZF**) are also shown in Fig. 2. They attach at one end to the lens (**Le**) margin and at the other end terminate in the pars plana (**Pl**) and pars plicata (**PP**) of the ciliary body as well as to a region anterior to the ora serrata (**OS**). The iris (**Ir**) can also be observed through the zonule filaments in Fig. 2. A portion of the lens (**Le**) is illustrated in Fig. 4, and many zonule filaments (**ZF**) extend from the ciliary processes (**CP**) to insert into the lens (**Le**) at its equator. [Fig. 2 courtesy of D. H. Dickson.]

Fig. 1, ×15; Fig. 2, ×35;
Fig. 3, ×80; Fig. 4, ×95

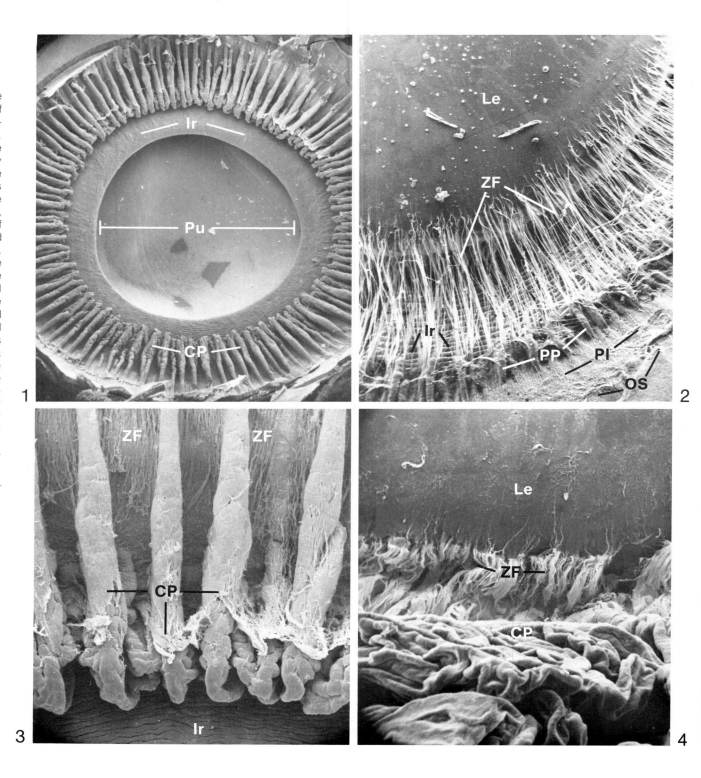

1

Lens fibers in cross section

Suture of anterior surface of lens

Lens capsule

Dividing cells

Cutaway view of adult lens showing embryonic lens inside

Suture on posterior surface of lens

Embryonic lens

Longitudinal section of lens fibers

2

C

B A

d

f

g

f

g

d

3

b

a

Suture

NERVOUS TISSUE
Eye—Lens

The lens is biconvex, elliptical in shape, and approximately 4 mm thick at its anteroposterior diameter. Its transparency is due primarily to the shape, arrangement, internal structure, and biochemistry of the lens cells. In the adult, the lens has no blood supply, and its nourishment must come from the aqueous and vitreous humor. The lens is surrounded by a transparent elastic capsule (see Fig. 1 and **d** in Fig. 2) that is actually a thick basement membrane. Although the capsule tends to compress the lens, so that its configuration is less spherical, tensions in the ciliary zonule filaments (**f** in Fig. 2) tend to oppose this action. The zonule fibers attach to the capsule to form the pericapsular or zonular lamella of the lens (**g** in Fig. 2). Directly beneath the lens capsule at the anterior (**A**), intermediate (**B**), and equatorial (**C**) zones (see Fig. 2) is a single layer of cuboidal lens epithelium. The cells at A and at B are diagrammed in cross section and in corresponding surface view. At the equator, the lens epithelium can divide, and the cells become highly elongated, sending processes anteriorly and posteriorly; the cells have a ribbon-like appearance. Because of their extensive length, lens cells are frequently called lens fibers. As additional lens cells are formed, older cells may lose their nuclei and become displaced to deeper parts of the cortex of the lens to surround the embryonal lens nucleus that is formed initially. By about midlife, a central lens nucleus has formed that consists of many older lens cells, surrounded by lens fibers of the cortex. The three-dimensional arrangement of the lens cells (drawn as bands) is diagrammed in Fig. 1. At areas where lens cells converge and meet, sutures are formed. The embryonal nucleus possesses a Y suture at both the anterior and posterior poles. Those cells that attach to the tips of the Y sutures at one pole attach to the fork of the Y at the opposite pole. In the adult lens cortex, the anterior and posterior organization of the sutures is more complex. The continuity of the suture throughout the thickness of the cortex is more accurately represented in Fig. 3 (see **a** and **b**). Complex interdigitations firmly interlock lens cells at the suture. [Figs. 2 and 3 redrawn and reproduced from M. Hogan, J. Alvarado, and J. Weddell, *Histology of the Human Eye*, W. B. Saunders Company, 1971.]

NERVOUS TISSUE
Eye—Lens

New cells are constantly added to the cortex from the lens epithelium at the lens equator. Cortical lens fibers may be 10–12 mm long (about 7 μm wide, 4.5 μm thick). Transverse sections of cortical lens cells, or fibers, in Figs. 1 and 2 illustrate their flattened hexagonal shape and the precise alignment of the cells in regular rows. In superficial layers of the lens cortex, interlocking interdigitations are evident at the edges along the length, as well as at the ends of the hexagonal lens fibers. The "ball and socket" connections (**arrows** in Figs. 3 and 4), the interdigitations between lens cells, are also thought to play a role in the transparency of the lens. In the deeper lens cortex, interdigitations are apparent along the long axis of the lens fibers, but they are not found at the short ends.

Fig. 1, ×810; Fig. 2, ×1220;
Fig. 3, ×1945; Fig. 4, ×6955

1

2

3

4

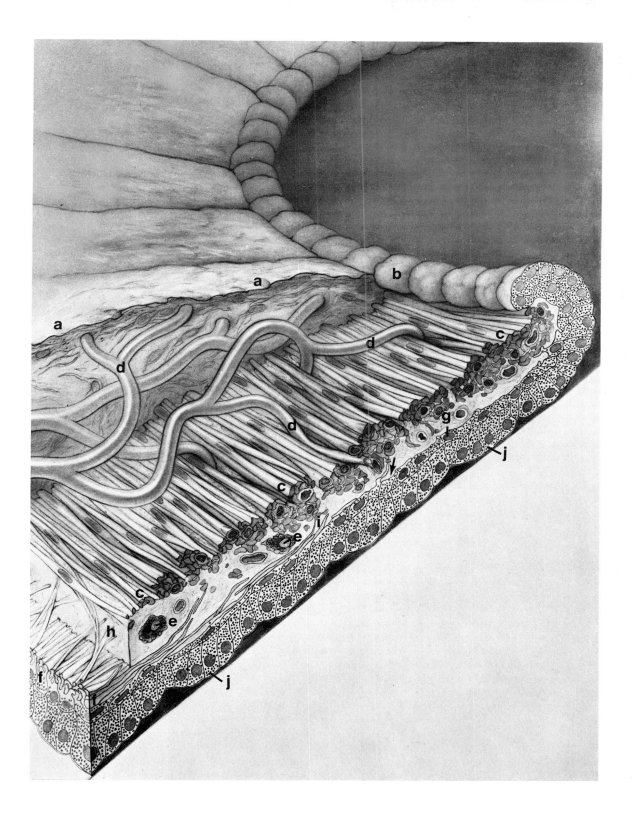

NERVOUS TISSUE
Eye—Iris

The pupillary portion of the iris is illustrated in this diagram. The cellular anterior border layer (**a**) terminates at the pigment ruff (**b**) in the pupil margin. The circularly arranged smooth muscle cells that form the pupil sphincter (constrictor) muscle are identified by (**c**). Vascular arcades (**d**) extend through the sphincter muscle toward the pupil margin. Capillaries, nerves, melanocytes, and clump cells (**e**) are found within and surrounding the smooth muscle. Several layers of dilator (**f**) smooth muscle are longitudinally (radially) arranged in the iris and terminate at the region of the **arrow**. Cuboidal epithelial cells (**g**) form the anterior epithelium of the pupillary margin. Spur-like extensions from the dilator muscle form Michel's spur (**h**) and Fuch's spur (**i**), which terminate among the smooth muscles of the sphincter muscle. The posterior iris epithelium (**j**) consists of columnar cells. Contraction of two layers of smooth muscle in the iris serves as an adjustable diaphragm to vary the size of the pupil. The muscles also serve to hold the iris against the lens surface so as to separate the anterior chamber from the posterior chamber. Contraction of the circular sphincter muscles decrease the pupil size, and the radially arranged dilator smooth muscle cells serve to increase the size of the pupil. The dilator muscle is innervated by sympathetic postganglionic neurons and responds to adrenergic compounds. The sphincter muscle is innervated by parasympathetic fibers of postganglionic nerves in the ciliary ganglion. The sphincter muscle thus constricts in response to cholinergic drugs. Both the sphincter muscle and the ciliary muscles are innervated by ciliary nerves and work in synchrony. Thus, in accommodating for close vision, ciliary muscles contract, as does the pupillary sphincter. [Reproduced by permission from M. J. Hogan, J. A. Alvarado, and J. E. Weddell, *Histology of the Human Eye*, W. B. Saunders Company, 1971.]

NERVOUS TISSUE
Eye—Iris

The anterior surface of the iris is shown from the rabbit (Figs. 1 and 2) and the rat (Figs. 3 and 4). Although a very thin sheet of cells (**EC**) forms this lining, there are extensive intercellular spaces (**IS**) present that may vary in size depending upon the degree of contraction of the smooth muscle cells within the iris. The cells lining the anterior surface of the iris were initially characterized as endothelium. Although endothelium may form a covering for the anterior surface of the human iris at birth, it is thought that it soon disappears, to be replaced by fibroblasts. The branched and irregularly shaped fibroblasts form an extremely thin covering on the anterior iris surface. Layers of branched melanocytes are found beneath the superficial layer of fibroblasts. Connective tissue fibers (**CF** in Fig. 2) of the iris stroma can be observed through the intercellular spaces (**IS**). These are predominantly collagenic fibers. Cells that have surface characteristics of macrophages (**Ma**) can also be found on the anterior surface of the iris (Figs. 3 and 4).

Fig. 1, ×1105; Fig. 2, ×3070; Fig. 3, ×2350; Fig. 4, ×5875

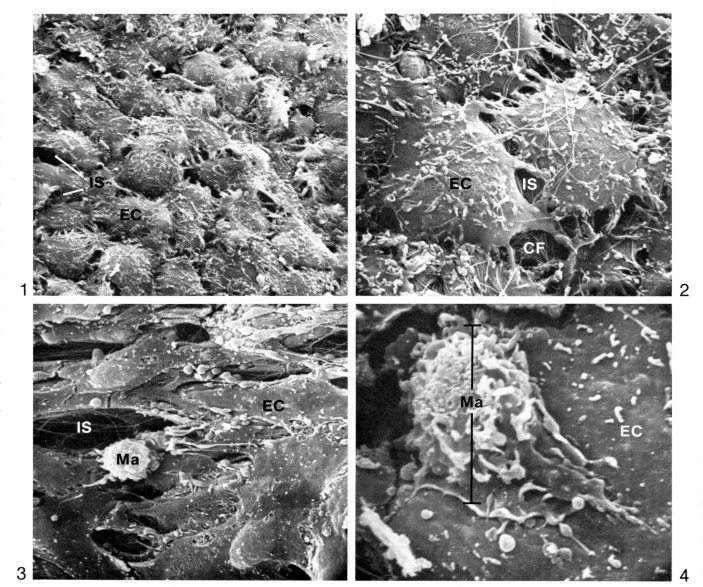

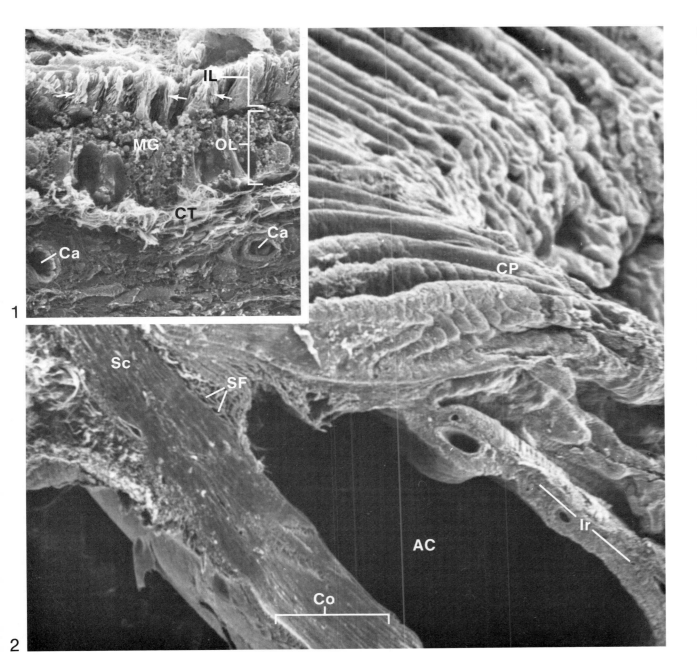

NERVOUS TISSUE
Eye—Anterior Margin

In the section of the anterior margin of the rabbit eye shown in Fig. 2, the continuity between the sclera (**Sc**) and the cornea (**Co**) can be observed. The anterior chamber (**AC**) contains aqueous humor and is bordered anteriorly by the cornea (**Co**) and posteriorly by the iris (**Ir**). The Iris represents an anterior extension of the ciliary body. The folded ciliary processes (**CP**) of the ciliary body are also identified, but the lens was removed in this preparation. The trabecular meshwork and spaces of Fontana (**SF**) have the appearance of a sponge-like tissue at the margin of the anterior chamber. The posterior surface of the pars plicata is lined by two layers of epithelial cells that take part in the production of aqueous humor. The inner layer (**IL** in Fig. 1) of epithelial cells, facing the interior of the eye, is in contact with the posterior chamber. The plasma membranes of these cells are extensively folded (**arrows**) along their lateral cell borders. The cells of the outer layer (**OL** in Fig. 1) are folded to a lesser extent and, in addition, contain numerous cytoplasmic melanin granules (**MG**). These two layers reside on a highly vascularized layer of connective tissue (**CT**) that contains many capillaries (**Ca**). A large amount of the fluid portion of the blood is forced through the fenestrated capillary walls and comes to lie beneath the double layer of epithelial cells. Ions, water, and small molecules from this tissue fluid are transported by both layers of epithelia into the posterior chamber. This fluid, known as aqueous humor, circulates into the anterior chamber and serves to nourish the lens and cornea.

Fig. 1, ×1335; Fig. 2, ×70

NERVOUS TISSUE
Eye—Anterior margin

The section of the anterior margin of the rat eye shown here illustrates the junction between the cornea (**Co**) and the sclera (**Sc**). Episcleral blood vessels (**EBV**) are transversely sectioned just external to the sclera. The corneal epithelium is continuous laterally with the epithelium of the bulbar conjunctiva (**BC**). The retina (**Re**) terminates just posterior to the ciliary body (**CB**) at a region called the ora serrata (**OS**). The ciliary body (**CB**), which is continuous with the choroid (**Ch**) layer, consists of folded ciliary processes (**CP**) to which many zonule filaments (**ZF**) are attached. Many of the zonule filaments serving to suspend the lens were broken during removal of the lens in this preparation. The iris (**Ir**), which extends anteriorly from the ciliary body, divides the anterior chamber (**AC**) and posterior chamber (**PC**). In this preparation the iris has been deflected against the inner surface of the cornea, so that the extent of the anterior chamber is not accurately depicted. Many blood vessels (**BV**) are sectioned in different planes in both the ciliary body and the iris. The location of the trabecular meshwork and spaces of Fontana (**TS**) are identified, as well as a portion of the canal of Schlemm (**CS**).

×290

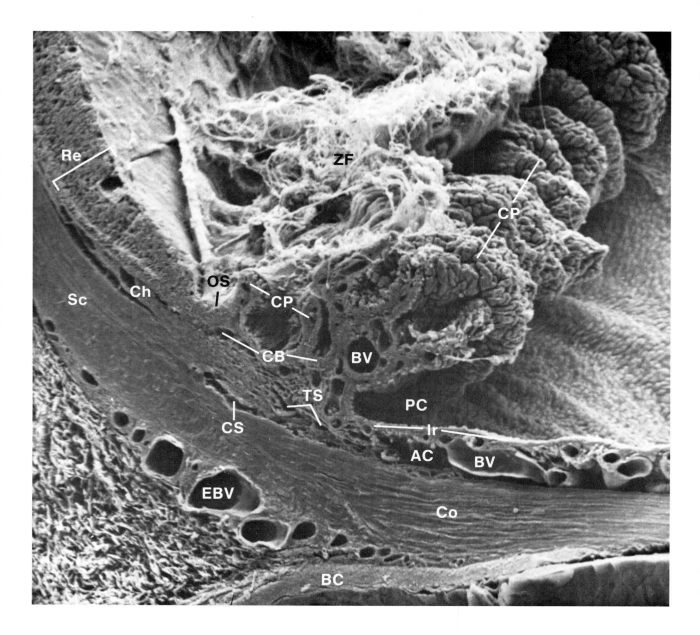

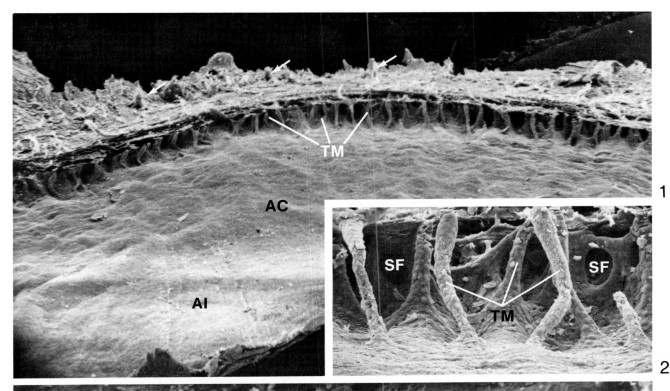

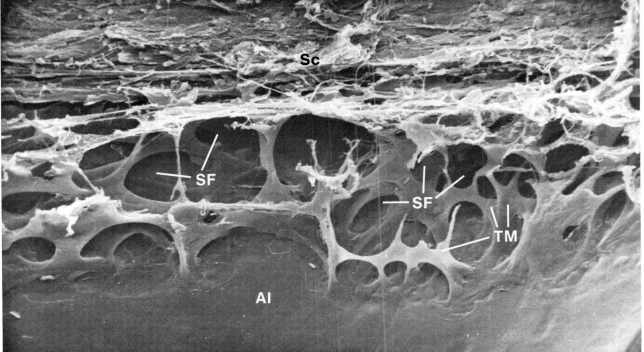

NERVOUS TISSUE
Eye—Limbus

The limbus is a region of transition between the cornea and the sclera. The structures contained within the limus are important functionally in providing a pathway for the removal and drainage of aqueous humor and in providing nourishment to the corneal periphery. The principal pathway by which aqueous humor exits from the anterior chamber is through a trabecular meshwork into the canal of Schlemm, from which aqueous humor eventually passes via aqueous veins into episcleral veins. Fig. 1 shows a view from within the anterior chamber (**AC**) of the rabbit eye toward the margin of the chamber (inner surface of limbus). In this region, there are a number of beams, or sheets, called trabeculae, which form a trabecular meshwork (**TM**). A considerable part of the anterior surface of the iris (**AI**) is still present, but the cornea and cornea-sclera junction have been removed, so that only remnants of their attachment remain (**arrows**). The trabecular meshwork (**TM**) of the rabbit eye is illustrated at higher magnification in Fig. 2. The trabeculae are covered with endothelial cells that are continuous with the endothelial lining of the cornea. Openings within and between the trabecular meshwork are called intertrabecular spaces of Fontana (**SF**). The trabecular meshwork (**TM**) and spaces of Fontana (**SF**) of the rat eye are illustrated in Fig. 3. In this region, the trabeculae are more flattened, and the cells covering the trabeculae have a smoother appearance. The size of the space may vary somewhat with the degree of contraction of smooth muscles in the ciliary body and iris. In Fig. 3, a part of the anterior surface of the iris (**AI**) is present, but the cornea was removed, leaving the cut surface of the sclera (**Sc**) exposed in the limbus. Thus, in three-dimensions, the trabeculae form a circular grid containing spaces through which aqueous humor passes from the anterior chamber to the canal of Schlemm. Obstruction in the normal drainage of aqueous humor can result in a rise in intraocular pressure and can lead to the disease called glaucoma.

Fig. 1, ×30; Fig. 2, ×85; Fig. 3, ×390

NERVOUS TISSUE
Ear

The relationships between the structural components of the external, middle, and inner ear are diagrammed here and in Fig. 1 on the following page. The external ear includes the pinna and external auditory meatus, which terminates as a canal within the temporal bone at the primary tympanic membrane, or eardrum. The eardrum constitutes an intervening "seal" between the middle ear chamber and the external auditory canal. The middle ear consists of a small tympanic cavity that contains three small auditory ossicles (malleus, incus, and stapes) and two small muscles (tensor tympani and stapedius). The upper part of the figure on this page diagrams the position of the ossicles with respect to one another. The middle ear cavity communicates with dead-end mastoid air cavities and extends as an auditory tube (eustachian tube) to the nasopharynx. The wall of the auditory tube is usually collapsed and closed except during the act of swallowing, when it opens to permit equalization of air pressure between the tympanic cavity and the exterior. Two potential openings exist between the middle and inner ear. The oval window is closed by the foot plate of the stapes, which is bound by an annular ligament, and the closely positioned round window is covered by a secondary tympanic membrane. Sound waves cause vibrations of the primary tympanic membrane, and these are in turn transmitted across the middle ear bonelets to produce pressure changes in the fluid (perilymph) that is present in the osseous labyrinth adjacent to the stapes foot plate. The internal ear, or labyrinth, is composed of an outer osseous, or bony, labyrinth and an internal membranous labyrinth. The osseous labyrinth consists of several small intercommunicating cavities within the temporal bone. The cavities take different forms, are lined by a thin periosteum, and contain a fluid called perilymph. The divisions of the osseous labyrinth include three semicircular canals, a vestibule, and a coiled cochlea. Another canal, called the perilymphatic duct, extends from the vestibule to the subarachnoid space around the brain. The perilymph within the osseous labyrinth is similar in composition to cerebrospinal fluid. The bony cavities and canals just described constitute the osseous labyrinth and contain within them thin walled membranous

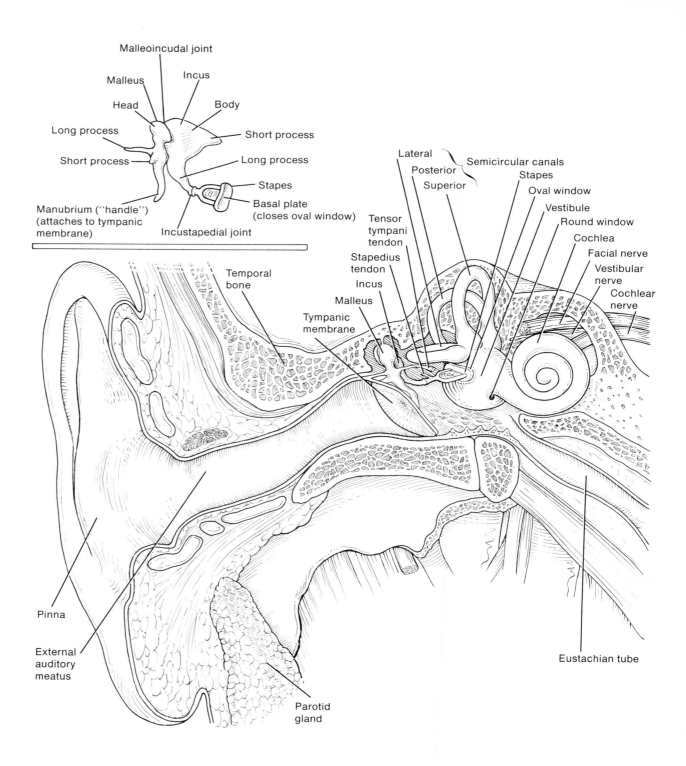

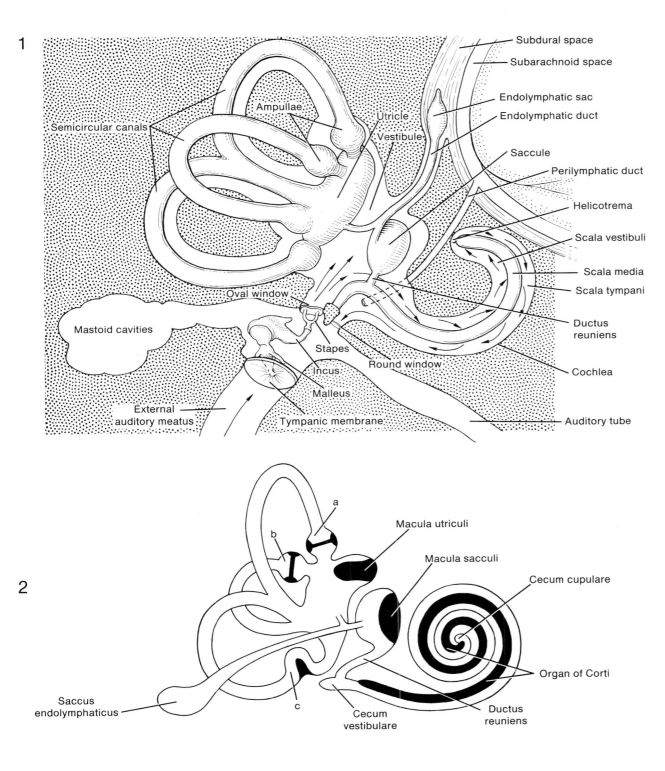

1

Semicircular canals

Ampullae

Utricle

Vestibule

Subdural space

Subarachnoid space

Endolymphatic sac

Endolymphatic duct

Saccule

Perilymphatic duct

Helicotrema

Scala vestibuli

Scala media

Scala tympani

Ductus reuniens

Cochlea

Auditory tube

Oval window

Mastoid cavities

Stapes

Incus

Malleus

Round window

External auditory meatus

Tympanic membrane

2

a

b

Macula utriculi

Macula sacculi

Cecum cupulare

Organ of Corti

Saccus endolymphaticus

c

Cecum vestibulare

Ductus reuniens

sacs or tubes which also intercommunicate and collectively make up the membranous labyrinth. These interconnected membranous sacs and tubes are suspended in perilymph but are filled with a fluid called endolymph. Each of the osseous semicircular canals contains a single membranous semicircular canal. The vestibule portion of the osseous labyrinth, however, contains two interconnected and elongated sac-like structures, the utriculus and the sacculus, which make up the membranous labyrinth in this portion of the inner ear. The membranous semicircular canals are dilated in regions called ampullae (**a,b,c** in Fig. 2) where they join the utriculus. The utriculus and the sacculus are also connected by a short tube that branches to form an endolymphatic duct. This duct extends to the subdural space of the brain, where it ends as a closed-ended endolymphatic sac (Fig. 1). A short, tubular structure called the ductus reuniens leads from the saccule and communicates with the membranous part of the coiled cochlea, called the cochlear duct or scala media. In many regions, the thin wall of the membranous labyrinth consists only of simple squamous cells and a small amount of connective tissue. Connective tissue fibers occasionally attach or anchor the membranous labyrinth to the wall of the osseous labyrinth. In some areas, however, the wall of the membranous labyrinth may be directly attached or adhere to the osseous labyrinth. In localized regions of the membranous labyrinth, the epithelium is thickened to form specialized vestibular and auditory sensory receptors. The position of the neurosensory areas in the membranous labyrinth are identified by shaded areas in Fig. 2. Neurosensory areas, called cristae ampullaris, are present in the wall of the ampullae. Neurosensory areas in the sacculus and the utriculus are termed the macula sacculi and macula utriculi, respectively. These vestibular sensory regions are concerned with maintaining equilibrium and with discerning movement and orientation in space. [Fig. 1 reproduced by permission from T. H. Bast and B. J. Anson, *The Temporal Bone and the Ear,* 1949. Courtesy of Charles C Thomas, Publisher, Springfield, Illinois; C. H. Best and N. B. Taylor, *The Physiological Basis of Medical Practice,* Copyright © 1971 by the Williams & Wilkins Co. Fig. 2 modified from von Ebner, reproduced by permission of W. Bloom and D. W. Fawcett, *A Textbook of Histology* (10th ed.), W. B. Saunders Company, 1975.]

NERVOUS TISSUE
Middle Ear

The delicate middle ear ossicles conduct vibrations from the eardrum to the oval window almost without distortion, and serve to amplify weak sound waves that move the eardrum. Since the malleus attaches to the inner surface of the eardrum and the tensor tympani muscle inserts via a ligament onto the malleus, contraction of the muscle results in the eardrum becoming tense for better transduction of higher frequency vibrations. The stapedius muscle inserts via a ligament into the stapes. Loud noise causes a reflex contraction of these muscles, which acts as a protective adaption to prevent damage to the neurons of the cochlea. Fig. 1 illustrates at low magnification a portion of the middle ear cavity, or tympanic cavity (**TC**), the temporal bone (**TB**), and part of the vestibule (**Ve**) of the inner ear. It is possible to observe part of the malleus (**Ma**), the incus (**In**), and the stapes (**St**). The foot plate (**FP**) of the stapes is flattened on its medial surface where it resides in the oval window. Fig. 2 is an enlargement of Fig. 1; only a small part of the annular ligament (**AL** in Fig. 2), which binds the foot plate into the oval window, remains in this preparation. In Fig. 3, the insertion of the ligament (**Li**) of the stapedius muscle onto the neck (**Ne**) of the stapes is denoted by the **arrow**. The two crura (**Cr**) of the stapes diverge from the neck, and when the body is in an upright position the two crura are directed anteriorly and posteriorly in the horizontal plane. The head of the stapes forms a depression, covered by cartilage, and articulates with the rounded lenticular process (**LP**) of the incus (Figs. 1 and 2). The lenticular process is continuous with the long crus of the incus. The short crus, hidden from view at this angle, is somewhat conical in shape and passes almost perpendicular to the long crus. The body of the incus possesses a saddle-shaped facet that articulates (**broken lines**) with the head of the malleus. The major process of the malleus, the manubrium, passes in a plane almost parallel to the long crus of the incus and is embedded at its distal end into the fibrous and mucous layers of the tympanic membrane.

Fig. 1, ×35; Fig. 2, ×45; Fig. 3, ×40

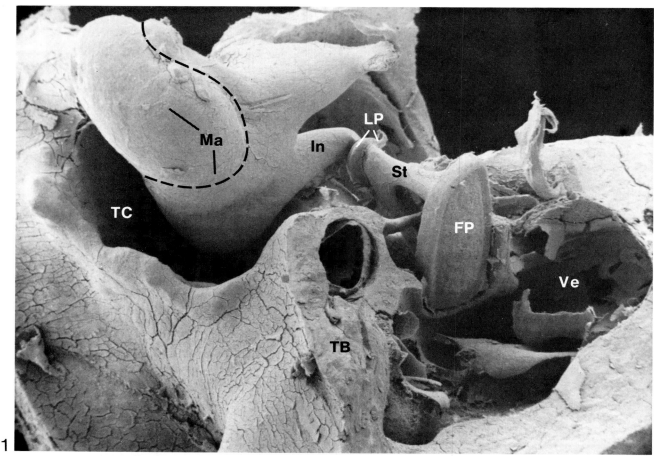

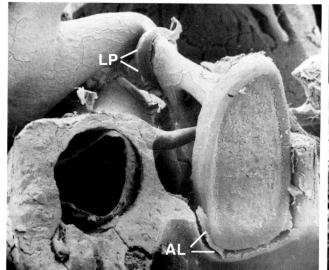

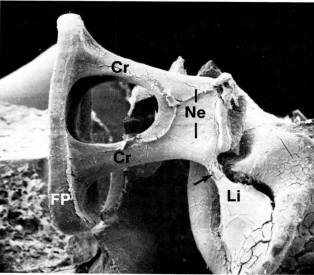

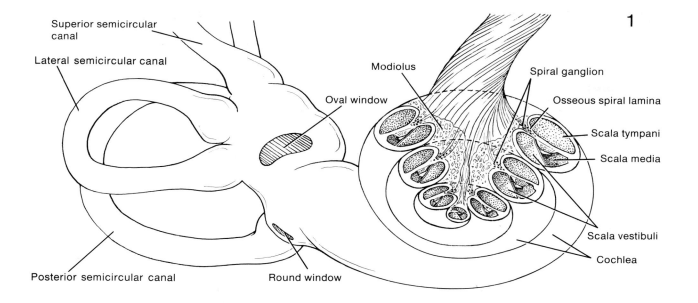

1

Superior semicircular canal

Lateral semicircular canal

Posterior semicircular canal

Oval window

Round window

Modiolus

Spiral ganglion

Osseous spiral lamina

Scala tympani

Scala media

Scala vestibuli

Cochlea

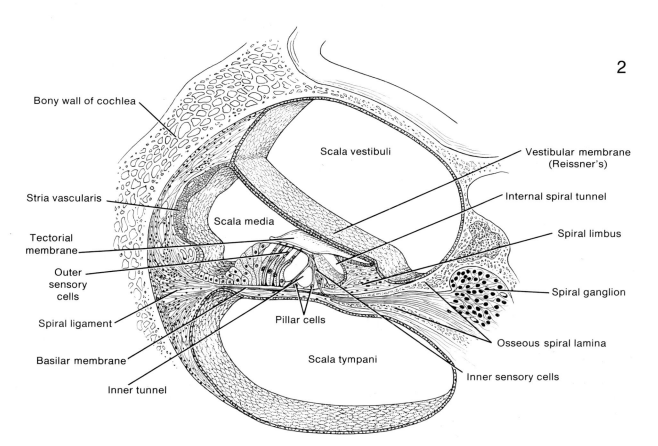

2

Bony wall of cochlea

Stria vascularis

Tectorial membrane

Outer sensory cells

Spiral ligament

Basilar membrane

Inner tunnel

Scala vestibuli

Scala media

Pillar cells

Scala tympani

Vestibular membrane (Reissner's)

Internal spiral tunnel

Spiral limbus

Spiral ganglion

Osseous spiral lamina

Inner sensory cells

NERVOUS TISSUE
Cochlea and Organ of Corti

The osseous cochlea is a bony canal that is coiled approximately $2\frac{3}{4}$ turns around a central core called the modiolus. In addition to a bony central core, bone also forms a complete outer covering. The modiolus is surrounded by a spiral lamina that is divided into an osseous spiral lamina and a membranous spiral lamina. The osseous spiral lamina consists of thin bony ridges or plates that extend laterally from the modiolus. Blood vessels, nerves, and the spiral ganglion are located between the bony plates of the osseous spiral lamina. The membranous spiral lamina is also called the basilar membrane. The spiral ganglion contains numerous cell bodies of afferent bipolar neurons. The dendrites of these bipolar neurons extend into the organ of Corti to innervate the base of the hair cells. The coiled cochlea is illustrated in section in Fig. 1. Details of the cochlear organization are shown in Fig. 2. Two large cavities in the osseous cochlea, the scala vestibuli and the scala tympani, contain perilymph. A central scala media or cochlear duct of the membranous cochlea contains endolymph. The floor of the scala media contains the organ of Corti, which is positioned on the basilar membrane. The organ of Corti consists of special sensory cells, called inner and outer hair cells, as well as supporting cells. A gelatinous tectorial membrane extends from the vestibular lip of the spiral limbus over the internal spiral tunnel to cover the surfaces of the inner and outer hair cells. A thin vestibular membrane (also called Reissner's membrane) separates the scala vestibuli from the scala media. The lateral wall of the scala media consists in part of a region called the stria vascularis. Specialized cells in this region are thought to form endolymph for the membranous cochlear duct. [Fig. 1 reproduced, with permission, from L. C. Junqueira, J. Carneiro, and A. N. Contopoulos, *Basic Histology* (2nd ed.), Lange Medical Publications, 1977. Fig. 2 reproduced, with permission, from W. Bloom and D. W. Fawcett, *A Textbook of Histology* (10th ed.), W. B. Saunders, Company, 1975.]

NERVOUS TISSUE
Cochlea

Since the cochlea is small and enclosed by bone, it is necessary to remove this bone to view the overall cochlear structure. In Fig. 1, a small portion of the temporal bone (**TB**) was removed to expose the apical turn (**AC**) of the coiled cochlea. In Fig. 2, more of the temporal bone was removed to expose all of the coils in the cochlea; here it is more apparent that the cochlea is shaped like a snail shell. The guinea pig cochlea has a larger number of turns than the human cochlea. Not only is the cochlea surrounded by bone, but it has a central bony axis called the modiolus. The plane of this axis is denoted by the **arrow** in Fig. 2. A spiral bony plate, the osseous spiral lamina (**OSL**) projects from the central modiolus. Communication between the vestibule and the osseous cochlea in the broad basal turn is illustrated in the lower right portion of Fig. 2. The thickened periosteum covering the outer wall of the spiral bony channel is called the spiral ligament (**SL**). In addition, the torn vestibular (Reissner's) membrane (**VM**) and the tectorial membrane (**TM**) can be observed at this magnification.

Fig. 1, ×35; Fig. 2, ×55

110

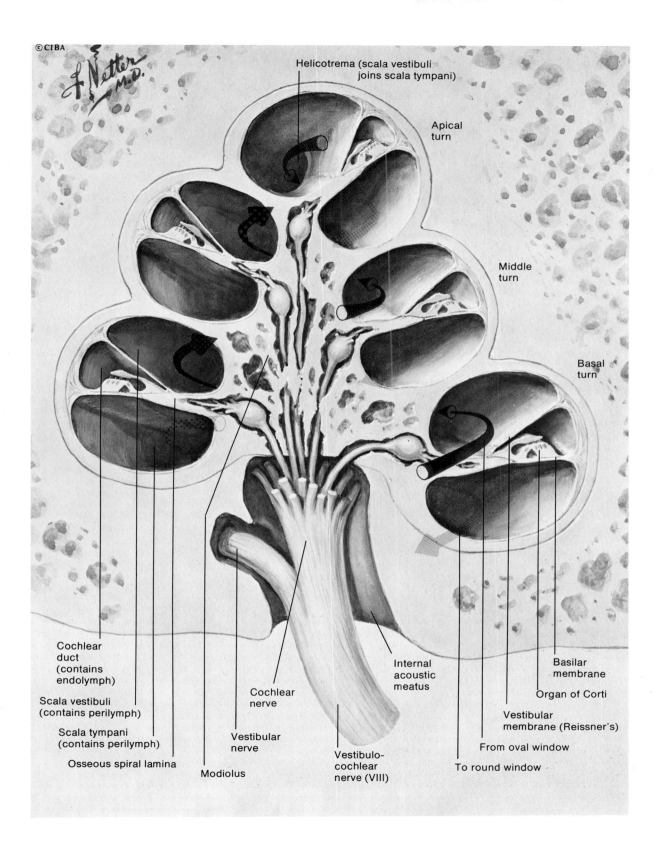

Helicotrema (scala vestibuli
joins scala tympani)

Apical
turn

Middle
turn

Basal
turn

Cochlear
duct
(contains
endolymph)

Scala vestibuli
(contains perilymph)

Scala tympani
(contains perilymph)

Osseous spiral lamina

Cochlear
nerve

Vestibular
nerve

Modiolus

Vestibulo-
cochlear
nerve (VIII)

Internal
acoustic
meatus

From oval window

To round window

Basilar
membrane

Organ of Corti

Vestibular
membrane (Reissner's)

NERVOUS TISSUE
Cochlea

This diagram serves to illustrate structural details within a section of the cochlea. The basal, middle, and apical turns of the cochlea are shown. The scala vestibuli is continuous with the scala tympani in a region of the apical end of the cochlea called the helicotrema. At this point the perilymph in the two scalae are continuous. The central bony modiolus is illustrated in the central axis of the cochlea, and the position of the spiral ganglion is denoted. The spiral ganglion contains bipolar neurons. The axons of these neurons extend as the cochlear nerve into the 8th cranial nerve to the dorsal and ventral cochlear nuclei in the pons. The dendrites of the bipolar neurons synapse with the hair cells of the organ of Corti. The cochlear duct, which contains endolymph, is separated from the scala vestibuli by the thin vestibular (Reissner's) membrane and from the scala tympani by the basilar membrane. The cochlear duct terminates blindly at the apical end of the cochlea. The organ of Corti and tectorial membrane project into the endolymph of the cochlear duct. The position of the organ of Corti on the surface of the basilar membrane is noted. The basilar membrane represents a specialization of perilymphatic tissues—that is, a portion of the lining of the scala tympani. It is a resonant structure that also supports the organ of Corti. It consists of many oriented filaments 80–100 Å thick that have been likened to collagen by some and called scleroprotein fibers by others. The filaments are embedded in a matrix, and may be organized into bundles. The fibers of the basilar membrane make up the so-called auditory strings. The basilar membrane terminates laterally by merging with the spiral ligament, which consists of periosteal connective tissue. [Diagram © Copyright 1970 CIBA Pharmaceutical Company, Division of CIBA-GEIGY Corporation. Reproduced with permission from *Clinical Symposia,* illustrated by Frank H. Netter, M.D. All rights reserved.]

NERVOUS TISSUE
Cochlea

Fig. 1 is a view into a middle turn of a sectioned cochlea, showing the scala vestibuli (**SV**), the cochlear duct (**CD**), or scala media, and the scala tympani (**ST**). A vestibular (Reissner's) membrane (**VM**) separates the scala vestibuli and the cochlear duct. In this section, the outer bony covering provided by the temporal bone (**TB**) is continuous with a small part of the osseous spiral lamina (**OSL**). A bony axis of the cochlea, the modiolus, is out of view to the left. Spiral ganglia (**SG**) contain bipolar neurons whose dendrites extend in a bundle (**Bu**) toward the organ of Corti to innervate the hair cells. The medial aspect of the cochlear duct (**CD**) is known as the spiral limbus (**SL**) and is the periosteal connective tissue of the osseous spiral lamina. The spiral limbus extends laterally and divides into a short tympanic lip (**TL**) and a short vestibular lip (**VL**), which forms a margin for a part of the internal spiral sulcus (**Su**). The tympanic lip is continuous with the basilar membrane (**BM**), but the vestibular lip is continuous with the tectorial membrane (**TM**). Because the gelatinous tectorial membrane is shrunken in this preparation and deflected from its normal position, the configuration of the internal spiral sulcus (**Su**) is not accurately represented. The spiral ligament (**Li**) is partially torn away from the surrounding temporal bone in this preparation. The position of the organ of Corti (**OC**) on the surface of the basilar membrane is denoted. The tympanic side (underside) of the basilar membrane adjacent to the scala tympani is covered with a single layer of cells (**BMC** in Fig. 2), characterized by long slender branches that form a network. The cells may form a compact layer on the basilar membrane in the basal turn of the cochlea. Toward the apex, however, the cells are more loosely organized. Fibrillae (**Fi**) of the basilar membrane can be viewed through the meshes of the cellular network. [Fig. 1 reproduced by permission from T. Tanaka, N. Kosaka, A. Takiguchi, T. Aoki, and S. Takahara, Observations on the cochlea with SEM. In O. Johari and I. Corvin (eds.), *Scanning Electron Microscopy / 1973*, IIT Research Institute, pp. 427–434 (1973).]

Fig. 1, ×355; Fig. 2, ×3660

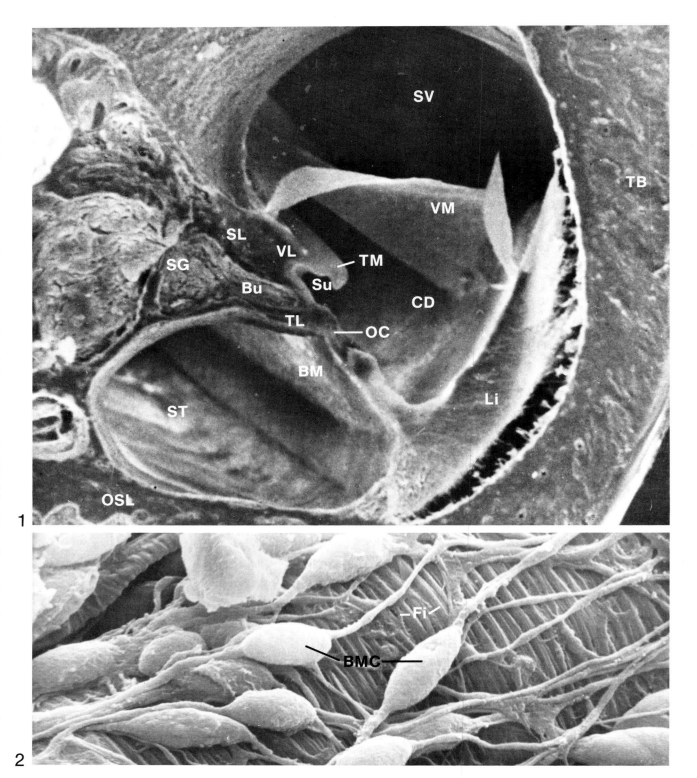

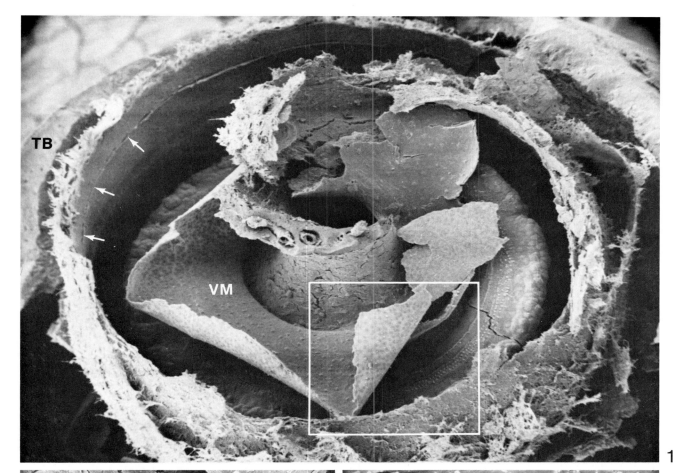

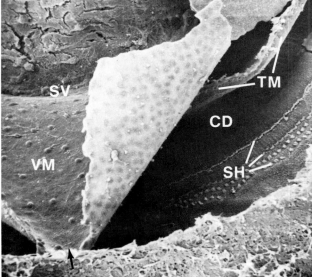

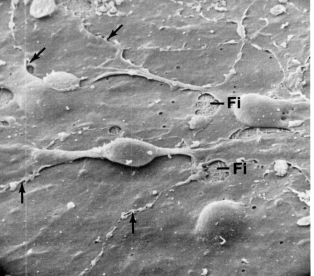

NERVOUS TISSUE
Cochlea

The temporal bone (**TB**) covering the apical end of the cochlea was removed in Fig. 1 to illustrate the vestibular membrane (**VM**). The lateral attachment of the vestibular membrane is broken over much of its extent, but the normal lateral attachment site is indicated by **arrows**. The region contained within the rectangle in Fig. 1 is enlarged in Fig. 2. The vestibular membrane (**VM** in Fig. 2) is attached in only a small region to the spiral ligament (**arrow**), which forms an internal lining of the outer temporal bone. The vestibular membrane separates the scala vestibuli (**SV**) from the cochlear duct (**CD**). Because the lateral attachment of the vestibular membrane is broken over much of its extent and folded over on itself, it is possible to view the two apposed layers lining both of its surfaces. Furthermore, the tectorial membrane (**TM**) is shrunken and retracted, making it possible to view the sensory hairs (**SH**) of the single row of inner hair cells and the three rows of outer hair cells projecting into the cochlear duct. The upper surface of the vestibular membrane adjacent to the perilymph in the scala vestibuli is illustrated at higher magnification in Fig. 3. The vestibular membrane is covered on each of its sides by highly attentuated simple squamous cells. The borders between cells are denoted by the **arrows** in Fig. 3, and the localized bulges probably indicate the positions of the cell nuclei. In small areas where the epithelium is incomplete, it is possible to observe delicate fibers (**Fi**) below the squamous cells.

Fig. 1, ×120; Fig. 2, ×220; Fig. 3, ×2045

NERVOUS TISSUE
Cochlea

The tectorial membrane normally extends into the cochlear duct to cover the stereocilia on the apical ends of the hair cells. In a small region in Fig. 1 the tectorial membrane (**TM**) is attached to the hairs of sensory cells (**arrow**), but it is detached and elevated in other regions. Cut edges of Reissner's vestibular membrane (**VM** in Figs. 1 and 2) and the spiral ligament (**SL**) are also identified. The central bony modiolus (**Mo**) has been broken in the upper portion of the illustration. An upper turn of the cochlea is exposed in Fig. 2. The tectorial membrane (**TM**) extends as a sheet from its attachment to the vestibular lip (**VL**) of the limbus. Since the tectorial membrane usually shrinks during fixation and dehydration, it is possible to view the hairs of the sensory cells in the organ of Corti (**Ha** in Fig. 2). The tectorial membrane is secreted by interdental cells of the spiral limbus. It consists primarily of a protein that possesses some characteristics common to epidermal keratin. A view of the upper surface of the tectorial membrane is shown in Fig. 2. In such a fixed preparation, the tectorial membrane has a fibrous appearance. It is composed of fibrils (Fig. 4) and ground substance and is sometimes divided into a limbal zone, a middle zone, and a marginal zone. The marginal zone ends laterally in what is termed a marginal net. The fibrillar appearance of the lateral edge of the tectorial membrane (**TM**) is illustrated in Figs. 3 and 4. This network is probably more extensive in fixed preparations than in living tissue, since the tectorial membrane is extremely prone to shrinkage. The undersurface of the marginal zone of the tectorial membrane can be observed in places to be in contact with the sensory hairs (**Ha**). The marginal net appears in some cases to terminate (**arrows** in Figs. 3 and 4) on the surface of supporting phalangeal (Deiter's) cells (**PH**) or at the boundary between phalangeal cells and Hensen's cells (**HC**).

Fig. 1, ×250; Fig. 2, ×355;
Fig. 3, ×1125; Fig. 4, ×980

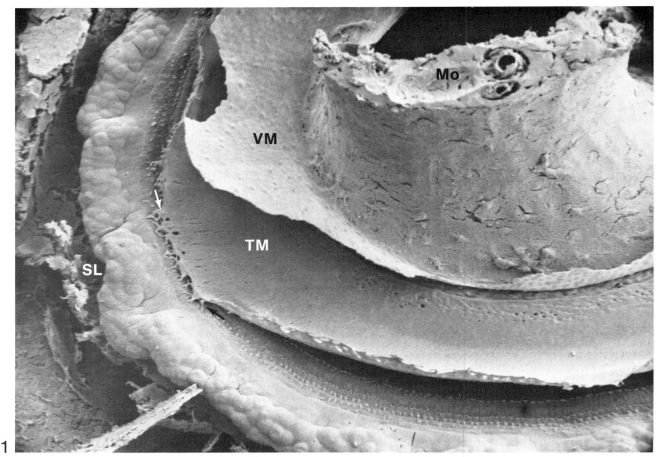

1

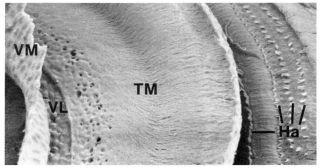

2

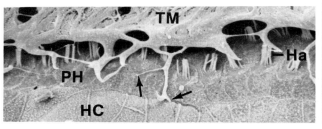

3

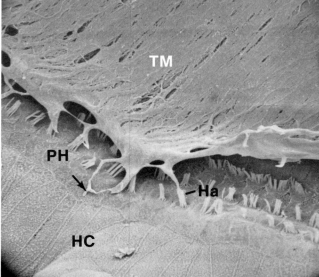

4

Fig. 1 labels:

Outer hair cells

Inner hair cell

Tectorial membrane

Outer tunnel

Cells of Hensen

Cells of Claudius

Cells of Boettcher

Basilar membrane

Outer phalangeal supportive cells of Deiter

Outer pillar cell

Spiral vessel

Tunnel of Corti

Inner pillar cell

Inner phalangeal supportive cell

Internal spiral tunnel

Myelinated nerve fibers

1

2

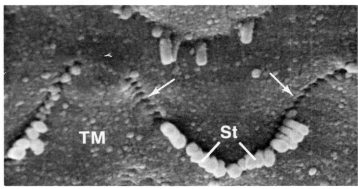

3

TM   St

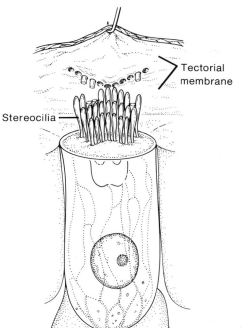

Tectorial membrane

Stereocilia

4

NERVOUS TISSUE
Organ of Corti

Fig. 1 diagrammatically illustrates the relationship between the tectorial membrane and the organ of Corti. The three rows of outer hair cells and the single row of inner hair cells are depicted in surface and cross-sectional views. The corresponding light photomicrograph in Fig. 2 also illustrates the organ of Corti in section. The tectorial membrane is partially retracted, an artifactual condition commonly resulting from shrinkage during specimen preparation. The undersurface of the tectorial membrane (**TM**) is shown in Fig. 3, where the imprints (**arrows**) of the tallest hairs (stereocilia) of the hair cells can be seen; some of the tallest stereocilia (**St**) are broken but remain attached to the tectorial membrane. Thus the tectorial membrane appears to be firmly attached to the tallest stereocilia on the hair cells. The resulting V-shaped pattern of indentations on the undersurface of the tectorial membrane (diagrammed in Fig. 4) corresponds to the arrangement of the tallest stereocilia on the hair cells. It appears that the hair cells are capable of transmitting sound of different frequencies and are located in different regions of the cochlea. Hair cells that transmit high frequencies are located at the base of the cochlea; those that transmit low frequencies are located at the apical end. Each hair cell is innervated by one or more nerve processes. The inner hair cells are supplied by a single neuron, but the outer hair cells may be supplied by several neurons. Electrophysiological evidence suggests that inner hair cells, which are unattached to the tectorial membrane, are velocity-sensitive and are stimulated by the drag of endolymph. The outer hair cells, however, are attached. They are stimulated by the shearing action of the tectorial membrane and basilar membrane and can thus be considered as "displacement detectors" (Dallos, 1973). [Fig. 4 supplied by T. Tanaka.]

Fig. 2, ×355   Fig. 3, ×10,770

NERVOUS TISSUE
Organ of Corti

The surface of the organ of Corti, after removal of the tectorial membrane, is illustrated in Fig. 1. A single row of inner hair cells (**IHC**) contains apical stereocilia that are linearly arranged. In contrast, the stereocilia associated with the three rows of outer hair cells (**OHC**) are arranged in a V or W configuration. The apical surface of the hair cells surrounding the stereocilia appear smooth and are termed cuticular plates (**CP** in Fig. 2). It is also possible to distinguish the free, or apical, surface of a number of other cells within the organ of Corti. These include the inner (spiral) sulcus cell (**ISC**), the heads of the inner pillar cells (**IPH**), the phalangeal plates (**PP**) of Deiter's (outer phalangeal) cells, and the surfaces of Hensen's cells (**HC**). Inner pillar cells reside basally on the basilar membrane. They narrow as they form the inner margin of the inner tunnel, but expand at their apical ends to form a flat plate that contacts adjacent pillar cells and the inner hair cells. The long outer pillar cells also reside on the basilar membrane. A part of the apical end, or head, of the outer pillar cells abuts on the plate of the inner pillar cells, but an extension of the outer pillar cell in this region, a phalangeal process, ends between adjacent outer hair cells in the first row. Surfaces of the phalangeal process of the outer pillar (**OP**) cells are illustrated in Fig. 2. Also shown are the surface of inner pillar (**IP**) cells, Deiter's cells, (**DC**), and the stereocilia (**St**) associated with row 1 and row 2 of outer hair cells.

Fig. 1, ×1460; Fig. 2, ×7750

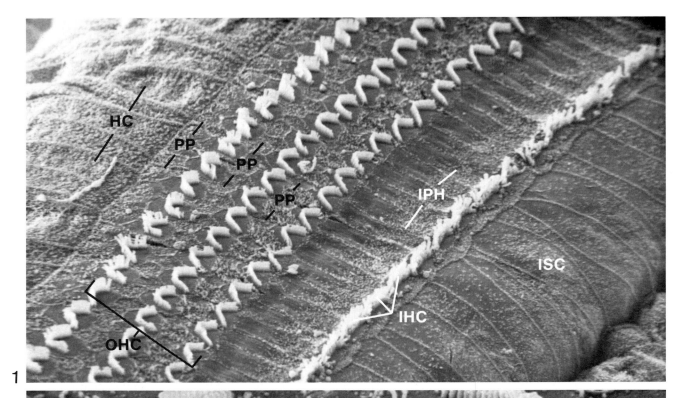

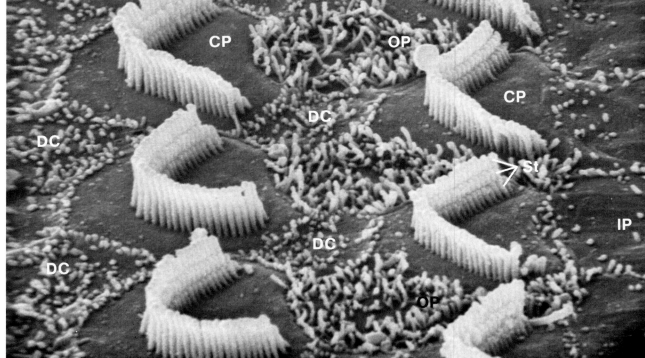

NERVOUS TISSUE
Organ of Corti

The single row of inner hair (**IH**) cells and the three rows of outer hair (**OH**) cells are shown in Fig. 1. It is also possible to distinguish in Fig. 1 the surfaces of the inner spiral sulcus cells (**SS**), the inner phalangeal cells (**IPh**), the inner pillar (**IP**) and outer pillar (**OP**) cells, and the three rows of Deiter's cells (**DC**), or outer phalangeal cells. Stereocilia (**St**) are arranged in a V or W configuration on the outer hair cells (Figs. 1 and 2). The stereocilia (**St**) of the outer row on each cell are the longest, and they are uniform in their length. Those of the inner row are the shortest, and those of the middle row are intermediate in length. The remainder of the hair cell surface is smooth and is called a cuticular plate (**CP** in Fig. 2). At higher magnification (Fig. 2), the three rows of stereocilia on a single outer hair cell are quite apparent, as is their arrangement into the form of a W. The tips of the stereocilia are often rounded. Stereocilia (**St** in Fig. 3) on the inner hair cells are more linearly arranged, but exhibit variation in their length from row to row, like those on the outer hair cells. Microvilli (**arrows** in Fig. 3) are associated with the exposed surfaces of the inner phalangeal cells. [Figs. 2 and 3 from T. Tanaka, "A scanning electron microscope study of the auditory organ," *Metabolism and Disease* 13(12):i–iii(19761.]

Fig. 1, ×2915; Fig. 2, ×24,920; Fig. 3, ×17,800

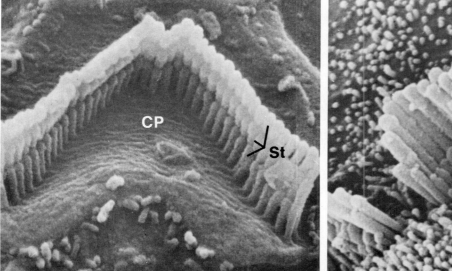

1

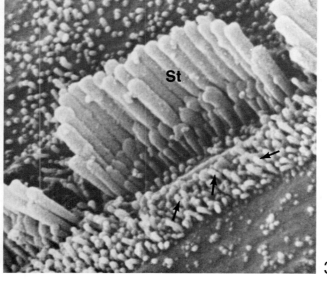

2

3

NERVOUS TISSUE
Organ of Corti

A fractured preparation illustrating the organ of Corti in radial section (Fig. 1) reveals the inner (**IP**) and outer (**OP**) pillar cells bordering the tunnel of Corti (**TC**). The basilar membrane (**BM**) forms the floor of the tunnel, and nerve fibers (**arrow**) traverse the tunnel. The stereocilia associated with the inner (**I**) and outer sensory hair cells can be observed in Fig. 1, as can the shape of the three rows (**1, 2, 3**) of outer hair cells. Cell bodies (**CB**) of the second row of outer hair cells and their stereocilia (**St**) are shown in Fig. 2. The cell bodies of Deiter's cells support the rows of hair cells, but each Deiter's cell extends a narrow phalangeal process upward between the hair cells. Several of these phalangeal processes (✱) are broken in this preparation. Both inner and outer hair cells are innervated at their base by afferent and efferent nerve fibers. In Fig. 3, it is possible to observe these nerve fibers (**arrow**) as well as the synaptic terminations (**Ne**) with the hair cell base (**OHC**). The cell bodies of the supportive Deiter's cells (**DC**) are located beneath the hair cell base and enclose the nerve fibers; their supportive nature can be visualized in Fig. 4. Deiter's cells (**DC**) reside on the basilar membrane and extend toward the base of the hair cells which they support. Hair cells do not reach the basilar membrane and are necessarily supported by Deiter's cells. Near this junction a thin phalangeal process (**PP**), containing microfilaments and microtubules internally, extends toward the surface of the organ of Corti (compare Figs. 2 and 4). A portion of the tectorial membrane (**TM**) overlying the stereocilia of the hair cells is also visible in Fig. 4. Several outer pillar cells (**OP**) are shown in Fig. 5. The base of these attenuated cells resides on the basilar membrane (**BM**), and the cells extend toward the surface of the organ of Corti, forming the lateral margin for the tunnel (**TC**) of Corti. [Figs. 1 and 3 reproduced by permission from H. Engström and B. Engström, *Acta Otolaryngol.* 83:65–70, 1977, Almqvist & Wiksell Periodical Co. Fig. 4 reproduced by permission from G. Bredberg, H. W. Ades, and H. Engström, *Acta Otolaryngol.* (Suppl. 301), 1972, Almqvist & Wiksell Periodical Co.]

Fig. 1, ×665; Fig. 2, ×2245
Fig. 3, ×4895; Fig. 4, ×1470; Fig. 5, ×970

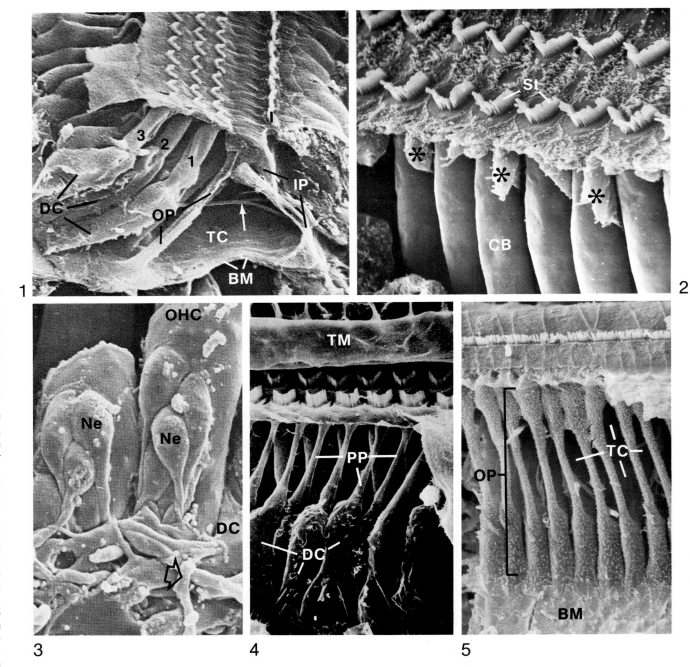

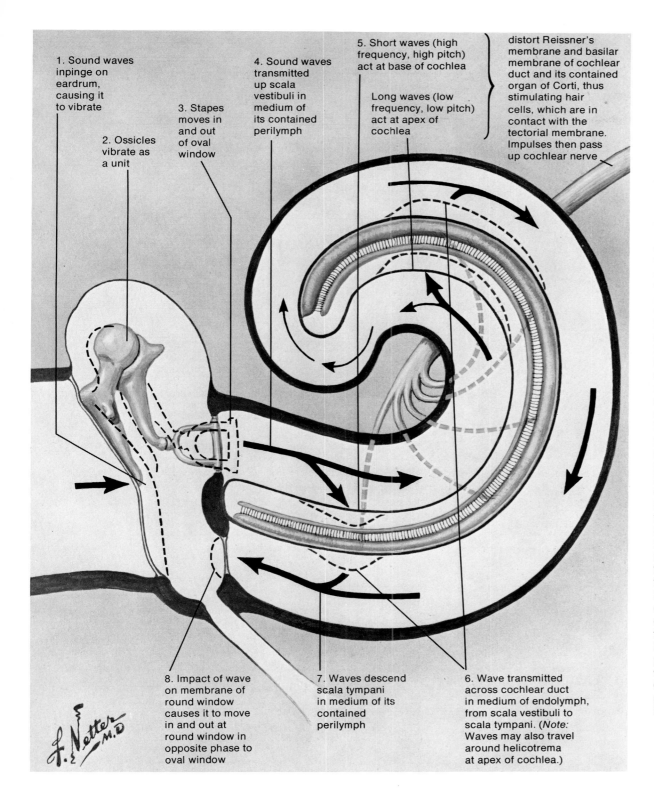

1. Sound waves inpinge on eardrum, causing it to vibrate

2. Ossicles vibrate as a unit

3. Stapes moves in and out of oval window

4. Sound waves transmitted up scala vestibuli in medium of its contained perilymph

5. Short waves (high frequency, high pitch) act at base of cochlea

Long waves (low frequency, low pitch) act at apex of cochlea

distort Reissner's membrane and basilar membrane of cochlear duct and its contained organ of Corti, thus stimulating hair cells, which are in contact with the tectorial membrane. Impulses then pass up cochlear nerve

8. Impact of wave on membrane of round window causes it to move in and out at round window in opposite phase to oval window

7. Waves descend scala tympani in medium of its contained perilymph

6. Wave transmitted across cochlear duct in medium of endolymph, from scala vestibuli to scala tympani. (*Note:* Waves may also travel around helicotrema at apex of cochlea.)

NERVOUS TISSUE
Organ of Corti

The mechanism by which the organ of Corti is involved in the process of hearing is only partially understood. Briefly, the sound waves that cause the eardrum to vibrate are transmitted to the ossicles, causing the stapes foot plate to move in and out of the oval window. As a result, pressure changes are set up in the perilymph, which is contained in the scala vestibuli. Vibrations generated within the perilymph are transmitted across the vestibular membrane to the endolymph of the cochlear duct. Here, a displacement occurs in the tectorial membrane–hair cell–basilar membrane complex. The movement of the stereocilia on the hair cells somehow results in a depolarization of these auditory mechanoreceptors, which is transmitted along afferent neurons that synapse with the base of the hair cells. The movement of the basilar membrane results in a shearing action on the hair cells, some of whose stereocilia are embedded in the tectorial membrane. The mechanical energy is thus transformed into electrical impulses by the hair cells. The basilar membrane is stiffer at the base than at the apex, and the fibers in the basal turn are relatively short. In the apex of the cochlea, the basilar membrane is broader, and more of its surface is covered with cells—a feature that apparently serves to dampen the oscillations of the membrane. The amplitude of sound waves along the basilar membrane from base to apex varies with their frequency. At high frequencies, there is maximum amplitude and maximum displacement of the basilar membrane in the basal turns, and the waves damp out before reaching the apical turns. In contrast, low-frequency waves have little displacement effect on the basal portion, but maximum amplitude is achieved toward the apex. These phenomena are the basis of pitch discrimination. Pitches are discerned by recognition of the point of maximum displacement of the basilar membrane. In addition, the rate of discharge in the hair cell–neuron complex is related to the magnitude of displacement. Pitch discrimination is thus based on the location of specific hair cells that are capable of initiating maximum action potential discharge in those afferent nerve fibers that synapse with them. The vibrations that are transmitted to the perilymph of the scala tympani cause the secondary tympanic membrane at the round window to move outward. [Copyright 1970 CIBA Pharmaceutical Company, Division of CIBA-GEIGY Corporation. Reproduced with permission from *Clinical Symposia,* illustrated by Frank H. Netter, M.D. All rights reserved.]

NERVOUS TISSUE
Vestibular Sensory Receptors—Maculae

That portion of the vestibular apparatus concerned with the detection of directional (positional) movement (i.e., tilting of the head) consists of two flattened, spot-like areas (maculae) located in the saccular and utricular cavities of the membranous labyrinth. The plane of orientation of the macula sacculi is vertical; that of the macula utriculi, horizontal. The macula utriculi is kidney-shaped (Fig. 1), but the macula sacculi resembles a hook (Fig. 4). In transverse section, the macula utriculi (Fig. 3) and macula sacculi (Fig. 6) are both organized into a single layer of cells in association with an overlying statoconial (otolithic) membrane. The single cell layer consists of supportive cells residing on a basement membrane and two types of sensory hair cells, type **I** and type **II** (Wersall, 1956). The type **I** sensory cell is bottle-shaped, possesses a larger free surface area, and is innervated by a nerve chalice that encloses almost the entire cell (Figs. 3 and 6). Type **II** cells are more cylindrical, possess a smaller surface area, and are innervated at their base by small terminations (often more than one per cell) of nerve fibers (Figs. 3 and 6). Each type **I** and type **II** cell is characterized by its 50–100 stereocilia, which project from the apical surface to varying heights, producing a pipe-organ configuration. Eccentric to each bundle of stereocilia is one long kinocilium (represented by the tallest and heaviest line projecting from each cell in Figs. 3 and 6). In contrast to the classification of hair cells based on innervation, another classification, resulting from recent observations with the scanning electron microscope, is based on the stereocilia-kinocilium height ratio. Hair cells designated as type **I** on this basis are those with comparatively short stereocilia, and cells designated as type **II** hair tufts are those with comparatively longer stereocilia. It is not known with certainty what, if any, correlation can be made between type **I** and **II** hair cells as defined on the basis of innervation and type **I** and **II** hair cells as defined on the basis of stereocilia length. It is now known, however, that the stereocilia length on vestibular hair cells can vary more than formerly thought, and may also vary between organisms. A more extensive classification of hair cells (based on kinocilium-stereocilia height ratios) is appearing in the literature (Lim, 1976). The sensory cells of the macula utriculi are morphologically polarized, so that the kinocilium of each cell faces toward an arbitrary line termed the striola (denoted by **broken lines** in Figs. 3 and 6; **S** in Figs. 2 and 5), which roughly divides the population of hair cells into two oppositely polarized groups. The hair cell polarization differs in the macula sacculi in that the kinocilium of each cell faces away from the striola. Cells defined as type **I** on the basis of innervation are twice as numerous as type **II** cells in the striola region. The overlying statoconial membrane consists of a gelatinous mass of calcium carbonate crystals. Regional variations in crystal size and membrane thickness are diagrammed in Figs. 2, 3, 5, and 6. [Figs. 1 to 6 courtesy of H. H. Lindeman, ''Studies on the morphology of the sensory regions of the vestibular apparatus,'' *Ergebn. Anat. EntwGesch.* 42(1):1–113 (1969). Springer-Verlag.]

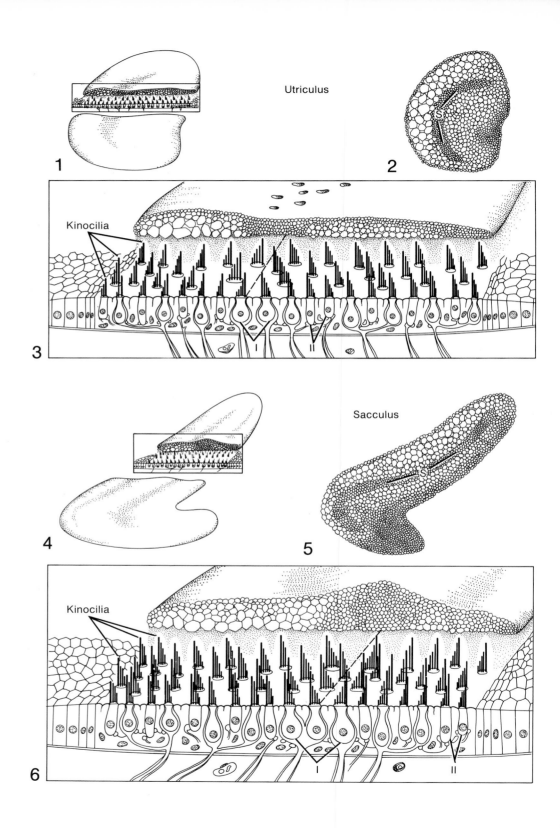

Utriculus

1

2

Kinocilia

3

Sacculus

4

5

Kinocilia

6

Vestibular Sensory Receptors—Maculae

A more detailed diagrammatic representation of the receptor hair cells (**RC**) and supportive cells (**SC**) that form the macula sacculi, as well as their relationship to the overlying otolithic membrane (**OM**) is illustrated in Fig. 1. The vestibular receptor of *Rana pipiens* is used here as a prototype because many of the electrophysiological measurements have been obtained in this amphibian and the functional properties of the sensory hair cells of this species are well characterized. In mammals, the vestibular apparatus contains two types of hair cells, as classified on the basis of differences in nerve innervation, cellular shape, and free surface area. In the frog, however, the cells have been classified alternatively into two types of hair tufts on the basis of a sterocilia-to-kinocilium height ratio (Hillman, 1976). The otolithic membrane (**OM**), containing calcium carbonate crystals or otoliths (**OL**), is supported by a filamentous base (**FB**) and possesses pores in areas overlying each hair cell. The kinocilium (**K**) is expanded distally into a bulb that is attached to both the otolithic membrane and the adjacent stereocilia (Fig. 2). Internally, the nonmotile kinocilium possesses an axial filament complex similar to cilia and flagella. The stereocilia (**S**) of each hair cell are variable in height and organized by rows in pipe-organ fashion. Bundles of intracellular filaments extend the length of the stereocilia into the apical cytoplasm to form a rigid, supportive cuticle (**C**), or cuticular plate (Figs. 1 to 4). The kinocilium originates proximally from a notch adjacent to the cuticle. Positional changes of the head are thought to cause a shift of the endolymph and otolithic membrane, which results in a shearing force on the underlying kinocilium–stereocilia complexes. When mechanical shearing forces displace the kinocilium–stereocilia complex toward the kinociliary side of the cells (Fig. 3), the unsupported base of the kinocilium plunges inward at the cuticular notch (**N**), forming a dimple (**arrow** in Fig. 3). This direction of displacement tends to deform the membrane mechanically, causing depolarization and increased activity of the vestibular nerve innervating the cell. Displacement in the opposite direction (Fig. 4), away from the kinociliary side of the cell, causes a raising up of the kinocilium and a convexity at its base, resulting in hyperpolarization and decreased activity of the innervating nerve. Because there are hair cells oriented (morphologically polarized) over a full 360 degrees, any directional movement of the statolithic membrane can be detected. Both afferent (**A**) and efferent (**E**) nerves synapse with the hair cell base. [Fig. 1 from D. E. Hillman, Morphology of peripheral and central vestibular systems. In R. Llinas and W. Precht (eds.), *Frog Neurobiology,* Springer-Verlag, 1976. Figs. 2 to 4 from D. E. Hillman, Observations on morphological features and mechanical properties of the peripheral vestibular receptor system in the frog. In A. Brodal and O. Pompeiano (eds.), *Progress in Brain Research* (Vol. 37), Elsevier Publishing Company, 1972.]

NERVOUS TISSUE
Vestibular Sensory Receptors—Maculae

Fig. 1 illustrates the otoliths (**Ot**) and otolithic membrane (**OM**) supported by the filamentous base (**FB**) that overlies the receptor epithelial surface (**RS**) in the frog utricular macula. The otolithic membrane is folded so that both surfaces can be viewed. The kinocilium–stereocilia complexes on the receptor cells in the frog are illustrated in Figs. 2 and 3. The distal expansion (**arrow** in Fig. 2) on the kinocilium (**Ki**) can be observed in addition to the organization of adjacent stereocilia (**St**). Microvilli (**Mv**) on the surface of surrounding supportive cells are also illustrated. In Fig. 3, the opposing polarity of hair cells on each side of the striola (denoted by **broken lines**) can be distinguished. Note that the kinocilia (**arrows** in Fig. 3) face away from the striola on opposing sides. The orientation of hair cells with respect to the striola region appears to be responsible for the capacity to detect multidirectional shifting of the otolithic membrane. Moreover, since the striola follows a semilunar path, there is a population of hair cells capable of detecting any movement over a full 360 degrees. In addition, a directional shearing force on the receptor surface will cause one population of hair cells to depolarize and those oppositely oriented on the other side of the striola to hyperpolarize. Important in the overall function of the sacculus is the attachment of the kinocilium to the stereocilia and the way in which the kinocilium is attached at the cell surface, as diagrammed on the preceding page. Because the stereocilia are inserted into a cuticular base but the kinocilium is not, displacement of the kinocilium occurs at its base, whereas the stereocilia slide in relation to each other. Precisely how a mechanical deformation of the cell membrane at the base of the kinocilium is transduced into a change in membrane potential has yet to be understood. [Figs. 1 and 2 from D. E. Hillman, Morphology of peripheral and central vestibular systems. In R. Llinas and W. Precht (eds.), *Frog Neurobiology*, Springer-Verlag, 1976. Fig. 3 courtesy of E. R. Lewis.]

Fig. 1, ×535; Fig. 2, ×10,235; Fig. 3, ×9345

122

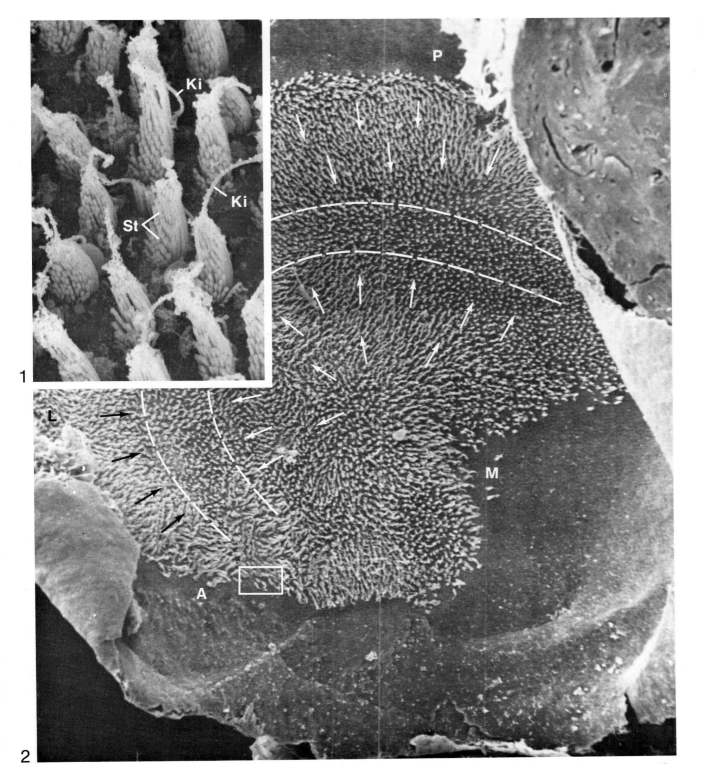

NERVOUS TISSUE
Vestibular Sensory Receptors— Maculae

The kidney-shaped macula utriculi of the guinea pig is shown at low magnification in Fig. 2. It is oriented in the horizontal plane when the head is in normal anatomical position. The gelatinous otolithic membrane and associated otoliths were removed to expose the utricular receptors. The surface of the utriculus can be divided into two areas (a pars externa and a pars interna), by an arbitrary strip that extends along the center of a semicircular area called the striola. The approximate course of the striola is indicated within the **broken lines**. The medial (**M**), lateral (**L**), anterior (**A**), and posterior (**P**) margins of the utricular macula are denoted. The macula utriculi consists of more than 7,000 hair cells that are less densely distributed in the striola region than peripherally. One prominent hair cell type in the guinea pig utricular macula, is shown in Fig. 1. Each sensory cell exhibits a surface polarity defined by a single kinocilium (**Ki**) that is positioned eccentrically relative to a group of stereocilia (**St**). The stereocilia are graduated in their height in pipe-organ fashion, but are arranged in parallel rows. The tallest stereocilia are adjacent to the kinocilium, and are approximately one-half the length of the kinocilium. There are more than fifty stereocilia on each of the cells shown in Fig. 1, but their number is variable. Whereas the kinocilium is attached to a basal body in the apical cytoplasm of the hair cell, the stereocilia emerge from a cuticular plate region on the hair cell surface. The position of the kinocilium with respect to the striola is denoted by **arrows**. Irregularities in the polarization pattern occur at the ends of the striola (**rectangle** in Fig. 2). When the kinocilium-stereocilia complex is deflected toward the kinocilium, there is an increased depolarization of the hair cells and a resulting increased rate of discharge in the afferent nerve fibers that synapse with the base of the hair cells. A displacement of the kinocilium toward and including the stereocilia tends to be inhibitory, and there is a decrease in impulse generation in the nerve fibers.

Fig. 1, ×3115; Fig. 2, ×200

NERVOUS TISSUE

Vestibular Sensory Receptors—Maculae

A portion of the guinea pig macular region is illustrated in Fig. 1 to convey the pattern of orientation of the hair cells in relation to the striola (**S**), and will serve as a reference in subsequent figures. The approximate boundary of the striola is demarcated by **broken lines.** The striola region is easily distinguished by the paucity of type II cells (defined on basis of innervation) in this region and the preponderance of type I cells. The difference in distribution is approximately two-fold in favor of type I cells. In the striola region (Fig. 2), the majority of cells possess shorter stereocilia compared to the hair cells outside of the striola, and no directional polarization is apparent here. A prominent cuticular plate (**CP**) is observed on one of the striola cells in Fig. 2. Figs. 3 to 6 illustrate details of the corresponding areas denoted by letters (**A, B, C, D**) in Fig. 1. The direction of orientation (polarization) of the kinocilium-stereocilia complexes with respect to the striola is designated by the **large arrows** in Figs. 3 to 6. The structural polarization of hair cells in the utriculus is such that the kinociliary side of the hair complex faces the striola. Figs. 5 and 6 show the relationship of the kinocilium (**Ki**) to the adjacent stereocilia (**St**). Cells with shorter stereocilia (✳ in Figs. 5 and 6) are occasionally interspersed among the majority of cells located outside the striola, and are sometimes oppositely polarized (✳ in Fig. 5) compared with the adjacent cells that possess longer stereocilia. The ability to sense directional motion (360 degrees) appears to depend on the differing orientation of the hair cells in relation to the striola. For example, a tilting of the head toward region **B** in Fig. 1 will cause the movement (and hence mechanical shearing) of the otolithic membrane in that direction. This direction of shearing would tend to hyperpolarize the hair cells in region **B** (Fig. 4) and depolarize the oppositely oriented hair cells in region **A** (Fig. 3). The integration of these contrasting impulses occurs in neurons of the vestibular nuclei, located in higher brain centers.

Fig. 1, ×170; Fig. 2, ×2405; Fig. 3, ×1555
Fig. 4, ×1555; Fig. 5, ×4360; Fig. 6, ×5650

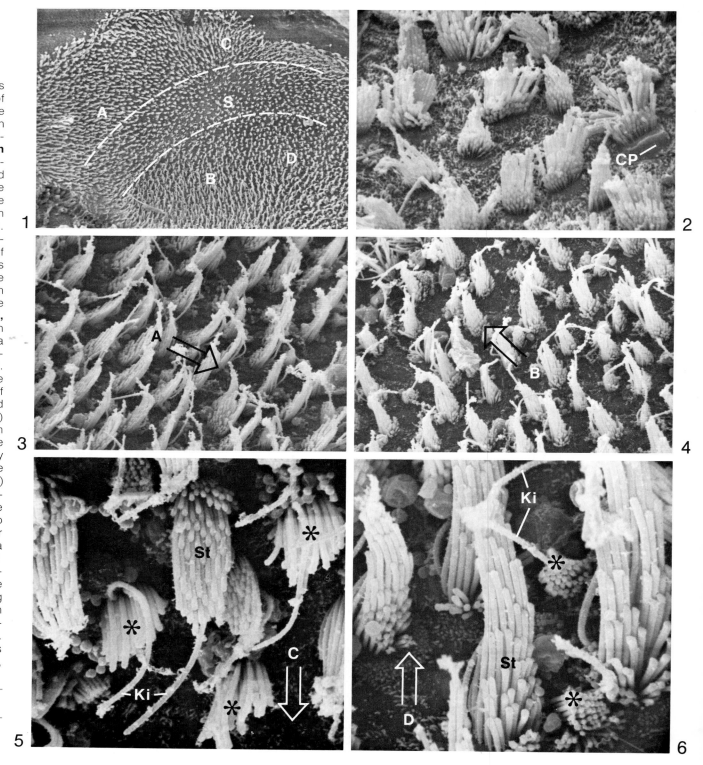

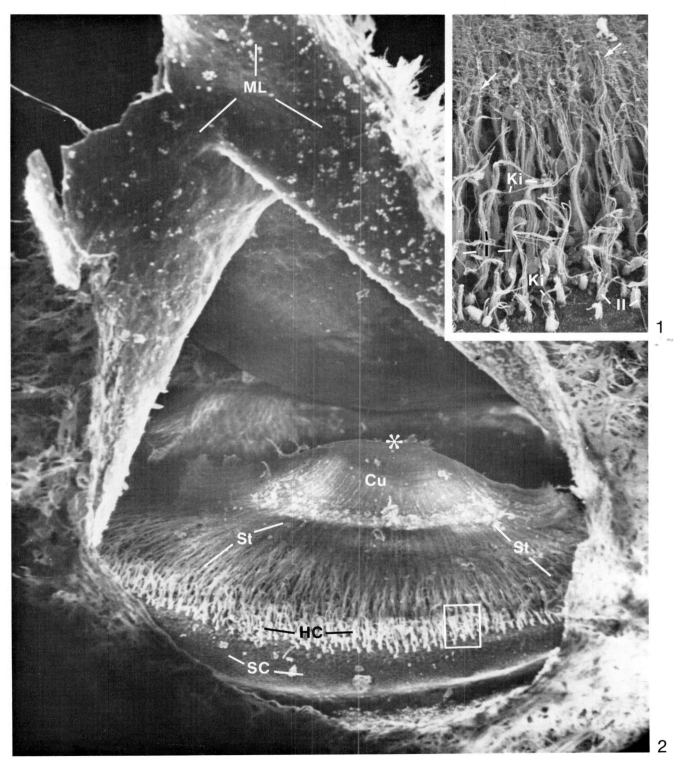

NERVOUS TISSUE

Vestibular Sensory Receptors— Crista Ampullaris

In Fig. 2 the membranous labyrinth (**ML**) of the ampulla has been opened to reveal the internally located sensory receptor, the crista. Hundreds of hair cells (**HC**), each possessing a number of sensory cilia, are concentrated in this area. Only polygonal supportive cells (**SC**) are located outside the boundary of the receptor area. A large, elevated gelatinous mass, the cupula (**Cu**), is attached at its base to the cilia of the hair cells. In the living state the cupula spans the entire height of the ampulla, but shrinkage often occurs during preparation, so that the apex (✳) of the cupula does not extend as high as it normally would. Toward its periphery, the cupula sends forth gelatinous strands (**St**) that radiate outward to contact the longest cilium (the kinocilium) of each hair cell. The crista region forms an elevated ridge that is shaped somewhat like a saddle, but the presence of the cupula obscures the shape in this micrograph. The area enclosed by the rectangle in Fig. 2 in enlarged in Fig. 1 to reveal the association (**arrows**) between the sensory hair cells and the gelatinous cupula. The cupula appears to consist of a delicate meshwork of fibrils that is loosely organized and overlies the hair cells. Two types of hair cells can be distinguished by the length of their stereocilia. Other types are defined on the basis of the kind of innervation at the base of the cell, as described previously for those of the macula. The type I cell (**I**), as defined on the basis of stereocilia length, possesses 5 to 6 long stereocilia that extend approximately three-fourths the height of the single kinocilium (**Ki**). Not all of the stereocilia of the same cell are of the same height; they often form a pipe-organ arrangement. The type II cell (**II**) possesses a long kinocilium, but the stereocilia are very short.

Fig. 1, ×915; Fig. 2, ×265

NERVOUS TISSUE
Vestibular Sensory Receptors— Crista Ampullaris

The epithelium of the membranous labyrinth in the region of the ampullae is elevated into a transverse ridge called the crista ampullaris. This low-magnification scanning electron micrograph of the ampulla from the anterior vertical semicircular canal illustrates the receptor area of the crista. The planar (**Pl**) ends of the crista have a dense concentration of receptor hair cells situated close to the ampulla wall (**AW**). A central isthmus (**Is**) portion of the crista is elevated such that the hair cells are more centrally located within the ampulla. The cupula has been removed in this preparation, but has the same general form as the receptor region. The planum semilunatum is identified at (**PS**). The epithelium contains receptor cells or sensory hair cells of different types as well as supporting cells. The supporting cells are continuous with the cells of the planum semilunatum. Each hair cell commonly bears a single long kinocilium about 50 μm or more in length and numerous adjacent stereocilia of variable length. The kinocilia extend into the base of the cupula and, internally, contain nine peripheral doublet microtubules and two central microtubules that end shortly after leaving the basal body. The stereocilia extend from a dense terminal web, or cuticular plate, in the apical end of the cell just below the plasma membrane. [Courtesy of D. E. Hillman, Morphology of peripheral and central vestibular systems. In R. Llinas and W. Precht (eds.), *Frog Neurobiology,* Springer-Verlag, 1976.]

×130

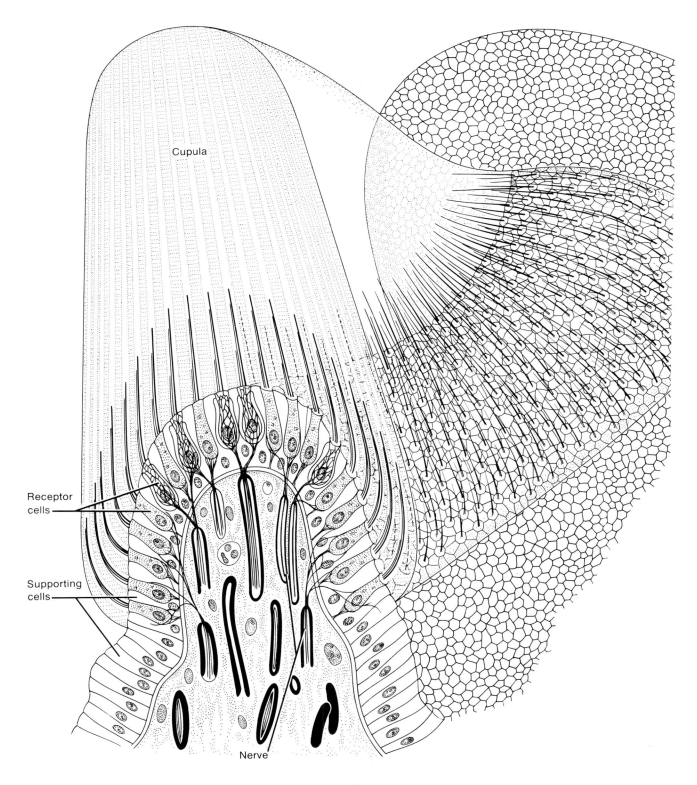

NERVOUS TISSUE
Vestibular Sensory Receptors—Crista Ampullaris

Shown here is a diagram of a crista ampullaris (canal receptor). The hairs of the receptor cells are covered by a gelatinous cupula that locally fills the space of the ampulla in this area. The cupula contains a high concentration of glycoprotein and is more viscid than the endolymph that surrounds the cupula. The receptor (hair) cells are innervated by both afferent and efferent nerve endings. In some of the hair cells, the afferent nerve endings form a chalice-like investment around the hair cell body. In other hair cells, a number of separate terminal nerve boutons terminate close to the basal portion of the hair cell plasma membrane. Supporting cells of the crista epithelium surround the hair cells. Other types of cells, called dark cells, are present at the base of the crista. It has been suggested that these cells may be responsible for maintaining the high potassium concentration of the endolymph. Movement of the head in space, or angular acceleration and deceleration, results in a displacement of the endolymph and cupula of the crista, which translates into a shearing action over the kinocilium–stereocilia complex of the hair cells. Within individual cristae, the polarity of the kinocilium-stereocilia complex is identical. A force directing the stereocilia toward the kinocilium causes increased activity in the vestibular nerve, which innervates the cells. Conversely, when a force directs the kinocilium toward the stereocilia, there is a decrease in activity in the vestibular nerve. The crista does not possess a well-defined striola. [Reproduced with permission from L. C. Junqueira, J. Carneiro, and A. N. Contopoulos, *Basic Histology* (2nd ed.), Lange Medical Publications, 1977.]

Cupula

Receptor cells

Supporting cells

Nerve

NERVOUS TISSUE
Vestibular Sensory Receptors— Crista Ampullaris

Sections of guinea pig crista ampullaris, with cupula removed, are illustrated in Figs. 1 and 2. The flask-shape form of the receptor hair cells (**RC**) within the ridge-like fold of neuroepithelum is shown. Nuclei of supporting cells (**SC**) are basally located in the neuroepithelium. Sections of many blood vessels (**BV**) in Figs. 1 and 2 are present in the connective tissue adjacent to the base of the neuroepithelium. Two types of hair cells can be distinguished in the crista ampullaris of the guinea pig on the basis of differences in stereocilia length. In the type I cell (**I**), several (5–6) of the stereocilia are long and extend with the kinocilium toward the cupula, but terminate shortly before reaching the kinocilium (Figs. 3 and 4). The remainder of the rows of stereocilia are much shorter (3–10 μm), but are arranged in a pipe-organ array. In the type II receptor cell (**II**), a single kinocilium (**Ki**) extends to the base of the cupula, but all adjacent stereocilia (**St**) are much shorter (5–10 μm) and are arranged in graded lengths (Figs. 3 and 4). In such an arrangement the tallest stereocilia are always adjacent to the kinocilium. In the cristae of the two vertical canals, the kinocilia of the hair tufts are directed away from the utricle. In the cristae of the horizontal canals, however, the kinocilia of the hair tufts are directed toward the utricle. The precise functional relationship between the cupula and the crista is not clear. The cupula has been compared to a resilient elastic diaphragm, bound circumferentially to the ampulla wall and more loosely bound across the subcupular space by filaments to the crista surface. In such an arrangement, displacement is greatest near the crista, and the amount of sensory hair displacement would be greatest for movement of a given volume of endolymph.

Fig. 1, ×460; Fig. 2, ×1470;
Fig. 3, ×1620; Fig. 4, ×3050

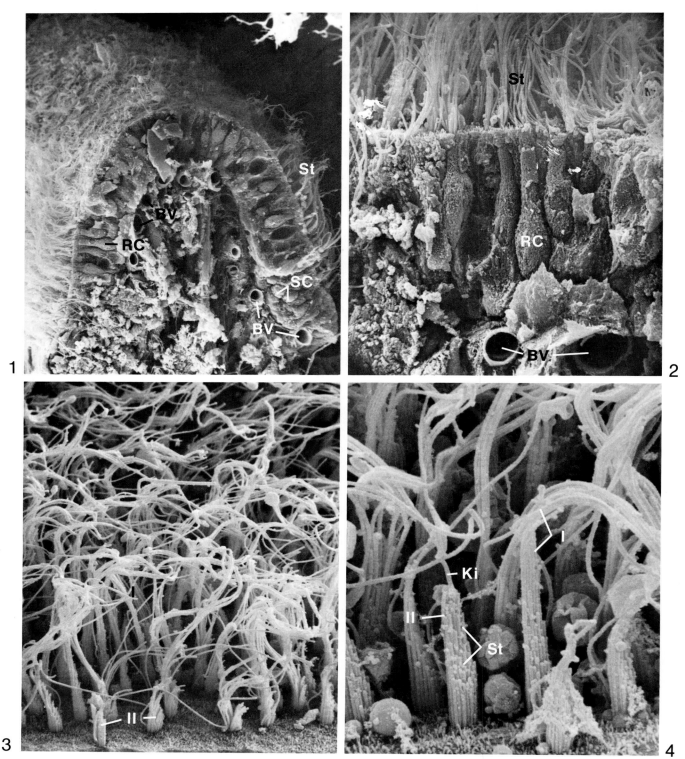

1

2

3

4

128

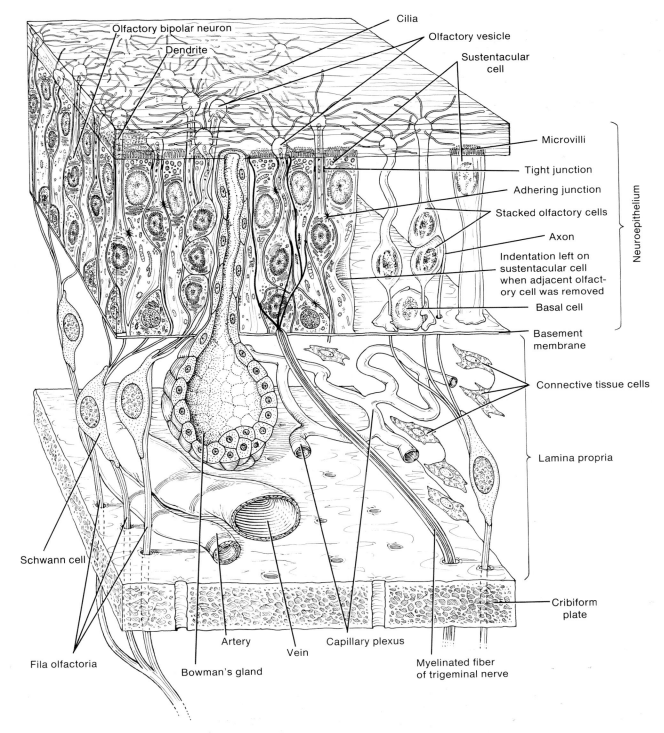

Cilia

Olfactory bipolar neuron

Dendrite

Olfactory vesicle

Sustentacular cell

Microvilli

Tight junction

Adhering junction

Stacked olfactory cells

Axon

Indentation left on sustentacular cell when adjacent olfactory cell was removed

Basal cell

Basement membrane

Neuroepithelium

Connective tissue cells

Lamina propria

Cribiform plate

Schwann cell

Fila olfactoria

Artery

Bowman's gland

Vein

Capillary plexus

Myelinated fiber of trigeminal nerve

NERVOUS SYSTEM
Olfactory Epithelium

The olfactory mucosa consists of a pseudostratified columnar epithelium that is taller than the surrounding respiratory mucosa. Olfactory cells, sustentacular cells, and basal cells make up the olfactory mucosa. Olfactory cells are actually bipolar neurons that taper into a thin dendrite at one end and into an axon at the other. The dendrite extends to the epithelial surface, where it expands into an olfactory vesicle. Basal bodies in the vesicle are continuous with cilia, which abruptly narrow to about half their normal width. These modified cilia are bathed in fluid on the surface of the olfactory epithelium and are intermeshed with cilia of other neurons as well as the microvilli of surrounding sustentacular cells. Although the histophysiology of the olfactory neurons is not well understood, stimulation of different olfactory cells by various odors appears to result from the depolarization of the plasma membrane that covers the olfactory vesicle cilia. In rare cases, in which olfactory cells lack cilia, depolarization apparently occurs on apical microvilli. At the opposite pole, the olfactory cell body narrows into a thin axon, which, together with those from other neurons, travels as the olfactory nerve to the olfactory bulbs of the brain, where synapses occur. The tall, columnar sustentacular cells are closely packed around the olfactory neurons and possess apical microvilli. Sustentacular cells separate the neurons and closely surround their axonal and dendritic processes. Although sustentacular cells possess many characteristics of secretory cells, their function in this regard is unclear. It has been suggested, however, that the function of the mucous surface of the olfactory mucosa may correspond to that of a gas chromatograph, separating gases as they flow past the olfactory surface (Mozell and Jagodowicz, 1973). The separation process is attributed to the solubilization of certain gases by the mucous secretion covering the surface. It is thought that the content of the mucus differs at specific regions on the olfactory mucosa, so that the separation and concentration of certain gases is regionalized on the surface. [Based on a drawing by S. P. Sorokin, ''The respiratory system.'' In R. O. Greep and L. Weiss, *Histology* (3rd ed.), McGraw-Hill Book Company, New York, p. 666, 1973.]

NERVOUS TISSUE
Olfactory Epithelium

A portion of the olfactory area was removed to expose the lateral surface of the epithelial cell sheet in Figs. 1 and 3. The lamina propria (**LP**) underlying the olfactory epithelium is visible in Fig. 1. Fig. 2 is a corresponding light photomicrograph of a similar area. Three different cell types constitute the olfactory epithelium: a sustentacular, or supportive, cell, a sensory olfactory cell, and a basal cell. The sustentacular cells (**SC**) are cylindrical in shape, becoming narrower at their base. Their nuclei are located in the upper third of the epithelial sheet. The free surface of the sustentacular cell has a number of short microvilli. Because of the yellow pigment contained within the sustentacular cell, the olfactory area has a yellow hue in the living condition. The olfactory cell (**OC**), a bipolar neuron, possesses a sensory dendrite (**De**) that transmits an impulse from the mucosal surface to the cell body (**CB**), which contains the nucleus. The nuclei of the bipolar neurons are positioned at different levels in the epithelium, giving it a pseudostratified appearance. From the soma (cell body) of the bipolar neuron, the impulse is transmitted via an axon (**Ax**) that penetrates the underlying connective tissue and joins other olfactory cell axons, forming bundles of neurons that constitute the olfactory nerve. As the dendritic processes reach the surface, they are wedged into crevices (**Cr** in Fig. 3) between sustentacular cells. Networks of long, thin sensory cilia (**Ci** in Fig. 3) can be observed to extend from the free surface of the olfactory cell dendrite. The small and relatively undifferentiated basal cells (**BC**), lying adjacent to the basement membrane, are thought to function as reserve cells for the replacement of the different types in the neuroepithelium.

Fig. 1, ×580; Fig. 2, ×685; Fig. 3, ×1440

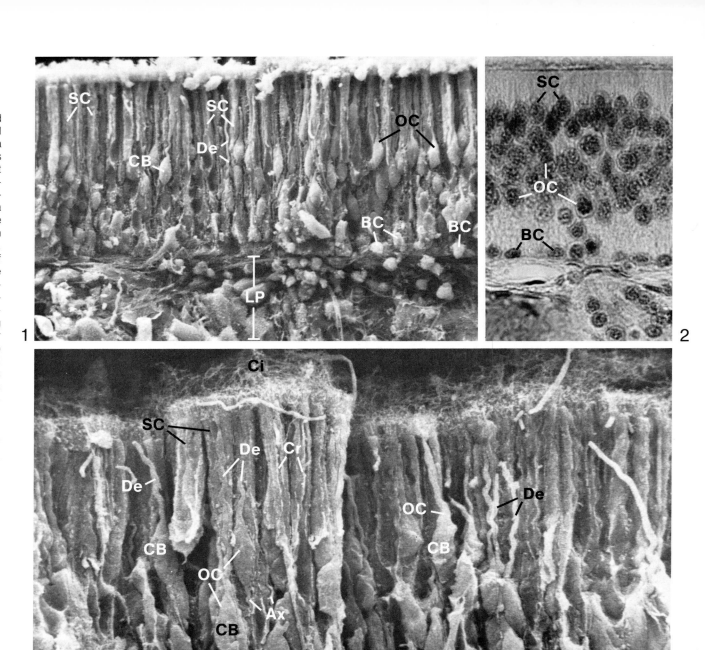

NERVOUS SYSTEM
Olfactory Epithelium

Details of the structural interrelationship between the bipolar neurons and sustentacular cells are revealed by the exposed lateral surface of the olfactory epithelium in these figures. The dendrite (**De**), cell body (**CB**), and axon (**Ax**) of each bipolar neuron are commonly invested by the surrounding sustentacular cells (**SC**). The dendrites are commonly curved, like a corkscrew. Areas of association between sustentacular cells and dendrites are denoted by **small arrows;** this is especially evident in Fig. 5, which is an enlargement of the area enclosed by the rectangle in Fig. 4. Thin cytoplasmic processes (**Pr**) that connect neighboring sustentacular cells are identified in Fig. 5; a dendrite present between the sustentacular cells also appears to be in contact with these processes. Areas of association between the sustentacular cell and cell body of bipolar neurons is denoted by **large arrows** in Figs. 2 and 4. In a number of locations a portion of the bipolar neuron has been separated away from the epithelium, revealing depressions (✱) on the surface of the sustentacular cells that correspond to the shape of the neuron cell body. Close packing and vertical stacking (Fig. 3) of the cell bodies of the olfactory neurons allows room for the extensive number of sensory dendrites that extend toward the epithelial surface. Such an arrangement causes the cell bodies to exist at different levels in the epithelium and contributes to the pseudostratified nature of the epithelium. It is apparent that the somas of these neurons could form synapses with one another in this arrangement, but such olfactory neuron interaction and possible informational processing has yet to be appreciated. The significance of the association between olfactory neurons and sustentacular cells is also not well understood. It is currently believed that sustentacular cells function to support and isolate individual bipolar neurons. Junctional complexes formed by membranes of adjacent sustentacular cells or between bipolar neurons and sustentacular cells are frequently observed. That the sustentacular cell may play a role in modulating the sensory function of the olfactory cell is an interesting speculation, but has not yet been investigated.

Fig. 1, ×1615; Fig. 2, ×2705
Fig. 3, ×3125; Fig. 4, ×3010; Fig. 5, ×9775

NERVOUS TISSUE
Olfactory Epithelium

The surface of the olfactory epithelium is illustrated in these figures. Figs. 3 and 4 also show a portion of the exposed lateral surface of the epithelium. Each dendrite (**De** in Fig. 3) extends toward the epithelial surface and ends in a bulbous expansion of cytoplasm referred to as the olfactory vesicle (**OV**). Approximately twelve to twenty-four (as high as one hundred in some species, such as the dog) cilia (**Ci**) originate from basal bodies located beneath the surface of each olfactory vesicle. In some cases the cilia taper into long, thin modified cilia (**MC**) that appear similar in diameter to microvilli (**Mv**) of the sustentacular cells. The length of the cilia is highly variable (2–100 μm) in different regions of the olfactory epithelium, as well as in different species. Cilia originating from the same olfactory vesicle may also vary in length. Olfactory vesicles are often located at different heights with respect to the plane of the surface epithelium. Many conspicuously protrude from the epithelial surface (see Figs. 1, 2, and 5); others appear to be located deeper in the epithelium and are only partially exposed (see Fig. 4 and **arrows** in Figs. 1, 2, and 5). Openings (**Op**) are also present on the epithelial surface. Some may represent the ducts of Bowman's glands opening onto the surface, but it appears that others are openings through which olfactory vesicles partially project with their respective cilia (❋ in Figs. 1 and 2). It is possible that the positioning of the olfactory vesicle is dynamic; the vesicle may exist at different levels with respect to the free surface according to the maturational or functional state of the olfactory bipolar neuron. Active movement of the olfactory cell dendrite has been observed (Vinnikov and Titova, 1957), and the phenomenon (called "olfactomotor reaction") may be partially responsible for the variable appearance of the olfactory vesicles. The corkscrew appearance of the dendrites observed on the previous page may have resulted from shortening of the dendrite during the olfactomotor reaction. The process of olfactory perception has been suggested to involve the olfactomotor reaction (Bronshtein, 1962).

Fig. 1, ×3240; Fig. 2, ×6155
Fig. 3, ×7035; Fig. 4, ×6595; Fig. 5, ×13,480

NERVOUS TISSUE
Olfactory Epithelium

The regional variation in the length of the olfactory cilia is demonstrated in these figures. The transition between an area possessing short cilia and an area exhibiting long cilia is apparent in Figs. 1, 3, and 4. That the regions with long, thin cilia (**Ci**) are olfactory areas is documented in Fig. 2, in which the lateral surface overlying such a region is exposed to reveal sustentacular cells (**SC**), dendrites (**De**), and cell bodies (**CB**) of olfactory cells. The longer cilia are densely distributed, and may travel in a parallel arrangement (**PA** in Fig. 1) along the surface of the epithelium or may be oriented in a haphazard, tangled fashion. The various arrangements present may be attributed to the reaction of the cilia to odoriferous molecules or to the manner in which the specimen was prepared (i.e., the effect of removing the mucous secretion from the epithelial surface). In the regions with longer cilia, the olfactory vesicles cannot be discerned. They may possibly be obscured by the density of the overlying cilia, allowing only those vesicles with short cilia to be observed (see adjacent areas with short cilia). Alternatively, these olfactory vesicles may be sunken among neighboring sustentacular cells, partially hidden from view (see **large arrows** in Figs. 1, 3, and 4). In the transition areas, the long cilia can be observed to originate from olfactory bulbs (**small arrows** in Figs. 3 and 4), and some appear to originate below the epithelial surface (✹ in Figs. 3 and 4), presumably from underlying olfactory vesicles hidden from view. The variations in the length of cilia from adjacent areas or from the same olfactory vesicle may reflect differences in the maturational or functional states of the olfactory cells, or the length of the cilia may distinguish different types of olfactory cells, capable of discriminating certain odors. A single ciliated respiratory cell (**RC** in Fig. 1) is sometimes observed in the olfactory epithelium, as well as large cells (**LC** in Figs. 1 and 3), which possess short cilia and whose identity has not been determined.

Fig. 1, ×2350; Fig. 2, ×1825;
Fig. 3, ×3400; Fig. 4, ×3240

NERVOUS TISSUE
Olfactory Epithelium

In humans, the olfactory mucosa is located in the superior portion of both nasal fossae, and forms the mucous membrane that covers the superior nasal conchae as well as the opposing septum. Three different views of the olfactory region are illustrated, and it is possible to compare the surface (Fig. 3) and lateral aspects (Figs. 1 and 2) of the olfactory epithelium. Each dendrite (**De**) originating from an olfactory cell body (**CB**) ends in a bulbous expansion of cytoplasm referred to as the olfactory vesicle (**OV**). Six to eight cilia originate from each vesicle and quickly taper into long, thin modified cilia (**Ci**) that appear similar in diameter to microvilli (**Mv**) of sustentacular cells (**SC**). The connection between a number of olfactory vesicles and cilia was removed in Fig. 2. At their proximal ends, the cilia exhibit the typical two central microtubules surrounded by nine peripheral doublet microtubules. But at the distal end of the cilium, few microtubules are present. This portion of the cilium is presumably the area that is sensitive to odors. The connective tissue under the epithelium contains tubulo-alveolar glands, called Bowman's glands, which synthesize a serous type of secretion that passes through a small duct onto the surface of the epithelium. This secretion assures that the epithelium is always moist, and serves as a solvent for various gases responsible for odors. Secretory droplets (**SD**) present on the epithelial surface are produced by sustentacular cells in some species and add to the serous secretion of Bowman's glands to produce a fluid that bathes the sensory cilia and microvilli.

Fig. 1, ×1280; Fig. 2, ×980; Fig. 3, ×1470

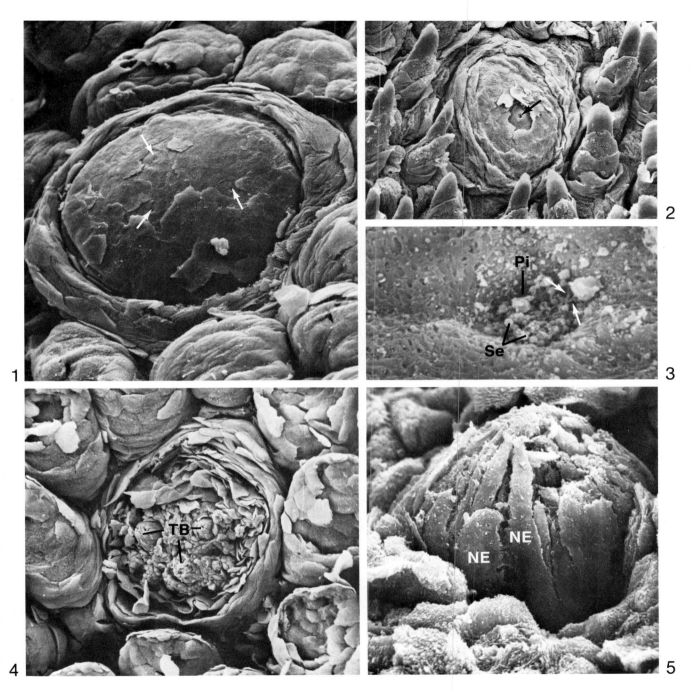

1

2

3

4

5

NERVOUS TISSUE
Gustatory Organs—
Taste Buds

Taste buds, as an example of neuroepithelium, are distributed in the mouth and throat, where they take the form of barrel-shaped structures in the epithelial layer (**TB** in Fig. 4). Only the tips of three taste buds are visible at the **arrows** in Fig. 1; the remainder are embedded within the epithelium covering a papilla of the tongue. Some surface epithelial cells have been removed from the papilla in Fig. 4, so that it is possible to observe the dome-shaped ends of three taste buds. One of those in Fig. 4 is enlarged in Fig. 5. Although taste buds are present on the exposed surfaces of both circumvallate (**arrows** in Fig. 1) and fungiform papillae (**arrow** in Fig. 2), they are particularly numerous along the sides of circumvallate papillae. Taste buds are oriented perpendicular to the epithelial surface. It is possible to observe a taste pore (**arrows** in Figs. 1 and 2), which is continuous below with a taste pit (**Pi** in Fig. 3) containing microvilli (**arrows** in Fig. 3) of neuroepithelial cells as well as secretory product (**Se** in Fig. 3). The taste bud consists of a number of elongate neuroepithelial cells (**NE** in Fig. 5) that are somewhat curved, tapering at their ends. There is a rather rapid turnover of cells, probably from a population of short basal cells. Some of the peripherally located, elongate neuroepithelial cells have apical secretory granules, perhaps a precursor of pit secretion. Branches of the chorda tympani division of the facial nerve and branches of the glossopharyngeal nerve penetrate the basal portion of the taste buds and terminate in close relation to a population of elongate cells. Such cells exhibit plasma membrane specializations (thickenings) comparable to presynaptic nerve endings and also contain many synaptic vesicles. In a manner not well understood, substances dissolved in saliva can affect the ends of neuroepithelial cells that project into the taste pit so as to generate nervous impulses in the nerve fibers that synapse with these cells.

Fig. 1, ×380; Fig. 2, ×355
Fig. 3, ×6075; Fig. 4, ×445; Fig. 5, ×2315

135

STRIATED MUSCLE TISSUE

Introduction

Muscle is a fundamental tissue type in which the basic protoplasmic property of contractility is most highly developed. The structural unit of striated skeletal muscle is a syncytial fiber that may be 1–40 mm long and 10–40 μm wide. The cytoplasm (sarcoplasm) of the multinucleate fiber contains sarcoplasmic reticulum (the smooth endoplasmic reticulum of the muscle fiber), a Golgi apparatus, mitochondria, polyribosomes, glycogen, lipid, and the pigmented protein myoglobin. Muscle fibers, in turn, consist of a variable number of myofibrils that often can be observed in photomicrographs, since they commonly measure 1–2 μm in diameter. The myofibrils consist of numerous myofilaments that, because of their small size, can only be observed with the transmission electron microscope. The specific and precise arrangement of myofilaments results in the presence of a number of bands, most of which can be observed in photomicrographs. Two types of myofilaments are present in striated muscle. The so-called thick filament is composed of the contractile protein myosin, and the myosin myofilament is approximately 1.6 μm long and about 140 Å in diameter. The myosin myofilaments that form the A-band of muscle are held in register at their centers by thin cross bridges that form the M-band (observed only by transmission electron microscopy). Each of the many myosin molecules making up the myosin myofilament is rod shaped and possesses two globular projections that form a "head" (the rod region is a double-stranded α-helix). The ability of myosin to bind adenosine triphosphate and to react with actin resides in the globular projections at the end of each myosin molecule. These projections can be observed with high-resolution transmission electron microscopy as slender cross bridges extending laterally from the ends of the myosin myofilaments toward the actin myofilaments. Myosin heads are arranged in pairs at intervals of about 143 Å along the length of the myofilaments, but each pair of myosin heads is rotated about 120° relative to the adjacent pair.

Thin filaments consist, in part, of the protein actin and are approximately one micrometer in length. Actin myofilaments are the principal constituent of the I-band of muscle. They are held in register by Z-filaments in the middle of the I-band, a region that is termed the Z-line, or Z-band, of muscle, where the protein α-actinin is localized. Globular actin molecules (G-actin) are arrayed in the thin filament in the form of a twisted double strand of beads, a double helix that exhibits functional polarity. That is, each actin bead, or molecule (about 55 Å in diameter), acts as if it had a distinguishable "front" and "back." This polarity is reversed on each side of the Z-line. Each thin filament contains 300–400 individual globular actin molecules.

Two regulatory proteins associated with actin myofilaments are tropomyosin and troponin. Tropomyosin is a long slender molecule (about 400 Å) that is in contact with seven actin molecules. Each actin strand has its own tropomyosin filament. Tropomyosin makes it possible for indirect communication between actin molecules, so that they respond as a cooperative unit. Troponin has a globular form. Each troponin molecule is associated with or attached to a tropomyosin molecule a short distance from its end.

A nerve impulse arriving at a motor endplate causes a change in permeability of the sarcolemma (muscle-fiber plasma membrane). This wave of depolarization is carried into the myofibrils of the muscle fiber by invaginations of the sarcolemma, called transverse tubules. The transverse tubules are closely positioned with respect to cisternal elements of the sarcoplasmic reticulum. An impulse arriving along the transverse tubule causes the release of calcium ions from the sarcoplasmic reticulum. The calcium ions then react with, or bind to, troponin. The binding of calcium ions to troponin results in molecular changes that cause the attached long tropomyosin molecules to shift their position or orientation with respect to the attached actin molecules so as to expose those sites on the actin molecules that are then capable of binding to the myosin head–ATP complexes. The force for contraction results from a cyclic process that occurs at the cross bridges formed by the globular heads, or projections, of the myosin molecules. Thus the myosin heads can attach to a succession of sites along the neighboring actin molecules in the actin myofilament. Myosin heads attach to the actin molecules at one angle and swivel and detach at another angle so as to cause the filaments to move past one another. The myosin heads on opposite sides of the A-band myofilaments swivel in opposite directions, which tends to bring I-band myofilaments into greater overlap with A-band myofilaments, thus decreasing the distance between adjacent Z-bands. Calcium ions are essential to this process, and the breakdown of ATP by myosin ATPase localized in the myosin heads (or cross bridges) plays an important role in energizing the process. The combination of myosin heads with ATP results in a "charged intermediate" that is capable of binding to those reactive

sites of actin molecules that are exposed by changes in position of tropomyosin molecules. The ATP then undergoes hydrolysis into ADP and P_i (inorganic phosphate), and energy is released. The massive repetition of these events results in a displacement of thick and thin filaments, designated by Hanson and Huxley as the "sliding filament mechanism" of muscle contraction. Since the myosin cross bridges on opposite sides of the A-band myofilaments "swivel" in opposite directions, they have been likened to oars pulled by rowers facing in opposite directions at the two ends of a racing shell (Murray and Weber, 1974). The result of this "rowing" is to pull the filaments into greater overlap, thus decreasing the distance between adjacent Z-bands and resulting in an overall shortening of the muscle. During relaxation, calcium ions are rapidly recaptured and sequestered within the cisternae of the sarcoplasmic reticulum. After contraction, the quick removal of calcium ions from the sarcoplasm around the myofilaments is achieved by a calcium pump located in the membranes of the sarcoplasmic reticulum. The absence of calcium ions prevents further cycling of the myosin cross bridges by producing a change in position of the tropomyosin molecules with respect to the actin molecules. This causes the tropomyosin molecule once again to cover the reactive sites on the actin mol-

ecules and prevents their interaction with the myosin heads.

Energy is essential to the contraction of muscle. It can be generated from the breakdown of ATP to ADP and free phosphate ions, but the amount of ATP available for muscle contraction is limited. While contraction of muscle is occurring, the phosphocreatine molecule can cleave to provide energy necessary for the recombination of ADP and inorganic phosphate into ATP. The availability of phosphocreatine is limited, so that under conditions of extensive exercise, further energy stores are provided by the combustion of glycogen, which is present in abundance in the sarcoplasm.

Muscle proteins, such as actin, myosin, alpha-actinin, are now known to be widely distributed in nonmuscle cells. These contractile and regulatory proteins are involved in a variety of cellular events, such as cell movement, the formation of cellular appendages (e.g., microvilli, lamellipodia, blebs, filopodia), the formation of the cleavage furrow during mitosis, and movements of molecules within membranes. The microvilli that form the brush border of intestinal epithelial cells contain actin filaments and other proteins (e.g., myosin and alpha-actinin), which are involved in the primitive motility observed in these microvilli (see Chapter 9).

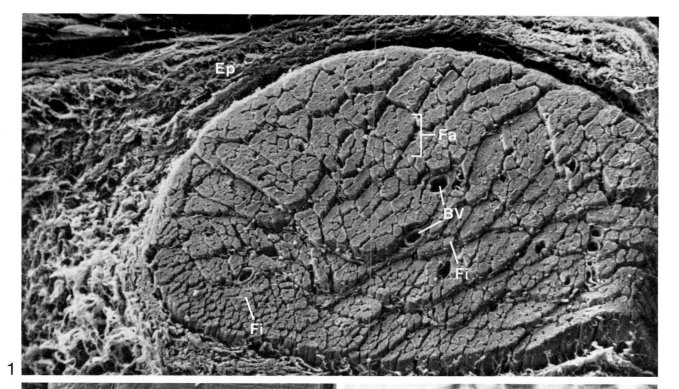

STRIATED MUSCLE TISSUE
Skeletal Muscle

A transverse section of rat extraocular muscle is illustrated by the scanning electron micrograph in Fig. 1. This muscle is surrounded by a connective tissue sheath called epimysium (**Ep**); inward extensions of this sheath, called perimysium, subdivide the muscle into fascicles (**Fa**), or bundles. Blood vessels (**BV**) contained within the perimysium appear in cross section in Fig. 1. Individual muscle fibers (**Fi**) are transversely sectioned in this micrograph. In the surface view of three muscle fibers shown in Fig. 2, it is possible to observe the transverse striations (**St**) of the muscle fiber in areas where the external membrane, or sarcolemma (**Sa**), of the muscle fiber has been sheared away (**arrows**). The connective tissue endomysium (**En**) that surrounds individual muscle fibers is also illustrated in Fig. 2. When longitudinally sectioned muscle fibers (**Fi**, Fig. 1) are stained with Heidenhain's iron hematoxylin, they exhibit an alternating light and dark banding pattern, as can be seen in Fig. 3. The dark-staining A-bands (**A**) are bisected by a thin, light H-band (**H**). Adjacent light-staining I-bands (**I**) are, in turn, bisected by a thin dark line termed the Z-line (**Z**). The distance between two Z-lines constitutes a unit of muscle organization known as the sarcomere. Nuclei (**Nu**) of the multinucleate skeletal muscle fiber are located just under the sarcolemma. Connective tissue elements of the endomysium (**En**) are present around individual muscle fibers. A single fiber, like that illustrated in Fig. 3, consists of many myofibrils, which, in turn, consist of numerous myofilaments. The myofibrils are arranged in register, which results in the precise banding pattern apparent in Fig. 3.

Fig. 1, ×215; Fig. 2, ×635; Fig. 3, ×710

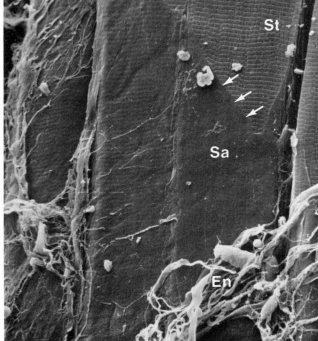

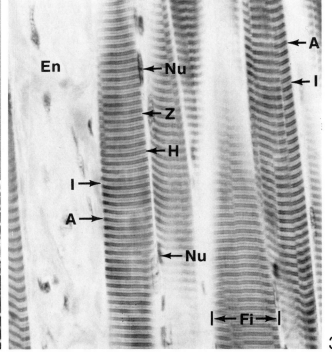

139

STRIATED MUSCLE TISSUE
Skeletal Muscle

Details of the organization and banding pattern of skeletal muscle myofibrils can be observed in the transmission electron micrographs in Figs. 1 and 3. Myofibrils (**Mf**) are made up of repeated units called sarcomeres (**Sa**), which are bounded by adjacent Z-bands. In the middle of the A-band (**A**) is an electronlucent region called the H-band (**H**). The width of the H-band is variable, and depends upon the extent to which muscle is contracted. A thinner band, the M-band (**M**), bisects the H-band at the middle of the A-band. This band results from numerous but thin connections that serve to hold the myosin myofilaments of the A-band in register. The width of the I-band (**I**) varies with muscle contraction, and is bisected by a dense Z-line, or Z-band (**Z**). Some elements of the membranous sarcoplasmic reticulum (**SR**) are present in Fig. 1. Also present in the micrograph are many small, dense granules of particulate glycogen. Details of myofibril organization are shown at higher magnification in Fig. 3. The scanning electron micrograph (Fig. 2) shows a number of myofibrils (**Mf**) that are traversed at regular intervals by transverse tubules (**TT**). Transverse tubules are tubular extensions of the sarcolemma into the fiber. The transverse tubules are located at the level of the Z-band in, for example, amphibian skeletal muscle and mammalian cardiac muscle. In human skeletal muscle, however, the transverse tubules are positioned at the junction of the A- and I-bands.

Fig. 1, ×8545; Fig. 2, ×11,500; Fig. 3, ×29,190

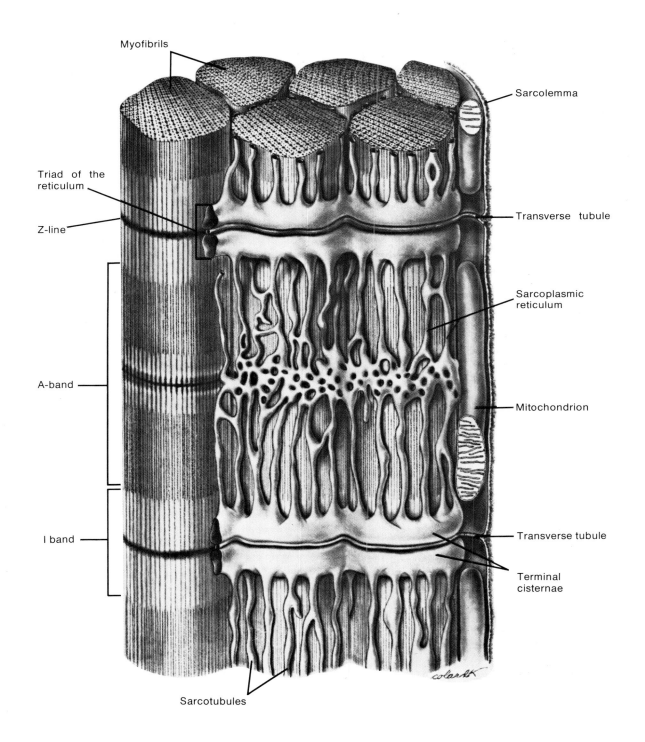

Myofibrils

Triad of the reticulum

Z-line

A-band

I band

Sarcotubules

Sarcolemma

Transverse tubule

Sarcoplasmic reticulum

Mitochondrion

Transverse tubule

Terminal cisternae

colarAk

STRIATED MUSCLE TISSUE
Skeletal Muscle

The sarcoplasmic reticulum, a specialized form of the smooth endoplasmic reticulum, is a membrane system of the muscle fiber that is structurally and functionally specialized for the rapid release and sequestering of calcium ions under appropriate conditions. It is widely distributed throughout the sarcoplasm around the myofibrils in the form of membranous cisternae, tubules, and vesicles. Flattened cisternae, also called terminal cisternae, are in contact with an invagination of the sarcolemma called the transverse tubule. The transverse tubule and two adjacent terminal cisternae constitute a triad. The transverse tubule and terminal cisternae lie over the middle of the I-band (at the Z-line) in some muscle (including frog skeletal and human cardiac muscle). Terminal cisternae communicate with tubular elements of the sarcoplasmic reticulum that extend longitudinally over the A-band. The tubular channels, sometimes called H-sacs, are continuous with an irregular branching form of the sarcoplasmic reticulum that is located at the level of the H-zone. The membrane of the transverse tubule adjacent to the terminal cisternae is specialized in the form of beaded units, similar to the subunits associated with gap junctions, which permit the passage of ions and small molecules. A finely granular material is frequently observed within the terminal cisternae, and it has been suggested that this material is capable of binding calcium ions. Terminal cisternae, when isolated by differential centrifugation, can, in fact, bind calcium. The wave of depolarization conducted along the sarcolemma and to the interior of the fiber by the transverse tubule causes the sarcoplasmic reticulum to release calcium ions, which can then diffuse among the myofilaments and initiate contraction by binding to troponin. This binding then causes the tropomyosin to change its association with actin so as to expose reactive sites capable of interacting with myosin heads. After contraction, the calcium moves back into the cisternae of the sarcoplasmic reticulum. Tropomyosin then covers the reactive sites on the G-actin molecules, and relaxation occurs. [Reproduced by permission from W. Bloom and D. W. Fawcett, *A Textbook of Histology* (10th ed.), W. B. Saunders Company, 1975.]

STRIATED MUSCLE TISSUE
Skeletal Muscle

It is possible to distinguish between the thin actin (**Ac**) myofilaments and thick myosin (**My**) myofilaments in the transmission electron micrograph shown in Fig. 1. In contracted muscle, such as that shown here, the actin filaments not only constitute the I-band, but extend into the A-band. The actin filaments branch as they approach the Z-line (**Z**), which bisects the I-band. Myosin myofilaments extend the entire length of the A-band (**A**). The narrow H-band (**H**) is also identified in Fig. 1. This micrograph also serves to illustrate the membranes of the sarcoplasmic reticulum (**SR**), the cisternae of which frequently contain a filamentous (**Fi**) material. The sarcoplasmic reticulum is differentiated into terminal cisternae (**TC**) immediately adjacent to the transverse tubules (**TT**), which are longitudinally sectioned. Electron densities (**ED**) are present at intervals between the transverse tubules and the terminal cisternae, and these densities are thought to represent a form of junctional complex involved in ion communication. The transverse tubule and the terminal cisternae of adjacent sarcomeres are termed a triad when transversely sectioned. Such a triad (**Tr**) is located in the upper left portion of Fig. 1. Fig. 2 illustrates myofibrils in transverse section. In Fig. 2, a section through the end of the A-band, it is possible to observe many transversely sectioned myosin myofilaments. The smaller actin myofilaments are arranged in such a way that six actin myofilaments surround a single myosin myofilament. Extremely thin filaments appear to interconnect (**arrows**) the actin and myosin myofilaments in some areas, and represent the globular heads of the myosin molecules. Mitochondria (**M**) and profiles of the sarcoplasmic reticulum (**SR**) are included in Fig. 2. The myofibril on the right is transversely sectioned near the end of the H-band; about 400 myosin myofilaments are present in this section. At this plane of sectioning, most of the black dots represent myosin myofilaments, and only a few actin myofilaments are visible in the lower left portion of the micrograph.

Fig. 1, × 87,220; Fig. 2, × 97,900

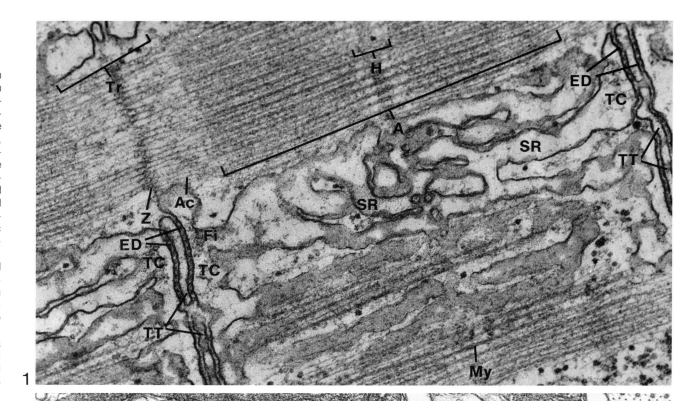

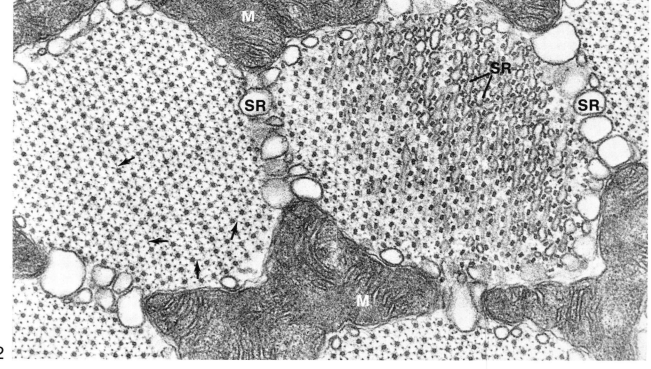

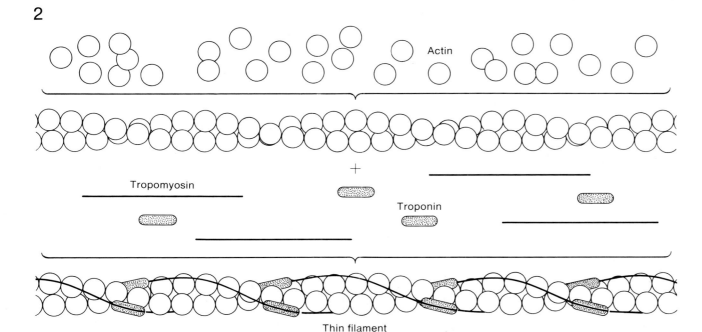

1

Myosin molecule

Cleavage site

|← 1200 Angstroms →|

Head

|← 429 Angstroms →|

Thick filament

2

Actin

+

Tropomyosin

Troponin

Thin filament

STRIATED MUSCLE TISSUE
Skeletal Muscle

A myosin molecule is long, rod-shaped, and has a double "head," or projection, at one end (Fig. 1). Several hundred such molecules are arranged in a sheath (Fig. 1) to form one long myosin myofilament, which measures approximately 1.5 μm in length by 140 Å in width. The myosin molecules are arranged with opposite orientation at each end of the myofilament (see Fig. 1 on the following page). The myosin heads bind adenosine triphosophate, which results in a so-called charged myosin-ATP intermediate (Murray and Weber, 1974). Myosin heads are arranged in pairs at intervals of about 143 Å along the length of the myofilaments, but each pair of myosin heads is rotated about 120° relative to the adjacent pair. As a result, there are six rows of myosin heads on a single myosin filament as viewed end-on. Actin molecules (Fig. 2) are globular or spherical and are arrayed in the thin myofilament in the form of a twisted double strand of beads (a double helix). Each actin bead (G-actin molecule) exhibits functional polarity: that is, it acts as if it had a distinguishable "front" and "back," and this polarity is reversed on each side of the Z-line. There are some 300–400 actin molecules in a single actin myofilament. Actin myofilaments are about one micrometer long. Two important regulatory proteins associated with the actin myofilament are tropomyosin and troponin. Tropomyosin (Fig. 2) is a long, slender molecule (about 400 Å), that is in contact with seven actin molecules. Each of the two actin strands in the myofilament has its own tropomyosin filament. The troponin molecule has a globular form and is attached to the tropomyosin molecule a short distance from its end. The molecular organization of the actin thin filament is diagrammed in Fig. 2. [Figs. 1 and 2 reproduced by permission from J. M. Murray and A. Weber, "The Cooperative Action of Muscle Proteins." Copyright © 1974 by Scientific American, Inc. All rights reserved.]

STRIATED MUSCLE TISSUE
Skeletal Muscle

The orderly interdigitation of myosin thick filaments and actin thin filaments in a myofibril is depicted in Fig. 1. Two sets of actin filaments extend from adjacent Z-lines into each end of adjacent A-bands. Actin filaments thus lie between and overlap the myosin myofilaments. Myosin heads are restricted from a central portion of the A-band myofilament. The portion of the A-band not occupied by actin filaments constitutes the H-band of muscle, the width of which is variable depending upon the extent to which the myofibril is contracted. During contraction (**left** part of Fig. 2) the myosin heads, or cross bridges, which are oriented in opposite directions on each end of the myosin myofilament, contact an actin molecule. The myosin heads then swivel to a new position and disconnect from the actin molecule. The myosin heads swivel in opposite directions on each side of the A-band. As a result, the thin actin myofilaments are propelled past the thick filaments, resulting in greater overlap between the myosin and actin filaments (**right** part of Fig. 2) and an overall shortening of muscle. The energy for this event is provided by the hydrolysis of adenosine triphosphate (ATP). The attachment or detachment of myosin heads and adjacent actin molecules in the presence of ATP and the regulatory muscle proteins does not occur unless regulatory calcium ions are also present. Contraction is initiated when calcium is released from the sarcoplasmic reticulum in response to a nerve impulse transmitted to the sarcolemma of the muscle fiber. Calcium is quickly removed from around the myofilaments by a calcium pump located in the membranes of the sarcoplasmic reticulum. When calcium is retrieved and stored in these membranes—an event requiring only a fraction of a second—further cycling of the myosin cross bridges or heads is prevented. [Figs. 1 and 2 reproduced by permission from J. M. Murray and A. Weber, "The Cooperative Action of Muscle Proteins." Copyright © 1974 by Scientific American, Inc. All rights reserved.]

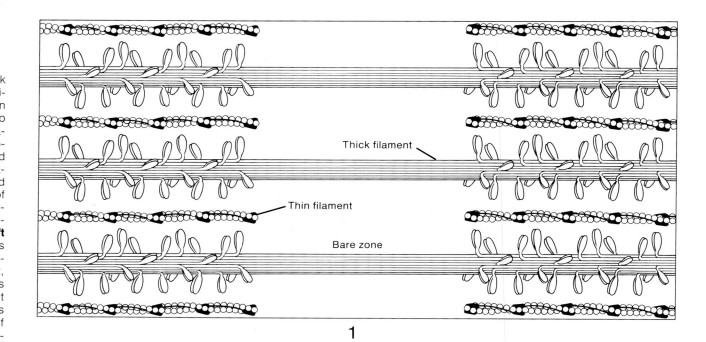

Thick filament

Thin filament

Bare zone

1

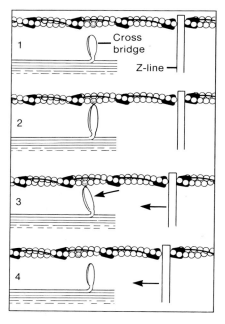

Cross bridge

Z-line

1

2

3

4

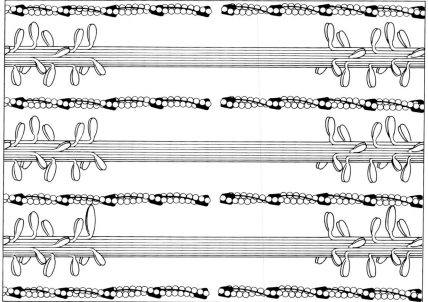

2

STRIATED MUSCLE TISSUE
Circulation of Skeletal Muscle

Muscle has an extensive blood vasculature that includes a rich capillary bed. These blood vessels are distributed within connective tissue that surrounds individual muscle fibers. The extensive vascularity of muscle cannot, however, be fully appreciated from the microscopic examination of tissue sections. Alternatively, casts of muscle vasculature viewed with the scanning electron microscope do convey the magnitude of the capillary networks associated with muscle fibers. A low-magnification scanning electron micrograph of large blood vessels (**BV**) associated with an extensive capillary plexus (**CP**) is shown from rat skeletal muscle in Fig. 1. At higher magnification (Figs. 2 and 3), it is apparent that many long capillaries (**single arrows**) follow linear courses with respect to the individual muscle fibers, but are tortuously folded along their length. The distance between capillaries approximates the distance between individual muscle fibers. Transverse capillary loops (**double arrows**) are also widely distributed and serve to interconnect the long capillaries. Thus the folding of the capillary networks is sufficient to enable them to change their length in response to changes in the length of muscle fibers during contraction and relaxation. An extensive vascular supply is a necessary requirement for a tissue as metabolically active as muscle. The muscle fibers are highly dependent for their activity on sources of nutrients (e.g., amino acids, monosaccharides, fatty acids) that supply energy for contraction.

Fig. 1, ×60; Fig. 2, ×155; Fig. 3, ×330

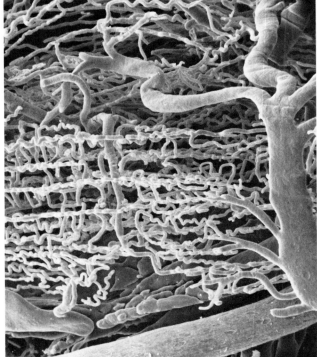

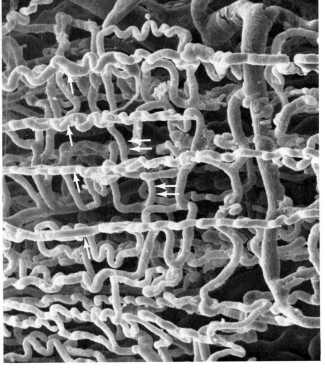

STRIATED MUSCLE TISSUE
Cardiac Muscle

Cardiac muscle, unlike skeletal muscle, consists of separate cellular elements, and is not a syncytium. Moreover, cardiac muscle is involuntarily controlled by innervation from both the sympathetic and parasympathetic divisions of the autonomic nervous system. Cardiac muscle fibers branch and anastomose, and the nuclei are centrally located rather than occupying an eccentric position, as in skeletal muscle. Fig. 1 illustrates a number of cardiac muscle fibers, and the cut edges of the myofibrils are denoted by **arrows**. The elevated transverse tubules (**TT**) traverse the myofibrils at the level of the Z-band. The capillary in Fig. 1 has been sectioned so as to expose an erythrocyte (**Er**). Sarcomeres (**Sa**) of cardiac muscle fibers are apparent in Fig. 2 because of the presence of overlying transverse tubules (**TT**). A nucleus (**Nu**) and mitochondria (**Mi**) of the sarcoplasm can also be observed in Fig. 2. Mitochondria are also present in the sarcoplasm around cardiac muscle fibers in Fig. 3. The sarcoplasm of cardiac muscle is generally more abundant in mitochondria, lipid, and glycogen than skeletal muscle. The sarcoplasmic reticulum of cardiac muscle is not as extensively developed as in skeletal muscle, and specialized terminal cisternae are usually absent. The cell boundaries between cardiac muscle cells are termed intercalated disks when viewed with the light microscope. In Fig. 4, the plane of fracture follows the intercalated disks (**ID**) of several cardiac muscle fibers. The intercalated disks between cardiac muscle cells have several junctional specializations associated with them. These include a fascia adherens (**FA**), which is illustrated in the transmission electron micrograph shown in Fig. 5. The actin myofilaments (**Ac**) insert into the fascia adherens of the intercalated disk. Although not clearly shown in this micrograph, desmosomes and gap junctions are also present at the periphery of the fascia adherens. [Figs. 1, 2, and 4 from H. D. Sybers and M. Ashraf, Preparation of cardiac muscle for SEM. In O. Johari and I. Corvin (eds.), *Scanning Electron Microscopy/1973*, IIT Research Institute, pp. 345–346.]

Fig. 1, ×3780; Fig. 2, ×7075
Fig. 3, ×7810; Fig. 4, ×3915; Fig. 5, ×28,035

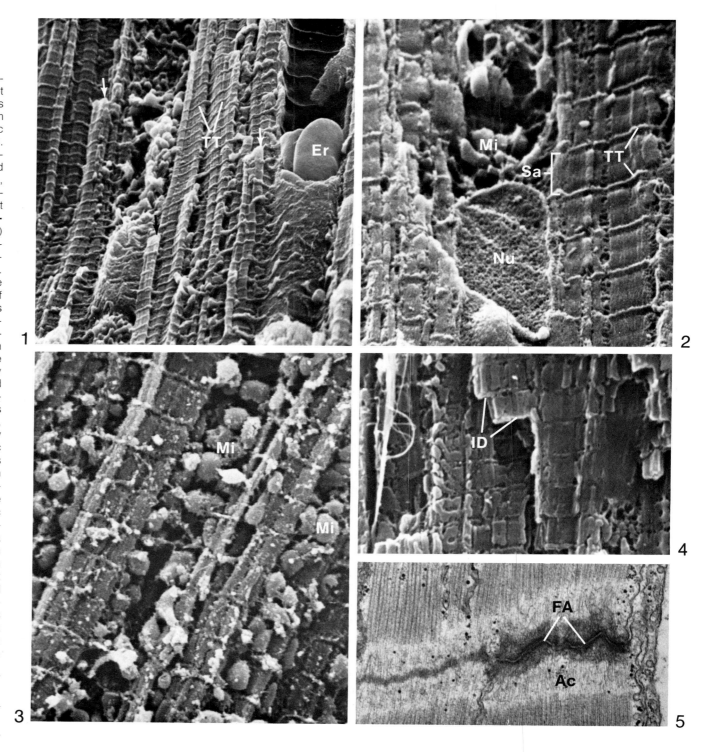

THE INTEGUMENTARY SYSTEM

Introduction

The skin, or integument, covers the body surface. It consists of a superficial epidermal layer attached to an underlying dermis or corium by a basement membrane and accounts for approximately 15% of the body weight in humans. Many infectious diseases and pathological processes are manifested by specific identifiable alterations of the skin, which are useful in diagnosis. The functions of the skin are closely related to its internal organization. Cells in the basal layer of the epidermis are capable of rapid proliferation over a long time period. Cells resulting from this proliferation undergo keratinization and form the most superficial layers of the epidermis. A distinction is commonly made between thick skin, which covers the palms of the hands and soles of the feet, and thin skin, which covers most of the remainder of the body. Thick skin has a thicker epidermis than thin skin; the stratum corneum (composed of keratin) and stratum granulosum (the cells of which contain keratohyalin) are especially thick. A layer called the stratum lucidum (containing eleidin) is absent in thin skin. The dermis of thin skin, however, is thicker than that of thick skin. Early in development the epidermis can, as a result of its ability to proliferate and differentiate, give rise to such appendages as hair, nails, sebaceous glands, and sweat glands.

Skin serves a variety of important functions. The tough protein keratin forms the outer layer of an uninterrupted cellular covering. The keratin serves as a waterproofing material and is impermeable to many disease organisms. Melanocytes derived from neural crest cells early in embryogenesis migrate into the basal layer of the epidermis and synthesize melanin pigment, which serves to protect against the harmful effects of ultraviolet light. Melanocytes can be recognized by the histochemical demonstration of tyrosinase, an enzyme required for the biosynthesis of melanin. Melanin granules may also be found in other cells of the epidermis, but they acquire the melanin by phagocytosis of pigment granules from the slender extensions of melanocytes that project into the narrow intercellular spaces between more superficial cells. The major portion of Vitamin D is synthesized in skin by the irradiation of ergosterol with ultraviolet light. Free nerve endings may be present in the epidermis, and various encapsulated nerve endings are present in the underlying dermis. Thus the integument represents an important interface in the reception of stimuli. Sweat glands are located in the dermis of the skin, the formation of which occurs early in development as an epidermal downgrowth. Through the activity of the sweat glands, the integument functions in thermoregulation and in the maintenance of water balance. Sweat glands are innervated primarily by the sympathetic nervous system. These sympathetic nerve fibers, unlike most sympathetic fibers, produce acetylcholine rather than norepinephrine at the synapse. Sebaceous glands develop from epidermal downgrowths that also differentiate into hair follicles. The cells synthesize a fatty substance called sebum and secrete it by a holocrine mechanism. Sebaceous glands are wedged between the hair follicle and arrector pili muscle, which attaches to the follicle and overlying epidermis. Cold serves as a stimulus to cause contraction of the arrector pili, and this results in compression of the sebaceous glands and the expression of sebum onto the surface of skin. Sebum serves to lubricate the skin surface, oils the hair, and reduces evaporation from the stratum corneum. Male sex hormone appears to increase the production of sebum by increasing proliferation, and hence turnover, of the cells in the gland.

An epidermal growth factor is now known to enhance the growth and keratinization of the epidermis and epithelial tissues, both *in vitro* and *in vivo*. This factor is a single-chain polypeptide that has been shown, after daily injection in mice, to cause precocious opening of the eyelids and precocious eruption of teeth. The epidermal growth factor also causes enhanced proliferation of epithelial cells in a variety of organ culture systems; for example, it induces proliferation of the corneal epithelium. In rodents, the growth factor appears to be produced by specific tubular cells of the submaxillary gland, and its secretion into serum is enhanced by α-adrenergic compounds and sympathetic stimulation.

THE INTEGUMENTARY SYSTEM
Skin

Skin consists of an avascular outer epidermis composed of stratified squamous keratinizing epithelium derived from ectoderm. The underlying dermis is derived from mesoderm and consists of connective tissue, blood vessels, and nerves. The dermis, which is firmly attached to the epidermis, also contains epidermal derivatives, including hair follicles, sebaceous glands, and sweat glands. The skin, primarily the dermis, contains free and encapsulated nerve endings that tranduce into electrical impulses such stimuli as touch (Meissner's corpuscle), pressure (Vater-Pacini corpuscle), heat (Ruffini corpuscle), cold (Krause's end-bulb), and pain. The skin is frequently attached to subcutaneous connective tissue, called the hypodermis, which commonly contains abundant fat cells, or adipocytes. The basal layer of the stratum germinativum, the stratum basale, continually proliferates to replace cells lost from the surface by desquamation. This layer of cells is also important in forming epithelial downgrowths into the dermis, which are capable of differentiating into hair follicles, sweat glands and sebaceous glands. As a result of mitosis in the basal layer of epidermis, cells are pushed toward the surface, but are anchored together by desmosomes. As the cells move closer to the surface and become further removed from the blood supply in the underlying dermal ridges, they undergo a transformation into the protein keratin, which forms the stratum corneum, the outer layer of skin. This transformation results in the death of cells and their subsequent loss from the skin surface. As the cells traverse the epidermis, there is a change in cell form from columnar to polygonal and, finally in the surface layers, to squamous. [Figure © Copyright 1967 CIBA Pharmaceutical Company. Division of CIBA-GEIGY Corporation. Reproduced with permission from *Clinical Symposia,* illustrated by Frank H. Netter, M.D. All rights reserved.]

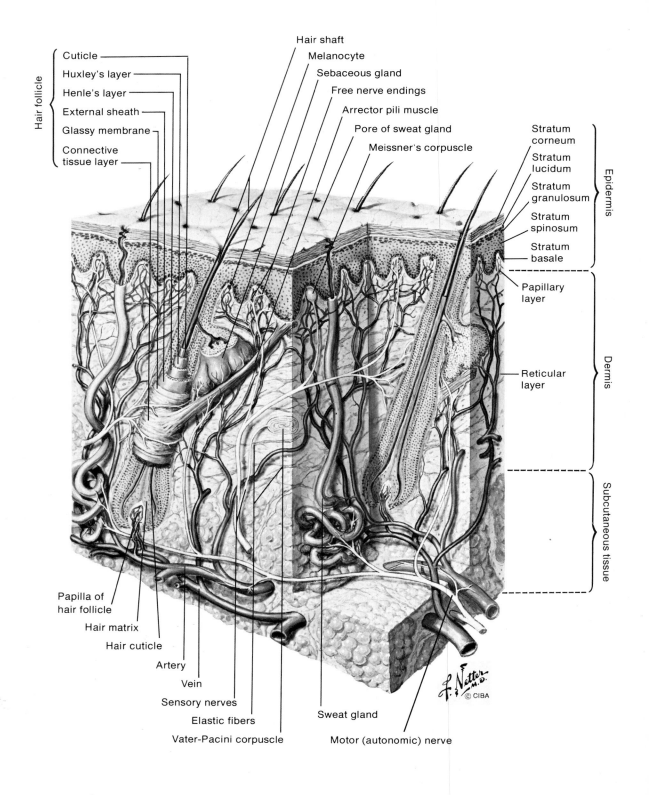

148

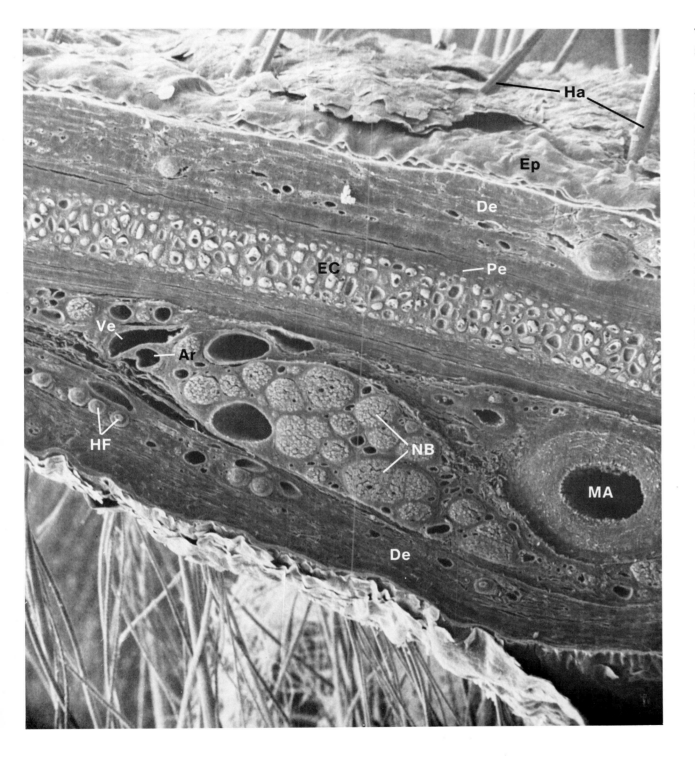

THE INTEGUMENTARY SYSTEM
Skin

A section through the rabbit ear is illustrated here. Numerous hairs (**Ha**) penetrate the epidermis (**Ep**). The underlying dermis (**De**) consists of densely arranged connective tissue fibers, blood vessels, nerves, and hair follicles (**HF**). Bundles of myelinated nerve fibers (**NB**) are transversely sectioned. A transverse section through a muscular artery (**MA**), an arteriole (**Ar**), and its companion venule (**Ve**) can also be observed. A plate of elastic cartilage (**EC**) surrounded by a perichondrium (**Pe**) is located in the center of the section and serves to provide flexible support for the external ear. As early as the third month of fetal life, the epidermis may grow down into the dermis to form a cellular shaft. The deepest portion of the downgrowth is called the germinal matrix and soon becomes arranged as a cap over a highly vascularized papilla of dermal connective tissue (see figure on previous page). The germinal matrix is connected to the overlying epithelium by a cellular external root sheath. Proliferation of cells in the germinal matrix produces hair growth. Initially, the proliferating cells of the germinal matrix are pushed up the center of the epithelial downgrowth. As they move further from the vascular supply in the papilla, they undergo transformation into keratin and eventually die. The initial cells participating in this process transform into a hair cortex (containing pigment) surrounded by a single layer of cells called the hair cuticle. A central hair medulla may or may not be present in hair follicles and hairs. Dividing matrix cells subsequently produce an internal root sheath that is pushed upward between the external root sheath and the hair cuticle.

×160

THE INTEGUMENTARY SYSTEM
Skin

Figs. 1 and 2 illustrate the skin of the rabbit and rat, respectively. Hairs (**Ha** in Figs. 1 and 2) penetrate the superficial squamous cells of the epidermis (**Ep** in Figs. 1 and 2). Blood vessels (**BV**) are present within the dense, irregular connective tissue of the dermis (**De** in Figs. 1 and 2). Hair follicles in the dermis and hair shafts piercing the epidermis are identified at **arrows** in Fig. 2. Also visible in the lower portion of the dermis and the adjacent hypodermis are many sectioned fat cells (**FC**) whose internal contents have been dissolved. A transverse section of a rabbit hair follicle is illustrated in Fig. 3; a central medulla (**Me**) is present in the hair and is surrounded by the cortex of the hair (**Co**) and the hair cuticle (**Cu**). The medulla consists of soft keratin, but the hair cortex and cuticle both contain hard keratin. Hard keratin, unlike soft keratin, lacks keratohyalin granules, contains more sulfur, and doesn't desquamate. The internal root sheath (**IRS**) consists of three layers; two are one cell thick. They include an inner cellular layer called the cuticle of the inner root sheath (**CI**), a middle Huxley's (**Hu**) layer, and an outer Henle's (**He**) layer. The cells in Huxley's and Henle's layers contain trichohyaline granules of soft keratin. The remaining layers of the hair follicle include the external root sheath (**ERS**) and a connective tissue sheath (**CTS**). A blood vessel (**BV**) and closely packed collagen fiber (**CF**) bundles of the dermis are visible around the hair follicle. Localized portions of the external root sheath also form sebaceous glands. A bundle of smooth muscle, the arrector pili muscle, extends from the base of the epidermis to insert on the lower exterior of the hair follicle (see the figure on p. 148).

Fig. 1, ×260; Fig. 2, ×100; Fig. 3, ×3845

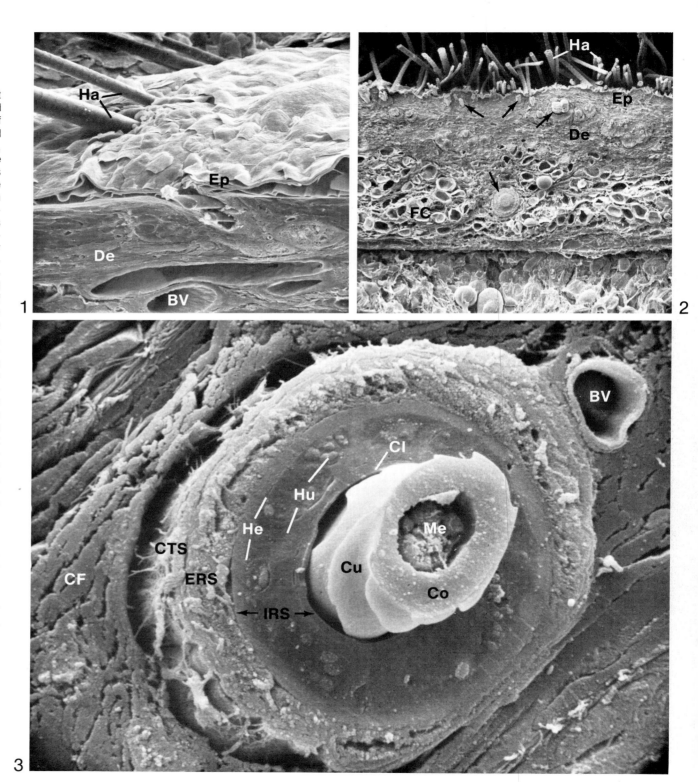

THE INTEGUMENTARY SYSTEM
Skin

Exposed hairs of the mouse (Fig. 1), rat (Fig. 2), and human (Fig. 3) are covered by a modified cellular layer called the cuticle of the hair. These cells are highly flattened and partially overlap one another. The palms of the hand and soles of the feet are covered with ridges and grooves that appear early in embryonic development and are determined by hereditary factors. Fig. 4 illustrates an acetate replica of a human fingerprint. The epidermal ridges (**ER**) occur over underlying ridges in the dermis. The epidermis grows partially down into the primary dermal ridges to subdivide them into secondary dermal ridges. Extensive capillary loops are present in the secondary dermal ridges. The circular apertures (**arrows**) located on the epidermal ridges in Fig. 4 represent the openings of sweat gland ducts.

Fig. 1, ×465; Fig. 2, ×4205;
Fig. 3, ×845; Fig. 4, ×35

THE DIGESTIVE SYSTEM

Introduction

Much of the digestive tract takes the form of a tube that consists of several layers. From inside to outside, the layers include: (a) a mucosa, or mucous membrane, (b) a submucosa, (c) a muscularis externa, and (d) a serosa. Of the various regions of the digestive tube, the inner mucosa layer shows the greatest structural variation. This layer consists, in turn, of three thinner layers: an epithelial layer, a connective tissue lamina propria, and one or more thin layers of smooth muscle that constitute the muscularis mucosa. The epithelial layer of the stomach is differentiated into numerous gastric pits and glands. The gastric mucosa functions to secrete hydrochloric acid, mucus, pepsin, gastric lipase, intrinsic factor, and the hormone gastrin.

Parietal cells in the gastric glands of the stomach mucosa are capable of producing 0.16 M hydrochloric acid, 0.07 M potassium chloride, and traces of other electrolytes in the stomach lumen. It is thought that chloride ions are derived from the blood and that hydrogen ions combine with them to produce acid in the secretory canaliculus of the parietal cell. The source of hydrogen ions is not clear, although they may be produced by ionization of water. The hydroxyl ions thus remaining in the cell may be neutralized by hydrogen ions formed by the dissociation of carbonic acid. Carbonic acid is formed from carbon dioxide and water, a reaction catalyzed by carbonic anhydrase present in the parietal cells. In many species, the polypeptide gastrin, histamine, and cholinergic nerve stimulation can each increase the production of hydrochloric acid. The gastric parietal cells in humans also produce and release a glycoprotein intrinsic factor that binds to vitamin B_{12}, thus facilitating its absorption by the intestine. Failure to produce the intrinsic factor can result in a deficiency in red blood cell production, a condition known as pernicious anemia. Since antibodies against parietal cell proteins may be present in the blood of individuals with pernicious anemia, it has been suggested that pernicious anemia is an autoimmune disease. In the rat, the intrinsic factor is apparently produced by chief cells. Chief (zymogenic) cells are present in the gastric glands and also produce the enzyme pepsin, which hydrolyzes proteins into smaller molecules. In calves, the chief cells can also produce rennin, which is capable of digesting milk protein, but pepsin serves this function in man.

The endocrine cells located in the stomach are called enterochromaffin cells (because of their special staining properties with silver or chromium salts), and are widely distributed along the entire length of the gastrointestinal tract. The enterochromaffin cells are commonly classified into two cell types: the argentaffin and argyrophil cells. Argentaffin cells in the mucosa of the digestive tract produce serotonin, or 5-hydroxytryptamine, which is a vasoconstrictor. Argyrophil cells, which are particularly abundant in the mucosa of the pyloric glands, produce the polypeptide hormone gastrin. Gastrin stimulates parietal cells to produce hydrochloric acid, and is secreted in response to distention of the stomach or the presence of partially digested proteins in the stomach lumen. Hydrochloric acid in sufficient quantity can inhibit secretion of gastrin. Argyrophil cells may contain histamine in some species, which can stimulate parietal cell activity. A glucagon-like substance that can stimulate insulin secretion by beta cells in the islets of Langerhans of the pancreas is also produced by a group of endocrine cells lining the digestive tract, including the stomach. These cells are responsive to stimuli from autonomic innervation, mechanical stimuli, and the presence of glucose in the stomach.

In the small intestine, the epithelial layer covers projections into the lumen of the intestine, called villi, and also lines the tubular intestinal glands, or crypts of Lieberkuhn. Villi are absent in the mucosa of the large intestine, but crypts of Lieberkuhn persist. The epithelial cells lining the intestinal glands are involved in the synthesis and secretion of digestive enzymes and some hormones. A number of different cell types have been described within the intestinal glands, including undifferentiated cells, absorptive cells, mucous cells, endocrine cells, and Paneth cells. The undifferentiated cells are mitotic and constitute a source of replacement for other cell types. The absorptive cells resemble those covering the villi, but are smaller, possess fewer cytoplasmic organelles, and usually have shorter microvilli.

The endocrine cells in the small intestine appear to be the same as those found in the stomach; they are identified by special stains that contain silver or chromium salts. The enterochromaffin cells (argentaffin and argyrophil cells) in the small intestine can be visualized in the lining of the intestinal glands and villi by using these stains. Although a number of hormones are probably produced by endocrine cells of the intestine, the cellular source of these hormones has not been unequivocally demonstrated. Those hormones produced by the intestine include serotonin (5-hydroxytryptamine), secretin (which stimulates the exocrine pancreas to produce a watery secretion rich in bicarbonates, but poor in enzymes), the catecholamine norepinephrine, cholecystokinin (promotes flow of bile), and a glucagon-like hormone. Paneth cells are located at the base of intestinal glands. These cells contain zinc and probably produce a peptidase (converting peptides to amino acids). Pancreozymin (CCK or cholecystokinin-pancreozymin) stimulates the pancreatic acinar cells to secrete enzyme-rich pancreatic juice. Some gastrin is secreted by the duodenal mucosa in man. The duodenal mucosa also produces the enzyme enterokinase, which converts inactive trypsinogen into the active enzyme trypsin. Lipase is produced by intestinal glands, but its source is unknown.

The submucosa consists of areolar connective tissue with many nerves, blood vessels, and lymphatic vessels. Postganglionic neurons of the parasympathetic division of the autonomic nervous system are located in the submucosa (Meissner's plexus) and between the two layers of the muscularis externa (Auerbach's plexus). The axons of these neurons, as well as postganglionic fibers of neurons in the sympathetic division of the autonomic nervous system, innervate the smooth-muscle cells of the muscularis mucosa and the muscularis externa. A number (20–30) of solitary lymphatic nodules (collectively called Peyer's patches) are closely packed in limited areas of the lamina propria in the ileum.

The muscularis externa typically consists of an inner circular and an outer longitudinal layer of smooth muscle. Alternate contraction and relaxation of these muscle layers, innervated by both the parasympathetic and the sympathetic divisions of the autonomic nervous system, produces peristaltic waves that move the contents of the digestive tube.

The outer serosal covering of the gut consists of a single layer of simple squamous (mesothelial) cells and an underlying thin layer of connective tissue.

Ingested food mixed with gastric chyme enters the duodenum (about 11 inches long) from the stomach and mixes with enzymes of the pancreatic juice and bile. Absorptive columnar cells possessing many microvilli are abundant in the intestinal mucosa, as are goblet cells (unicellular mucous glands). The vast surface area provided by the apical microvilli of the absorptive cell facilitates the absorption of many ingested substances. Water, electrolytes (e.g., sodium, chloride, iron, and calcium ions), vitamins (e.g., vitamin B_{12}), bile, monosaccharides, amino acids, and fats are examples of the numerous substances absorbed along the duodenum, jejunum, and ileum of the small intestine. Monosaccharides and amino acids absorbed into the intestinal epithelial cell pass into the intercellular space and into the nearby capillaries. Absorption of fats requires energy, and emulsified fats pass via pinocytotic, tubular, or vesicular structures through the plasma membrane of the absorptive cells as micelles. Within the absorptive cells, the micelles are coated with a lipoprotein material before they pass to the lymphatic capillary (lacteal) of the intestinal villus.

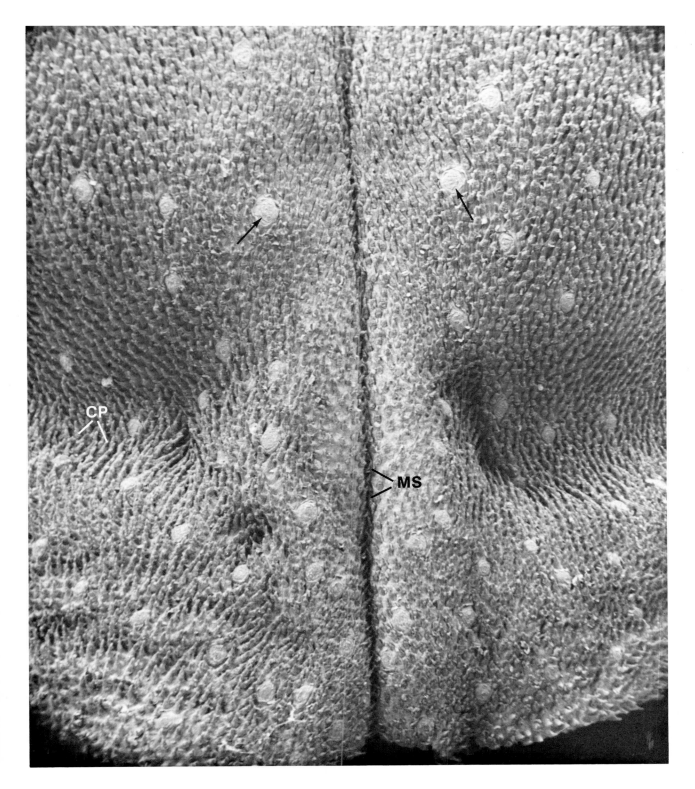

THE DIGESTIVE SYSTEM
Tongue

The central mass of the tongue consists of bundles of striated muscle arranged in three different planes. The muscle is closely surrounded by connective tissue of the lamina propria, and the tongue is covered with stratified squamous epithelium. Many cores of connective tissue project into the epithelial layer to form a variety of surface elevations called papillae that cover the anterior part of the dorsum of the tongue. The anterior dorsal surface is divided into symmetrical halves by a longitudinal furrow called the median sulcus (**MS**), illustrated here from the rat. The extensive anterior oral portion of the tongue is separated from a narrow posterior portion, called the root, by a V-shaped groove termed the sulcus terminalis. The apex of the V-shaped groove projects posteriorly. Large circumvallate papillae, 7–12 in number, are distributed along this groove. The dorsal surface of the tongue posterior to the sulcus terminalis contains a number of nodular projections produced by underlying lymphatic nodules. These are the lingual tonsils. The many conical projections (**CP**) of the mucous membrane covering the anterior dorsal surface of the tongue give it a characteristic raspy appearance. In addition, the large rounded papillae (**arrows**) are less numerous and appear red in the living state because of the numerous blood vessels in the underlying connective tissue.

×50

THE DIGESTIVE SYSTEM
Tongue

The conical projections illustrated on the dorsal surface of the tongue are called filiform papillae (**Fi** in Fig. 1). From this angle it is easy to see how these papillae could collectively provide a rough surface to aid in feeding. Filiform papillae may range from 2–3 mm in length. They also contain nerve endings for tactile sensation. Each papilla (Fig. 2) is shingled over its entire surface with several layers of flattened cells that form part of the stratified squamous epithelium. Dead cells of the top layer are constantly being shed by the process of desquamation. These cells are replaced by cells from underlying layers. In certain regions of the tongue of the rat and man, the superficial cells at the tips of the filiform papillae become dry and are transformed into hard scales consisting of a nonliving, hard coating called keratin. This process, termed keratinization, acts to provide a sharp reinforced tip for the papillae, creating an abrasive surface that is useful in feeding. Keratinized or cornified cells (**arrows**) may also be located at the base of filiform papillae. When digestion is disrupted during illness, the normal desquamation of the keratinized surface epithelial cells may be delayed. The cells may accumulate on the tongue surface together with bacteria and form the gray layer that is characteristic of the "coated" tongue.

Fig. 1, ×420; Fig. 2, ×1215

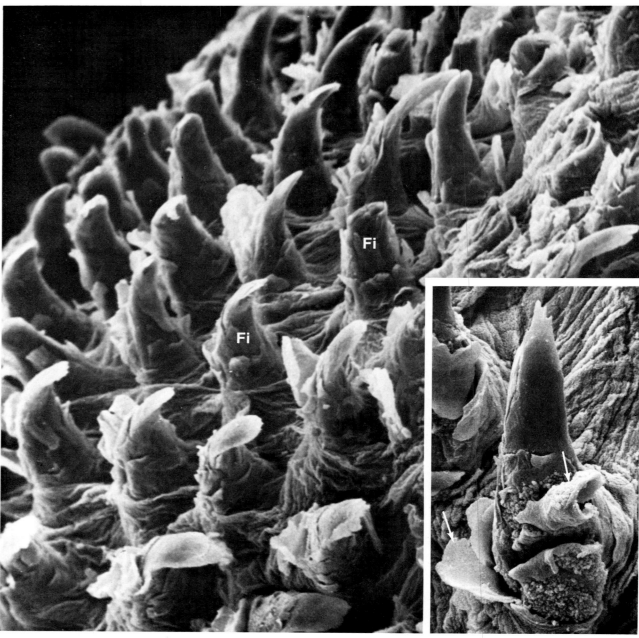

1

2

THE DIGESTIVE SYSTEM
Tongue

Filiform papillae on the rabbit tongue are illustrated in Figs. 1 and 2. These papillae taper distally into thin fila, hence the name filiform papillae. The shape of filiform papillae is variable, not only in different regions of the tongue, but among different species as well. Papillae with two or three tips (Fig. 2) owe their form to the shape of the internal connective tissue core underlying the epithelium. The tips of some filiform papillae, such as those shown from the rabbit in Figs. 3 and 4 may be nonkeratinized or only slightly keratinized. The filiform papilla illustrated in Fig. 3 closely resembles those present in man. The desquamating squamous cells give the papilla surface a scaly appearance. An enlargement of the rectangle labeled **A** in Fig. 3 gives a view of the papilla tip, which is enlarged in Fig. 4. **Arrows** indicate certain types of dividing bacteria that normally inhabit the oral cavity. At this magnification, a pattern of folds (**Fo**) (called microplicae) on the cell surfaces is apparent. Fig. 5 is an enlargement of the rectangle labeled **B** in Fig. 3. Borders (**B**) between adjacent squamous cells can be delineated, as well as the intricate pattern of microfolds, or microplicae, on the cell surfaces. These specializations of the plasma membrane of nonkeratinized squamous cells are illustrated in more detail, and their possible significance discussed, in the next section, which deals with the hard palate. Additional magnification of the microfolds (Fig. 6 reveals that many of them are continuous with irregular, branching processes that appear free at one end (**arrows**). Although the significance of these processes is unknown, they may be remnants of surface folds of adjacent cells that were shed during desquamation. Since the folds on the surface of one cell may interdigitate with the folds on the surface of adjacent or overlying cells, the arrangement could function in part to anchor the thin cells together into sheets.

Fig. 1, ×215; Fig. 2, ×160; Fig. 3, ×535
Fig. 4, ×3565; Fig. 5, ×5340; Fig. 6, ×12,815

157

THE DIGESTIVE SYSTEM
Tongue

Fungiform papillae (**Fu**) are illustrated in Figs. 1 (rat) and 2 (rabbit). They are larger and less numerous than filiform papillae (**Fi**), but are interspersed among them. Fungiform papillae have a short, somewhat narrowed stalk, but have a smooth free surface (**Fu** in Fig. 2). Several of the surface squamous cells (**SC**) were removed from the fungiform papilla in Fig. 1. Such a preparation, however, illustrates the layered arrangement of the squamous cells covering the papilla. On many fungiform papillae, neuroepithelial cells form taste buds. One such taste bud is located (**arrow**) on the fungiform papilla in Fig. 1 (see Gustatory Organs, Chapter 6). The circumvallate papilla is a third type of papilla present on the dorsal surface of the tongue. Numbering from 8–12, they are found at the junction of the anterior and posterior portions of the tongue. The circumvallate papilla, as viewed from the surface in Fig. 3 consists of a wall (**W**), or vallum, surrounding the central papilla (**CP**) and an intervening furrow, or moat (**arrows**). Taste buds are concealed beneath the papilla epithelium. The opening, or pore, of one such taste bud (**OTB**) is apparent on the surface of the circumvallate papilla in Fig. 3, but the majority of taste buds are positioned along the lateral surface of the central papilla, so that they open into the furrow. In addition, serous glands located below the papillae empty a secretion into the furrow to keep it cleansed and moistened.

Fig. 1, ×615; Fig. 2, ×245; Fig. 3, ×910

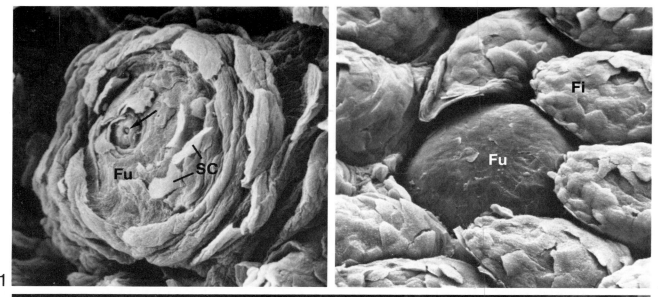

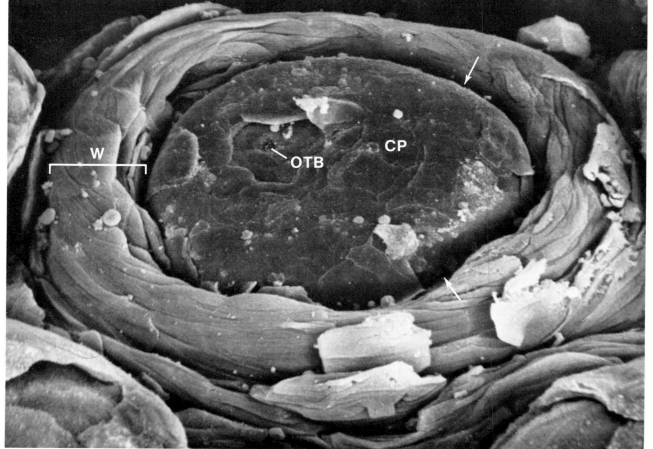

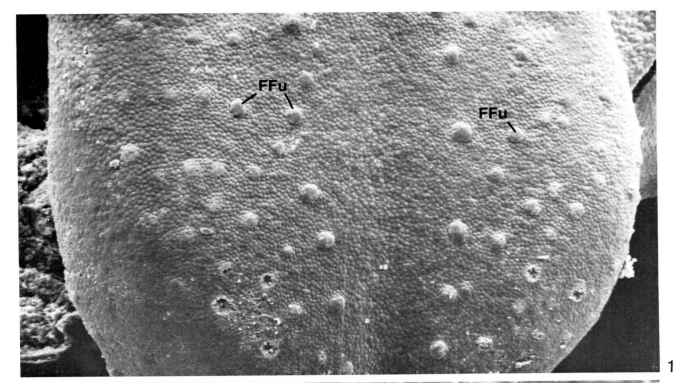

THE DIGESTIVE SYSTEM
Tongue

The dorsal surface of a rat tongue from a fetus near term is illustrated in Figs. 1 and 2. This surface differs markedly from the specialized dorsal surface of an adult tongue. Many individual surface epithelial cells can just be resolved in Fig. 1. Only a few rounded tubercles representing forming fungiform papillae (**FFu**) are present on the tongue at this developmental stage. Filiform papillae are not yet differentiated. Fig. 2 illustrates at higher magnification one forming fungiform papilla (**FFu**) as well as the surface of a number of epithelial cells surrounding it. Since the epithelial cell borders are distinct, it is possible to note that the cells covering the forming fungiform papilla are larger than surrounding cells. The plasma membrane at cell borders (**arrows**) appears to be folded and to possess short microvilli. The exposed surfaces of the epithelial cells may contain a single central cilium (**Ci**) and short microvilli (**Mv**) that are variable in their number and distribution.

Fig. 1, ×140; Fig. 2, ×705

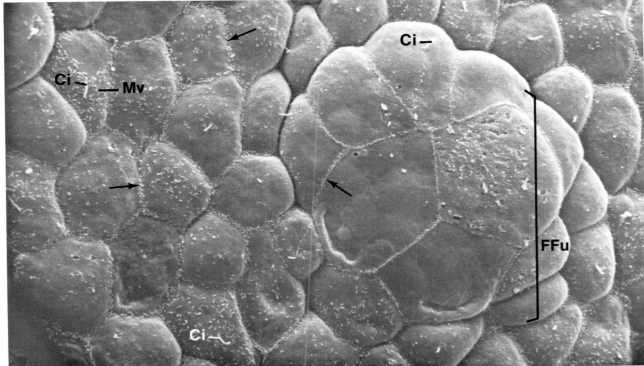

THE DIGESTIVE SYSTEM
Tongue—Linguil Tonsils

The posterior portion of the tongue is devoid of papillae and is termed the root of the tongue. Numerous rounded projections, or nodules, populate the dorsal aspect of the tongue and are collectively called lingual tonsils (**LT** in Figs. 1 and 2). The lingual tonsils are projections of the epithelium that are produced by nodular infiltrations of lymphocytes below the surface. The tonsils are advantageously situated at points of potential entry for bacteria, viruses, and other foreign material. Thus lymphocytes in the tonsils are assured easy access to invading foreign substances. The openings (**Op**) illustrated on the free surface of each lingual tonsil result from the deep invaginations of the surface epithelium, called crypts. Mucous glands, situated in the underlying muscle of the tongue, empty their secretion through ducts leading into the bottom of the crypts. The secretion can then pass up the crypt and onto the tongue surface.

Fig. 1, ×120; Fig. 2, ×340

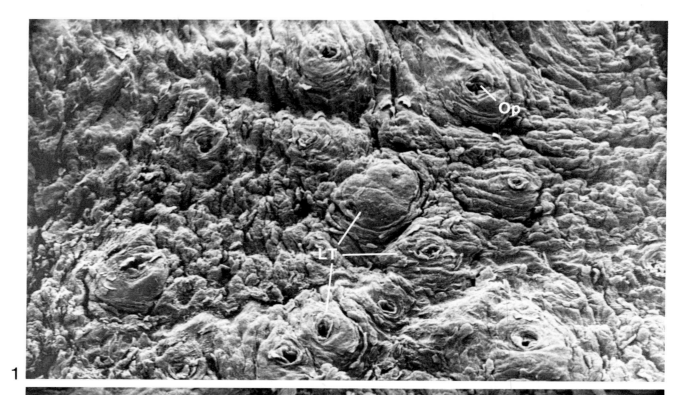

1

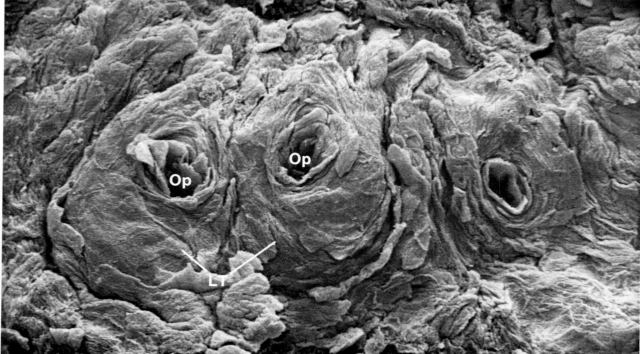

2

160

THE DIGESTIVE SYSTEM
Hard Palate

The anterior portion of the roof of the mouth has a bony framework and is called the hard palate. The oral surface of the hard palate is covered by mucoperiosteum, which consists of a mucous membrane and the periosteum of underlying bone fused together. The epithelium of the mucous membrane is stratified squamous in type. The mucoperiosteum in the rabbit is folded so as to form lateral ridges (**LR** in Fig. 1). The area enclosed within the rectangle in Fig. 1 is enlarged in Fig. 2 to show a number of surface squamous epithelial cells in the process of being shed by desquamation (**arrows**). Since the epithelium of the hard palate is usually kept moist, it is nonkeratinized or only partially keratinized. The continual loss of surface cells from the epithelium necessitates replacement by generative cells located in the basal layer of the epithelium. As new cells divide they tend to move into intermediate layers of the epithelium and firmly adhere to neighboring cells by means of numerous desmosomes, which differentiate as specializations of the plasma membrane of adjacent cells. As will be illustrated subsequently, the free surfaces of nonkeratinized, stratified squamous epithelial cells, like those in the hard palate, possess numerous microscopic ridge-like folds of the plasmalemma that are called microfolds, or microplicae.

Fig. 1, ×35; Fig. 2, ×995

1

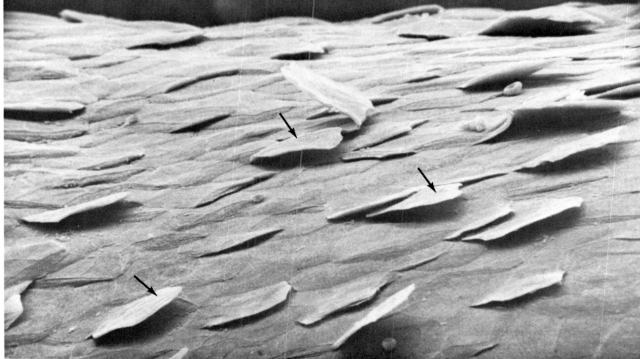

2

THE DIGESTIVE SYSTEM
Hard Palate

Microplicae are approximately 0.1–0.2 μm wide and variable in height (about 0.2–0.8 μm) and length. They may be straight or curved, often branched, and variable in their density. The plasmalemmal specializations are rather uniformly present on the moist surfaces of stratified squamous epithelial cells, which are only partially keratinized or nonkeratinized and subject to abrasive action. The distribution of microplicae, or microfolds (**Mf**), is apparent on the superficial desquamating cells (**SC**) visible in Fig. 1. The borders between cells possessing microplicae are distinct and characterized by ridge-like elevations (Fig. 1). In those regions in which cells have been detached (Fig. 2) or lost, it is evident that both the underlying cell surface and the free surface of the cells possess microplicae. At the **arrow** in Fig. 2, the microfold of a superficial cell appears to be continuous with a microfold (**Mf**) of another cell just below. Interdigitations of microfolds from two abutting cells is suggested at the **arrows** in Fig. 3. In addition, small finger-like extensions (**Ex**) project between the microfolds (**Mf**). The microplicae present on superficial cells in stratified squamous epithelium are similar to those associated with cells deeper in the epithelium that form intercellular interdigitations which increase adhesion between the cells. In deeper layers of the epithelium, desmosomes are associated with many of the microplicae. In the more superficial layers of the epithelium, microplicae persist, as shown, but desmosomes are fewer in number and consequently adhesion is considerably reduced. As a result, superficial cells in the epithelium can more easily desquamate than underlying cells. Since stratified epithelia are subject to wear and tear, the microplicae may also serve to reduce the frictional area between the plasma membranes of adjacent cells and to trap lubricating fluids such as mucin. In this regard, when the outer layer of epithelial cells is exposed to air and loses its coating of fluid, the cells become keratinized and are devoid of microplicae.

Fig. 1, ×4050; Fig. 2, ×10,125; Fig. 3, ×12,290

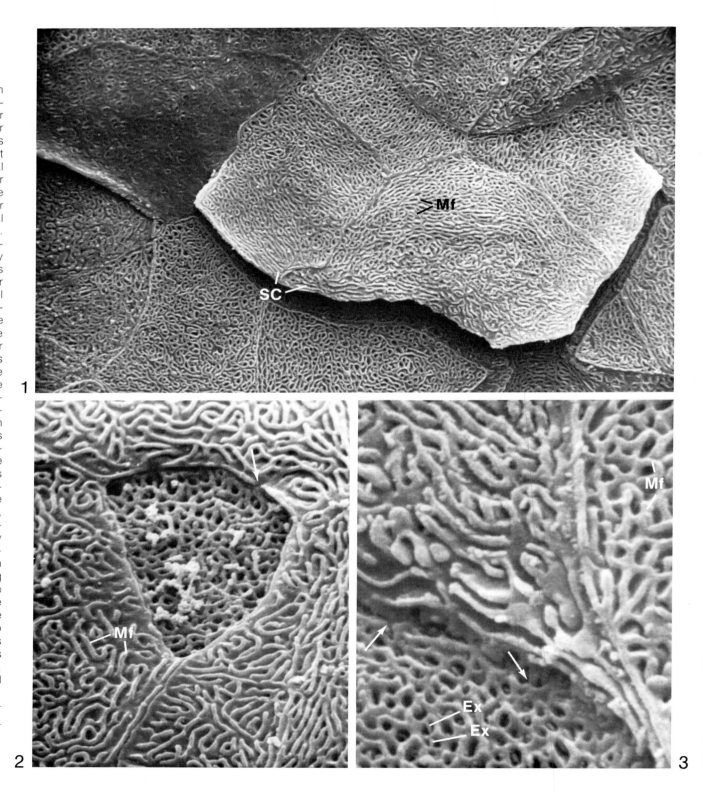

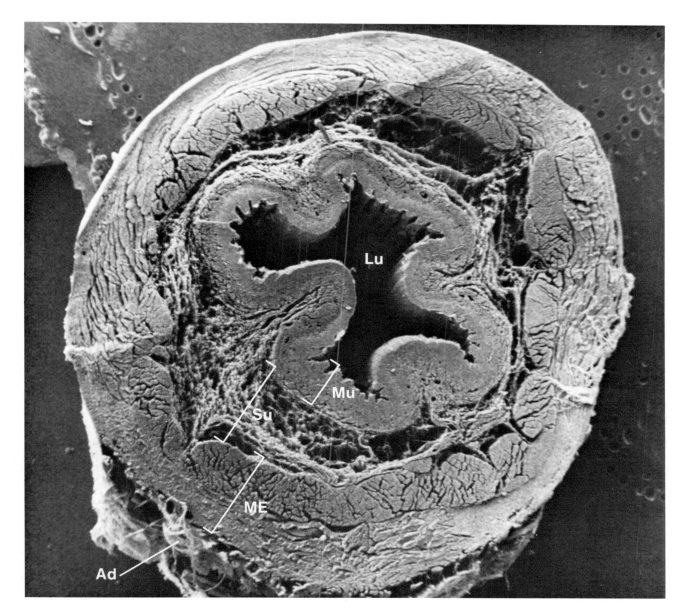

THE DIGESTIVE SYSTEM
Esophagus

The esophagus is a muscular tube that functions to transport food from the pharynx to the stomach via muscular contractions. Little or no absorption of materials occurs in the esophagus. As this figure shows, the central lumen (**Lu**) of the tube is surrounded from inside to outside by a mucous membrane, or mucosa (**Mu**), a submucosa (**Su**), a muscularis externa (**ME**), and a thin adventitia (**Ad**), or serosa. The mucosa, or mucous membrane, as will be illustrated in Fig. 1 on the next page, consists of three layers: an epithelium, a connective tissue lamina propria, and a muscularis mucosa. The mucosa may exhibit longitudinal folds produced by contraction of the muscularis mucosa. The submucosa consists of collagenous and elastic fibers. Blood vessels, nerves, and lymphatic vessels are present. Mucous glands, called esophageal glands, may be present in some regions of the submucosa. The muscularis externa consists of an inner circular and an outer longitudinal layer, although some variation in this pattern has been observed. Striated muscle predominates in the upper third of the esophagus. Both smooth muscle and striated muscle are found in the middle third, whereas only smooth muscle is found in the muscularis externa of the lower third of the esophagus. In this regard, there seems to be considerable variation in the distribution of smooth and striated muscle between different animal species. The outer adventitia layer is composed of loose connective tissue, and a serosa surrounds only that portion of the esophagus below the diaphragm. Above the diaphragm, the connective tissue adventitia serves to anchor the esophagus to surrounding structures.

×60

THE DIGESTIVE SYSTEM
Esophagus

The epithelial portion of the mucosa consists of stratified squamous (**SS**) epithelium. This epithelial type protects the esophagus against abrasion from coarse food in the lumen (**Lu**). In some mammals, including the rat, the epithelium tends to be cornified, but in man the epithelium does not become cornified. The thin connective tissue lamina propria (**LP**) of the mucosa is apparent in Fig. 1. Near the junction with the stomach, mucous glands called cardiac glands may be present in the lamina propria. The outermost layer of the mucous membrane, or mucosa, is a layer of muscle called the muscularis mucosa (**MM** in Fig. 1). This longitudinally oriented band of smooth muscle is present in much of the wall of the esophagus, although it is absent in the upper portion. The muscularis mucosa permits independent movements within the width of the mucous membrane. A portion of the submucosa (**Su**) is illustrated in Figs. 1 and 2. In addition to collagenous and elastic fibers, blood vessels (**BV**) fibroblasts, macrophages, and lymphocytes may be present. The muscularis externa (**ME**) and adventitia (**Ad**) are shown in Fig. 2. Striated muscle fibers (**arrows**) are apparent in the outer portion of the muscularis externa in Fig. 2.

Fig. 1, ×485; Fig. 2, ×125

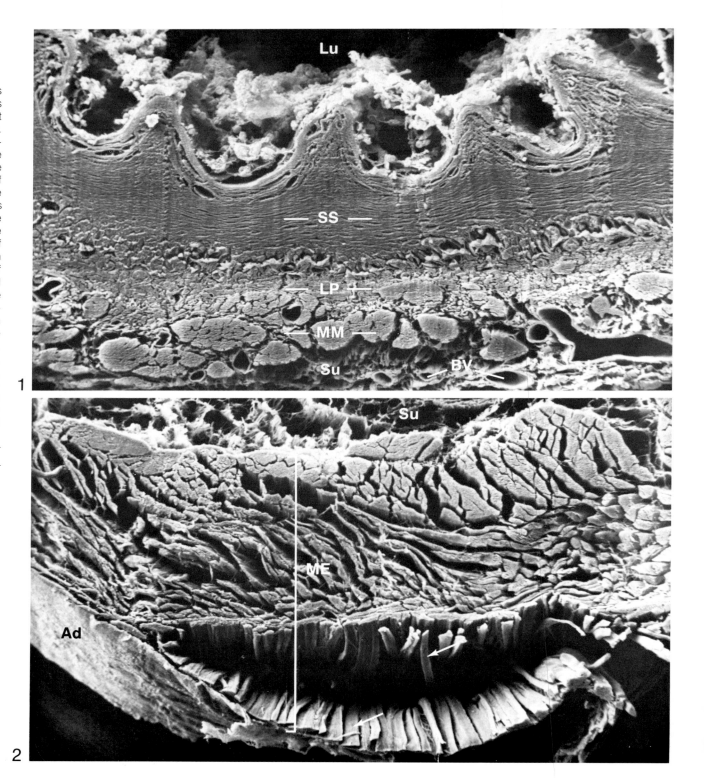

164

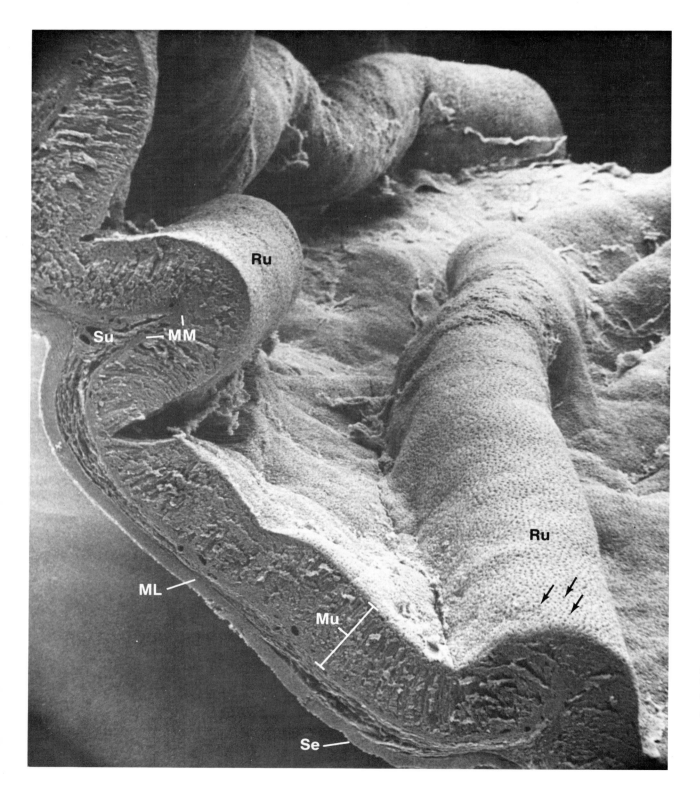

THE DIGESTIVE SYSTEM
Stomach

The stomach is a dilated segment of the digestive tract that is divided into a narrow region adjacent to the esophagus, called the cardia; a superior region above the cardiac orifice, called the fundus; a large region called the corpus, or body; and a pyloric region that communicates with the duodenum. The stomach functions in the production of hydrochloric acid, mucus, digestive enzymes, and hormones. The inner layer of the stomach is called the mucosa, or mucous membrane. The lining epithelium of this layer consists of simple columnar cells that synthesize and release abundant amounts of mucus. Goblet cells are not differentiated in the gastric epithelium. The gastric mucosa is further characterized by numerous invaginations of the lining epithelium into the lamina propria layer of the mucosa. Each gastric gland is thus a tubular downgrowth of epithelium into the lamina propria. These invaginations result in the presence of numerous unbranched or branched tubes consisting of two major regions: gastric pits and gastric glands. The luminal surface and the wall of the rat stomach are illustrated in the low-magnification scanning electron micrograph shown here. The following layers are identified: (1) the mucosa (**Mu**), or mucous membrane, which consists predominantly of numerous gastric pits and glands, (2) the thin muscularis mucosa (**MM**), consisting of smooth muscle and also constituting a portion of the mucosa, (3) the connective tissue of the submucosa (**Su**) layer, and (4) the smooth muscle band that forms the muscularis layer (**ML**). The position of a thin outer serosa (**Se**) is also denoted in the figure. The stomach wall thus consists of the same four layers as the esophagus, but the mucosa and muscularis externa are thicker in the stomach. In the empty stomach, the mucosa and submucosa may be folded. These folds, called rugae (**Ru**), disappear when the stomach is full. The numerous small apertures (**arrows**) on the surface of the ruga in the right portion of the figure are openings into gastric pits.

×55

THE DIGESTIVE SYSTEM
Stomach

The inner mucosa contains many long, tubular gastric glands (**GG**) oriented perpendicular to the stomach wall. The upper, or superficial, portion of the tube is called the gastric pit and is lined by a single layer of mucus-producing columnar cells similar to and continuous with the cells that form the noninvaginated portion of the epithelium. These epithelial cells also secrete electrolytes. The gastric pits vary in length; they may extend for at least half the length of the tube in the pyloric region of the stomach, but the pits are shorter in the fundus and body of the stomach. In the cardiac and pyloric regions, the glandular portion of the tubes (the gastric glands proper) are lined principally with mucus-producing cells. In contrast, the gastric glands of the body and fundus contain many parietal cells, which produce hydrochloric acid, and chief, or zymogenic, cells, which produce proteolytic enzymes. A small amount of connective tissue, the lamina propria, forms a packing between the gastric pits and glands. A layer of smooth muscle, the muscularis mucosa (**MM**), forms the third layer of the mucosa, and assists in moving mucus secreted within the gastric glands into the stomach lumen. The submucosa (**Su**) consists of loosely packed connective tissue cells and fibers together with blood vessels, nerves, and lymphatic vessels. An artery (**Ar**) and a vein (**Ve**) are visible in this section. The muscularis externa (**ME**) consists of spirally arranged smooth muscle bundles oriented in three major directions: an external longitudinal layer (**LM**), a middle circular layer (**CM**), and an inner oblique layer. The middle layer is particularly thick in the region of the cardiac and pyloric orifices and acts as a sphincter in these regions. The serosa (**Se**) is very thin and consists of a small amount of connective tissue covered by a single layer of squamous mesothelial cells.

×155

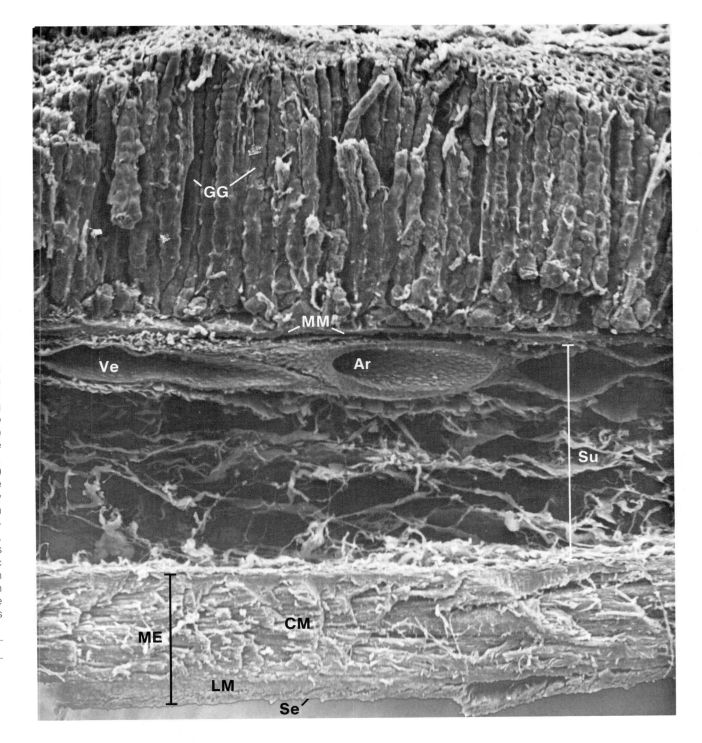

166

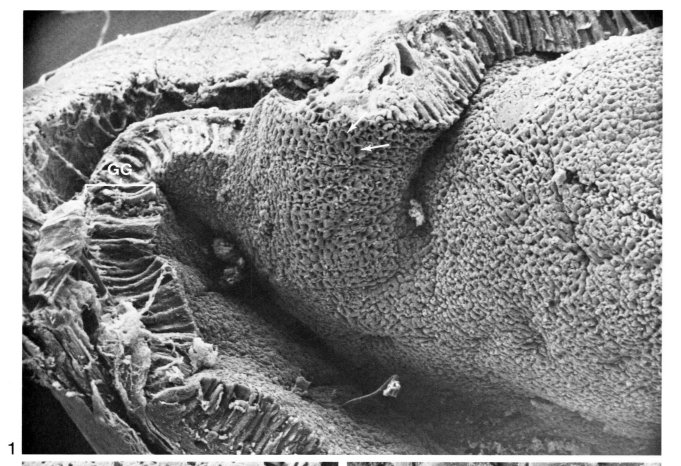

THE DIGESTIVE SYSTEM
Stomach

The tubular glands of the gastric mucosa are usually divided into a superior pit region, a middle neck region, and a lower glandular region. Thus the gastric gland is joined via a neck to the more superficial gastric pit, or foveola. Sometimes an isthmus region is distinguishable between the gastric pit and neck. Generative cells are present in addition to the mucous cells in this region. The cells have a renewal rate of about 2–3 days and can differentiate to produce either replacement cells that move superfically into the gastric pits or cells that migrate basally into the gastric glands where they slowly differentiate into parietal and zymogenic cells. Several glandular tubes may communicate with a single gastric pit. Differences in terminology exist with respect to specific regions of the gastric gland, as do structural variations in the gastric glands of different species. Fig. 1 shows a surface view of many gastric pits and longitudinal sections of the simple or branched tubular glands. The entrances to the many gastric pits, or foveolae, are denoted by **arrows** in Figs. 1 and 2. The peripheries of several gastric glands (**GG**) are shown in Figs. 2 and 3. The individual cells (✳) and surrounding lamina propria are illustrated in Fig. 3. The lamina propria (**LP**) of the gastric mucosa is largely restricted to a thin packing among the gastric mucosal glands. It consists of loosely arranged connective tissue fibers, fibroblasts, and macrophages. There is an extensive network of blood capillaries (**Ca** in Fig. 3) and lymphatic vessels. The blood vessels provide those raw materials required for synthesis and secretion by the cells of the gastric glands, and also facilitate absorption of nutrients.

Fig. 1, ×95; Fig. 2, ×140; Fig. 3, ×695

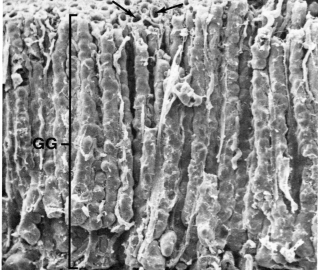

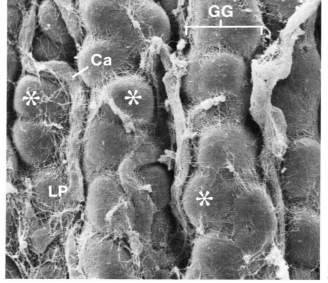

THE DIGESTIVE SYSTEM
Stomach

Casts of the gastric blood vascular system are illustrated in all figures. Fig. 1 illustrates at low magnification the blood vascular system in a region spanning the entire stomach wall. In Fig. 1 corresponding layers of the stomach are identified, including the mucosa (**Mu**), submucosa (**Su**), and muscularis externa (**ME**). The luminal side of the stomach wall is located at the top of the figure. Arteries supplying the stomach enter the serosa, divide into large branches that penetrate the muscularis externa, and enter the submucosa. A plexus of blood vessels forms in the submucosa, from which branches originate to supply the mucosa layer. An extensive capillary network traverses the entire mucosa and forms a plexus around each of the many gastric glands. The plexus of blood vessels associated with the gastric glands is illustrated in Figs. 1 and 2. This capillary network is especially well developed around the gastric pits, or foveolae, of the gastric mucosa as viewed from the luminal surface in Fig. 3.

Fig. 1, ×100; Fig. 2, ×105; Fig. 3, ×165

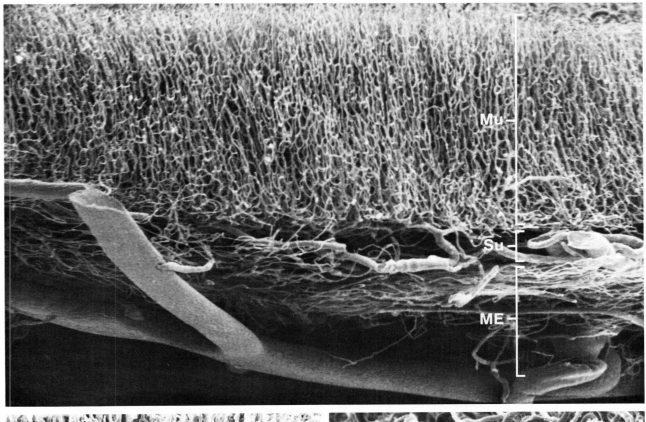

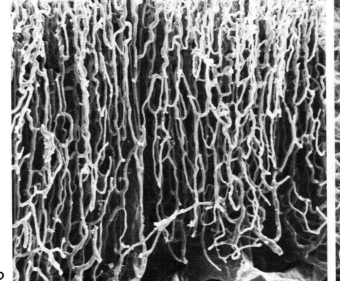

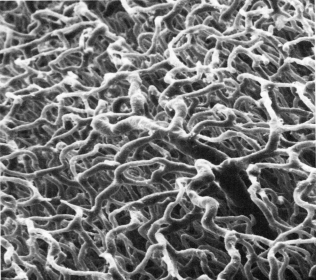

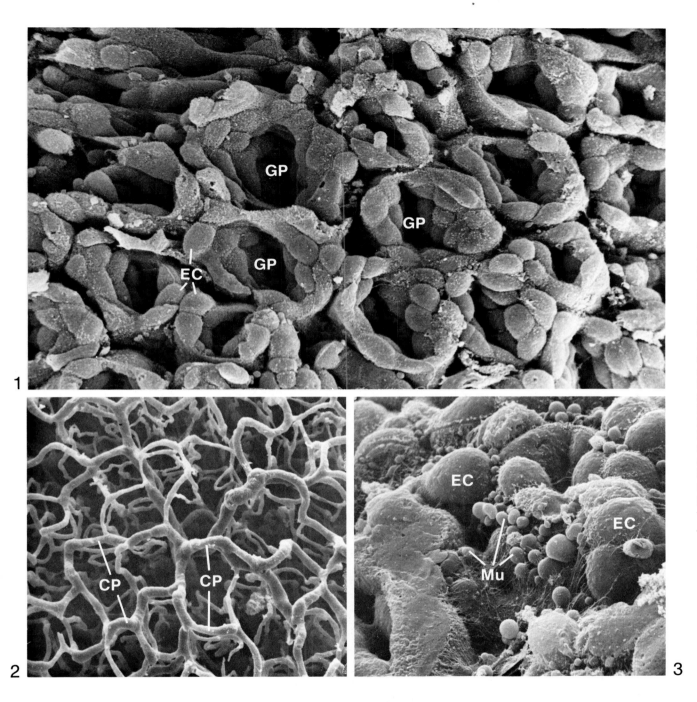

1

2

3

THE DIGESTIVE SYSTEM
Stomach

The surface of the gastric mucosa is shown in Fig. 1. As viewed from the gastric lumen after removal of the mucous coat, several gastric pits (**GP**), or foveolae, are apparent. The cell boundaries of the surface epithelial cells (**EC**) are distinct. The epithelial cells extend downward to line each gastric pit, which is continuous below with a gastric gland. The cells secrete mucus and electrolytes, such as calcium phosphate, sodium and potassium chloride, and sodium and potassium bicarbonate. A cast of the microcirculation associated with the gastric pits is shown in Fig. 2. As viewed from the luminal surface of the stomach, it is possible to observe many loops forming a capillary plexus (**CP**). Each capillary loop surrounds a gastric pit that would extend through its center in the intact stomach. The surfaces of several gastric epithelial cells (**EC**) are illustrated in Fig. 3. The rounded apical surface of these cells is either completely devoid of microvilli or possesses only a few short projections at the margin of the cell surface. These epithelial cells produce and secrete mucus directly without differentiating into goblet cells as do those in the intestines. Various forms of mucus (**Mu**) adhere to the gastric surface in Fig. 3. The concentration of hydrochloric acid is potentially high enough to damage cells and tissues, but it is thought that the large quantity of mucus normally produced by the stomach protects the gastric mucosa from damage.

Fig. 1, ×940; Fig. 2, ×535; Fig. 3, ×2135

THE DIGESTIVE SYSTEM
Stomach

The transmission electron micrograph in Fig. 1 shows a portion of a gastric gland in longitudinal section. A capillary (**Ca**) appears within the connective tissue of the lamina propria layer. The parietal cells (**PC** in Figs. 1 and 2) are widely distributed in the gastric glands of the body of the stomach. The plasma membrane of this cell is highly specialized to form what was first described as an ''intracellular canaliculus,'' which may be lined by many microvilli. The intracellular canaliculus (**IC**), an invagination of the cell membrane, communicates with the lumen (**Lu**) of the gastric glands. Since the canaliculus is not actually intracellular, as once supposed, a more appropriate term would be secretory canaliculus. Numerous mitochondria (**M**) are present in the cytoplasm and provide much of the energy for transcellular transport of hydrogen and chloride ions for the production of hydrochloric acid. Pepsinogen is synthesized by chief, or zymogenic, cells (**ZC** in Fig. 1) and released in response to nervous stimulation. The low pH of the gastric juice, due to the presence of hydrochloric acid, results in the conversion of pepsinogen into pepsin, the active form of the enzyme that cleaves peptide bonds and thus contributes to the hydrolysis of ingested proteins. Enterochromaffin cells (**EC** in Fig. 3), which have an affinity for chromium and silver salts, are located at the base of gastric glands and may contain dense cytoplasmic granules (**arrows**). Enterochromaffin cells are often divided into two populations: argentaffin and argyrophil cells. The argentaffin cells produce serotonin, or 5-hydroxytryptamine, which is a vasoconstrictor. Argyrophil cells, particularly abundant in the pyloric glands, produce the polypeptide hormone gastrin.

Fig. 1, ×3230; Fig. 2, ×6230; Fig. 3, ×8455

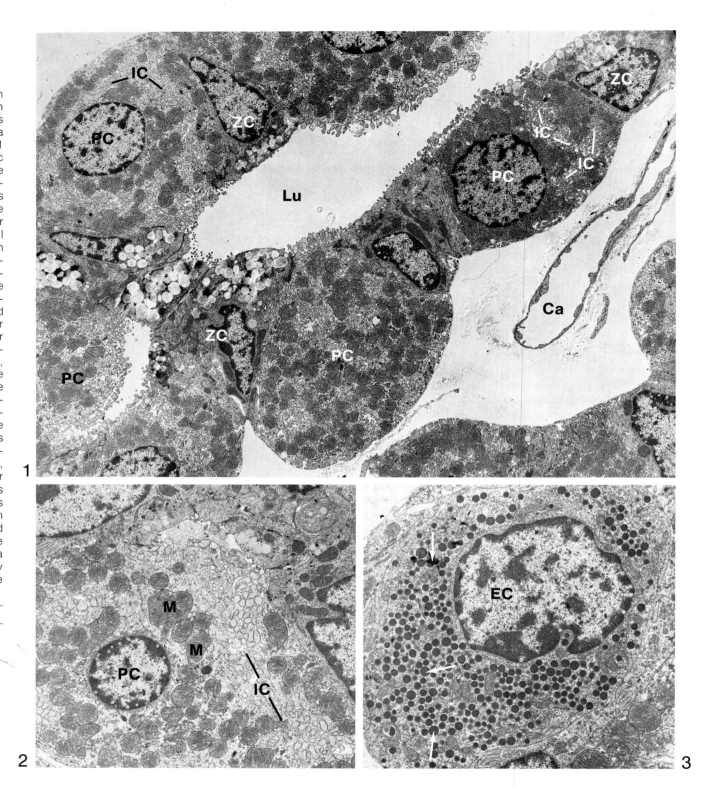

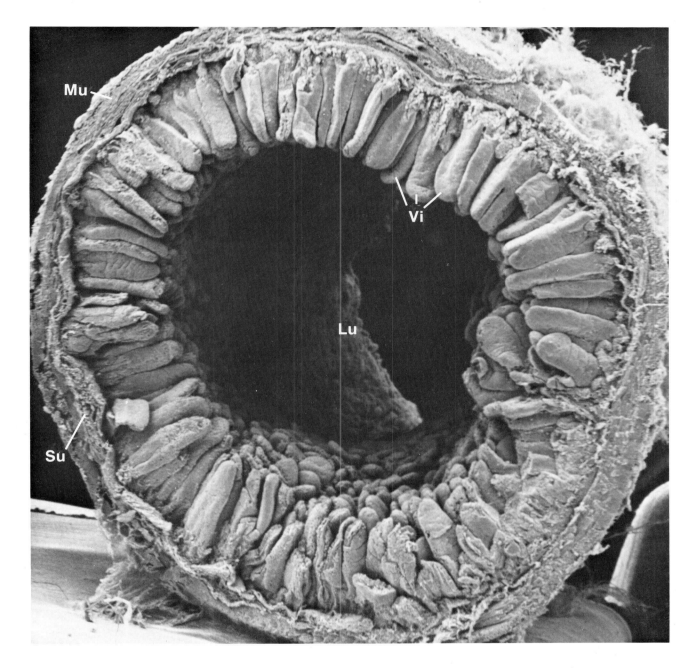

THE DIGESTIVE SYSTEM
Small Intestine

The tubular form of the small intestine is illustrated here. The products of digestion are absorbed in the small intestine, which in humans is approximately 6 m long. The small intestine is spatially divided along its length into the duodenum, jejunum, and ileum. Permanent folds of the mucosa and submucosa layers of the small intestine are present in some animal species, including humans. (*Note:* these folds, called plicae circulares, or valves of Kerckring, are absent in the rat; all of the scanning electron micrographs of the small intestine are from the rat.) Many villi (**Vi**), actually folds of the mucosa layer, project 0.5–1 mm into the intestinal lumen (**Lu**). Villi are most widely distributed in the duodenum and least numerous in the ileum. They are long, broad, or leaf-like in the duodenum, whereas in the ileum they are more finger-like in shape. Clearly, the villi are a structural modification that serves greatly to increase the effective absorptive and secretory surface of the mucosa. The villi, in turn, are covered by a layer of simple columnar epithelial cells that have a striated apical border comprised of microvilli. Unicellular mucous gland cells, called goblet cells, are interspersed among the absorptive cells. Goblet cells are fewest in number in the duodenum, but increase greatly in number toward the ileum, and their secretion (glycoproteins and sulfated mucopolysaccharides) protects the epithelial layer, lubricates the digestive tube, and participates in the formation of feces. The villi contain a connective tissue core, or lamina propria. Large blood vessels, capillaries, a lymph vessel (lacteal), and some smooth muscle cells of the muscularis mucosa are located in the lamina propria. The submucosa (**Su**) and muscularis (**Mu**) layers of the small intestine can also be distinguished in this low-magnification scanning electron micrograph.

×45

THE DIGESTIVE SYSTEM
Small Intestine—Duodenum

The figures on the page show sections of the wall of the rat duodenum, prepared by cutting. Fig. 1 shows both the surface and the interior of the broad villi (**Vi**). The intestinal glands (**CL**), the submucosa (**Su**), the muscularis externa (**ME**), and the thin outer serosal (**Se**) covering can also be distinguished. In the section depicted in Fig. 2, it is possible to observe the base of a villus (**Vi**) and a number of intestinal glands, or crypts of Lieberkuhn (**CL**). The connective tissue lamina propria (**LP**) forms a packing among the intestinal glands and extends into the cores of the villi. The position of the muscularis mucosa (**MM**) is denoted in Fig. 2. The muscularis mucosa of the small intestine, although thin, usually consists of an inner circular and an outer longitudinal layer of smooth muscle. Contraction of these smooth muscle cells permits independent movements of the mucosa and assists in expressing the secretion of intestinal glands into the lumen of the tube. The submucosa (**Su**) layer contains connective tissue elements as well as a number of Brunner's glands (**BG** in Fig. 2). Brunner's glands are present in the submucosa of the duodenum, especially in the proximal and middle regions. They are coiled, tubular glands that open into the base of intestinal glands through short ducts that penetrate the muscularis mucosa. Simple columnar cells make up the secretory portion of Brunner's glands. Although the function of these cells is not completely understood, they produce a mucus with a high bicarbonate content that may function to protect the intestinal epithelium from the gastric juice and pancreatic enzymes. The secretion may also include enterokinase and a proteolytic enzyme activated by gastric hydrochloric acid. Enterokinase is important in the conversion of pancreatic trypsinogen (inactive) to trypsin (active). Sections through a number of tubular intestinal glands (crypts of Lieberkuhn) are shown in Fig. 3. It is possible to distinguish the columnar cells (**CC**) that line the lumen (**Lu**) of the intestinal glands. Connective tissue elements of the lamina propria (**LP**) and blood vessels form a packing around the intestinal glands.

Fig. 1, ×220; Fig. 2, ×230; Fig. 3, ×420

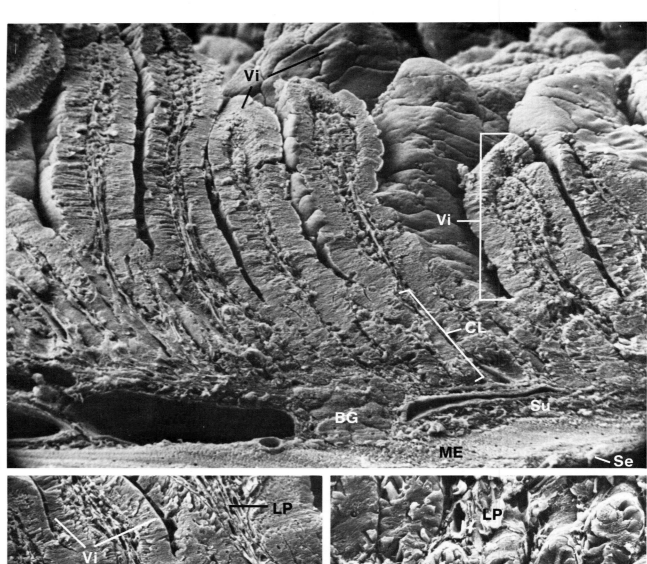

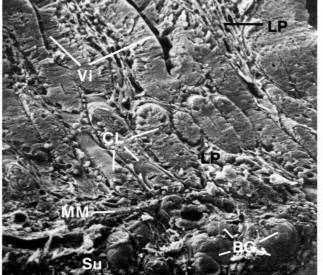

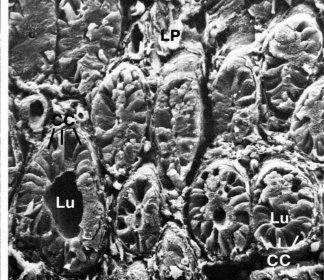

172

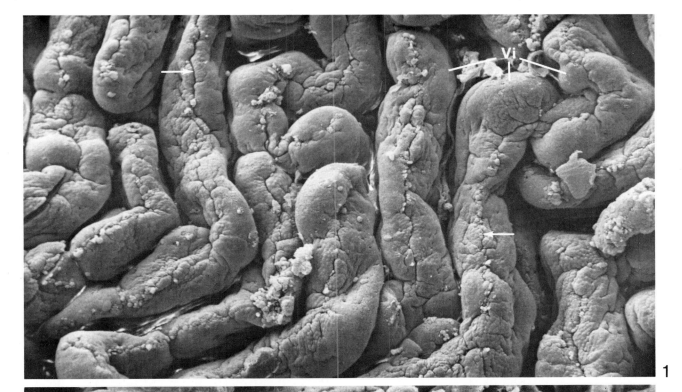

1

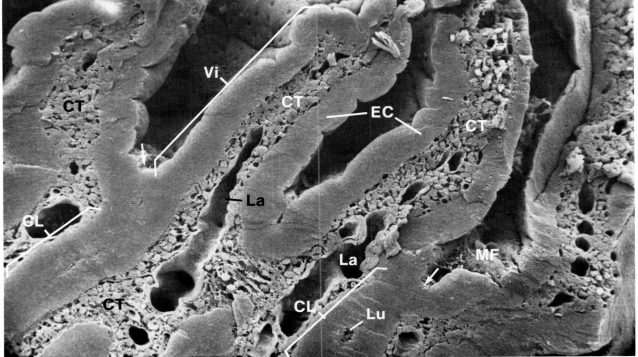

2

Villi (**Vi**) are broadest in the duodenum, as illustrated from the lumen in Fig. 1. Fig. 2 is a cryofractured preparation that illustrates the interior of a part of the intestinal mucosa. In this figure, it is possible to discern several villi (**Vi**), which are elevations, or evaginations, of the mucosal floor (**MF**). Tubular invaginations of the mucosal floor occur at the bases of villi and are called intestinal glands, or crypts of Lieberkuhn. The approximate location of the orifices into two crypts of Lieberkuhn (**CL**) are denoted by the **arrows** in Fig. 2. Intestinal glands in the duodenum are longer than those found in the jejunum and ileum, and possess a narrow lumen (**Lu**), which is partially revealed in Fig. 2. Intestinal villi are covered with a single layer of epithelial cells (**EC**) that is continuous with the layer of epithelium that lines the intestinal glands. Connective tissue (**CT**) elements of the lamina propria form a packing between the intestinal glands and the central core of each villus in the intestinal mucosa. Within this packing, many cells such as plasma cells, granular leukocytes, and lymphocytes may be found. Blood vessels, nerves, and centrally located lacteals (**La**), or lymph vessels, are also contained within the connective tissue of the lamina propria. The replacement rate of the absorptive cells of the mucosal epithelium is approximately 2–4 days. This process involves the active proliferation and movement of undifferentiated cells located in the lower portion of the intestinal glands. These cells differentiate as they migrate from the crypt onto the mucosal floor and up the villi. Finally, the cells are extruded from the crest of the villus (**arrows** in Fig. 1) at the end of their migratory cycle. Consequently, the size of villi may vary depending upon the rate of proliferation and desquamation of the surface epithelial cells.

Fig. 1, ×185; Fig. 2, ×375

THE DIGESTIVE SYSTEM
Small Intestine—Jejunum

The tongue-shaped villi of the jejunum (as viewed from its lumen) are illustrated at different magnifications in Figs. 1 to 3. In Figs. 1 and 2 the intestinal wall was mechanically stretched so as to illustrate the distribution of numerous villi (**Vi**) in relation to the mucosal floor (**MF**). At higher magnification (Fig. 2) it is also possible to discern the openings, or orifices (**arrows**), of the tubular intestinal glands, which are oriented below and at right angles to the mucosal floor. Fig. 3 shows the apical surface of individual absorptive cells (**AC**) and goblet cells (**GC**) covering the villus. A cast of the blood vascular system of a single villus is illustrated in Fig. 4. The cast of the blood vessels roughly conforms to the shape of a villus, even though all cellular and fibrous elements were digested. This micrograph shows clearly the extensive network of capillaries (**Ca**) that supply each villus. In the intact organ the capillary plexus is closely associated with the base of the covering epithelium. Such a close relationship is clearly advantageous for the absorption of digestive products into the blood vascular system. Each villus receives one or more arterial blood vessels, and capillaries converge at the tips of the villus to form one or more venules, which extend into the submucosal plexus.

Fig. 1, ×80; Fig. 2, ×235; Fig. 3, ×725; Fig. 4, ×255

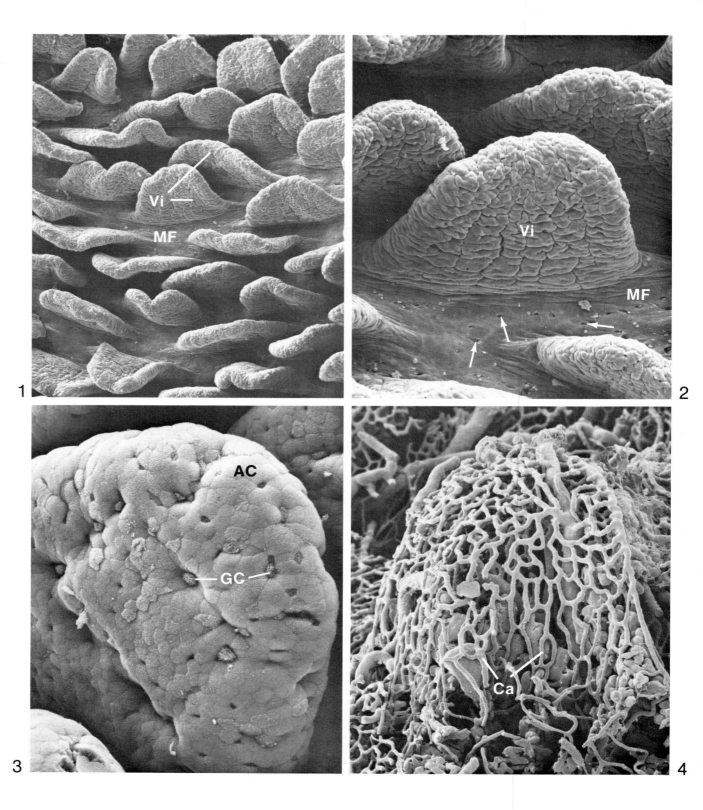

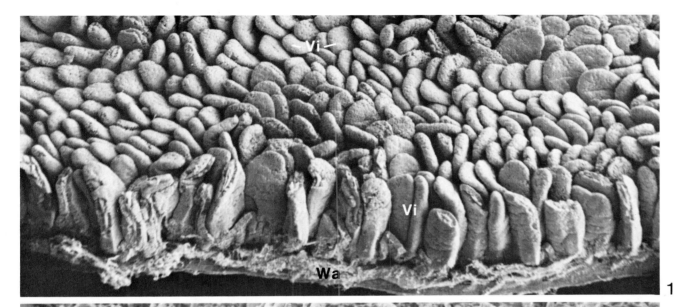

1

THE DIGESTIVE SYSTEM
Small Intestine—Jejunum

When the intestinal wall is not mechanically stretched during preparation (Fig. 1), the mucosal floor cannot be visualized; nevertheless, this preparation provides a more realistic view of the distribution and close packing of the many villi (**VI**) present in the jejunum. The angle of observation is such that a cut portion of the jejunal wall (**Wa**) can be seen in addition to the luminal view of villi. For comparison, a cast of the blood vascular system of the jejunal wall is shown in Fig. 2. As would be expected, the blood vascular system is most extensive in the region of the mucosa, especially in relation to the many villi. Although it is not possible to identify all regions of the cast with certainty, the capillary plexus (**CP**) associated with each villus is clear. This capillary plexus is supplied by arterioles (**AI**) that originate from arteries (**Ar**) in the submucosa. The capillaries drain into one or more venules (**Ve**), which converge into medium-sized veins (**Vn**) in the submucosa. Small capillary vessels (**CV**) are also present in the mucosa surrounding the approximate location of intestinal glands. The bundles of straight vessels present in the lower portion of the micrograph are probably those associated with the smooth muscle cells of the muscularis externa. The distinction between arterial and venous components of the replica is based on differences in vessel diameter and on their location with respect to the intestinal wall.

Fig. 1, ×50; Fig. 2, ×180

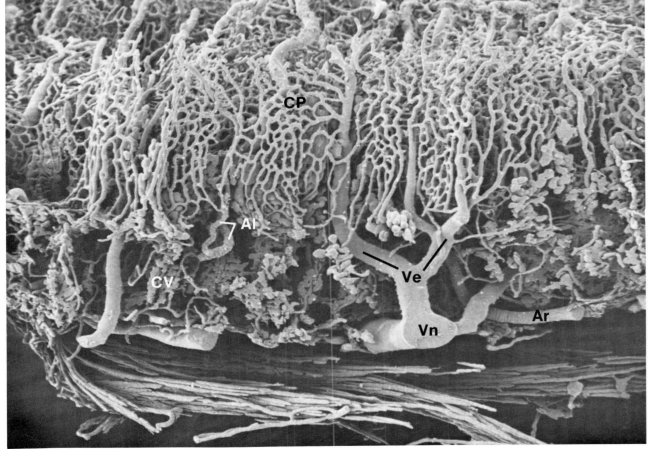

2

THE DIGESTIVE SYSTEM
Small Intestine—Jejunum

Illustrated here are the two most prevalent epithelial cell types that cover intestinal villi. The free, or apical, surface of a goblet cell is usually smaller in area than that of the absorptive columnar cells. The goblet cell surface (Fig. 1) often has microvilli (**Mv**) associated with it, and although their number and length appear to be variable, they are larger and wider than those microvilli (**MA**) associated with adjacent absorptive cells. Occasionally, epithelial cell microvilli have expanded tips (**arrows** in Fig. 1). In Fig. 2, the microvilli of absorptive cells (**MA**) are observed to form an orderly array because of their close packing. The microvilli clearly function to increase the cell surface area available for absorption. Adjacent epithelial cells are covered with small, discrete droplets (**arrows** in Fig. 2) that are probably mucigen granules of goblet cells. A goblet cell (**GC**) is illustrated in a cryofractured preparation in Fig. 3; note the number and arrangement of internal mucigen granules (**MG**). Goblet cell microvilli (**arrows** in Fig. 3) may be much smaller and more sparsely distributed than the microvilli (**MA**) of adjacent absorptive epithelial cells (**AC**), depending upon the secretory state of the cell. The absorptive epithelial cells (**AC**), illustrated from surface and side views in Fig. 4, reside on a basal lamina (**BL**), and their apical microvilli (**MA**) vary somewhat in their packing.

Fig. 1, ×13,770; Fig. 2, ×7510; Fig. 3, ×7340; Fig. 4, ×3000

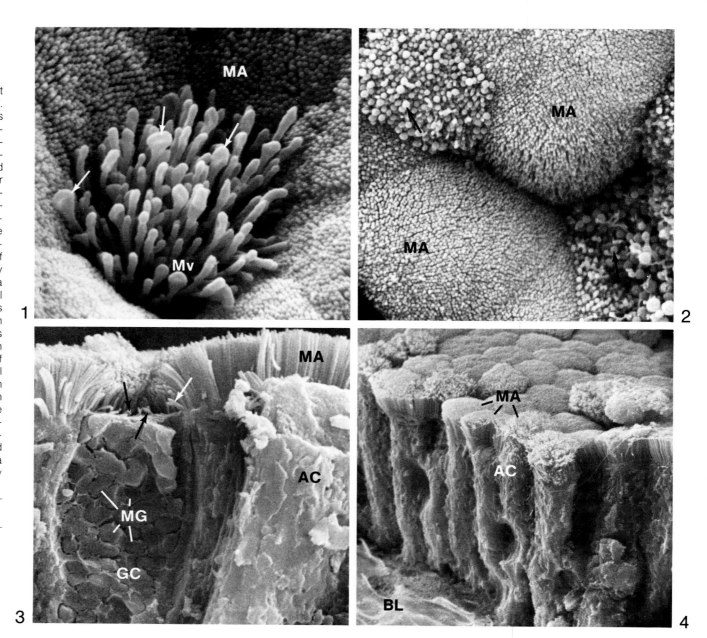

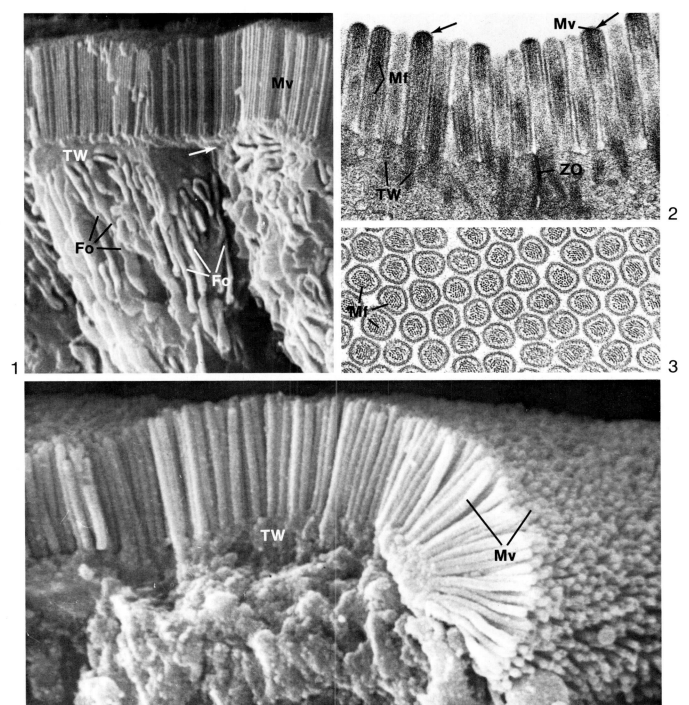

The size, number, and arrangement of the closely packed microvilli (**Mv**) associated with intestinal epithelium are depicted in the scanning electron micrographs shown in Figs. 1 and 4. Transverse and longitudinal sections of the microvilli are shown in the transmission electron micrographs (Figs. 2 and 3). The core of each microvillus contains 40 to 50 microfilaments (**Mf**) that have been demonstrated to contain the contractile protein actin. The filaments (about 60 Å in diameter) extend into the apical cytoplasm and constitute a component of the terminal web (**TW** in Fig. 2). Another contractile protein, myosin, is associated with that portion of the microfilaments located in the terminal web. The presence of the contractile proteins, actin and myosin, confer a primitive motility to the microvillus border of the intestinal epithelial cells. Between adjacent epithelial cells, specializations of the lateral plasma membrane exist in the form of junctional complexes. One type of junctional complex, the zonula occludens (**ZO** in Fig. 2), seals off the narrow intercellular space and prevents material in the lumen of the intestine from passing between cells. Other types of junctional complexes that form between intestinal epithelial cells include the macula adherens, zonula adherens, and the gap junction (or nexus). Some of the junctional complexes serve to anchor the cells in the epithelial sheet and, together with the terminal web, stabilize the form of the apical end of the cells. The lateral portions of the epithelial cells are highly folded (**Fo** in Fig. 1), and these folds of the lateral plasma membrane of one cell interdigitate with those of adjacent cells. This arrangement probably adds stability to the epithelial sheet. Both inner and outer leaflets of the plasma membrane covering the microvilli are resolved in Fig. 3. Carbohydrates make up the outer portion of the plasma membrane, providing a thin filamentous coating to the tips of the microvilli (**arrows** in Fig. 2). This thin layer, called a glycocalyx, together with the plasma membrane of the microvilli, contains enzymes that can dephosphorylate and cleave disaccharides.

Fig. 1, ×9345; Figs. 2 and 3, ×61,765; Fig. 4, ×17,910

THE DIGESTIVE SYSTEM
Appendix

The appendix is a tubular evagination of the cecum portion of the large intestine. All illustrations used are of the rabbit appendix. The appendix is constructed of the same basic layers present in the remainder of the digestive tract. The inner mucosa, or mucous membrane, consists of simple columnar epithelium, which is invaginated to form many tubular crypts of Lieberkuhn, but villi are absent. The connective tissue lamina propria surrounding the epithelium is heavily infiltrated with lymphatic nodules, and the muscularis mucosa is poorly developed. The submucosa forms a thick layer around the mucosa and consists of connective tissue with blood vessels, nerves, and some fat cells. The muscularis externa is thinner than in the intestine, but consists of an inner circular and outer longitudinal layer of smooth muscle. A serous coat, or serosa, forms the outermost layer of the appendix. The mucosal surface, as viewed from the lumen of the appendix, is shown in Figs. 1 and 2. The horizontal fold of the mucosa at the top of Fig. 1 marks the junction between the appendix and cecum. The lumen of the appendix is usually small, sometimes obliterated, and frequently contains dead cells and detritus. Large rounded projections (**Pr**) are apparent in the folded mucosa in both figures. These projections are folds of the epithelial layer over lymphatic nodules, which occur in large numbers in the underlying lamina propria and submucosa. The smaller rounded or slit-like apertures (**arrows**) are the openings into the tubular crypts of Lieberkuhn of the mucosa.

Fig. 1, ×45; Fig. 2, ×135

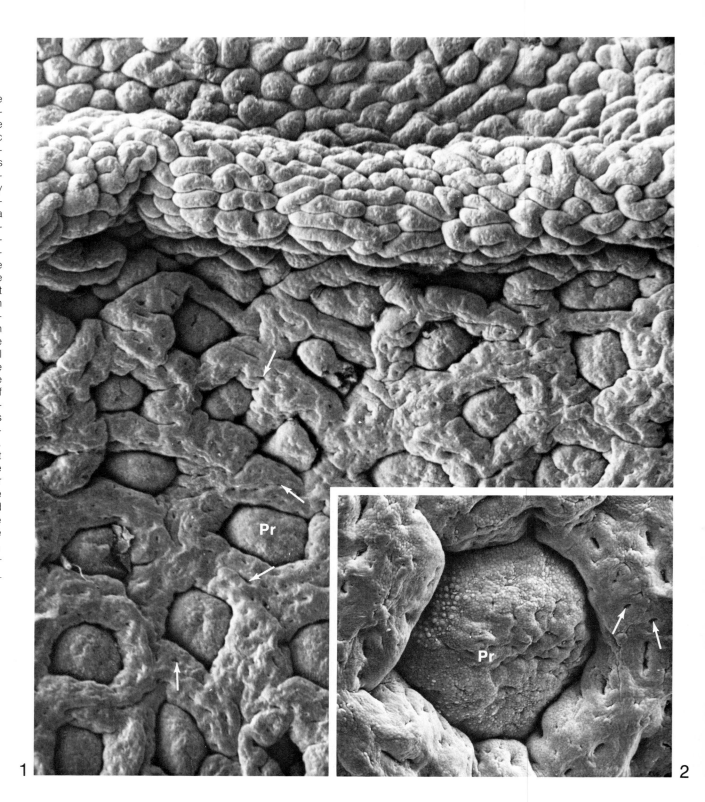

1

2

178

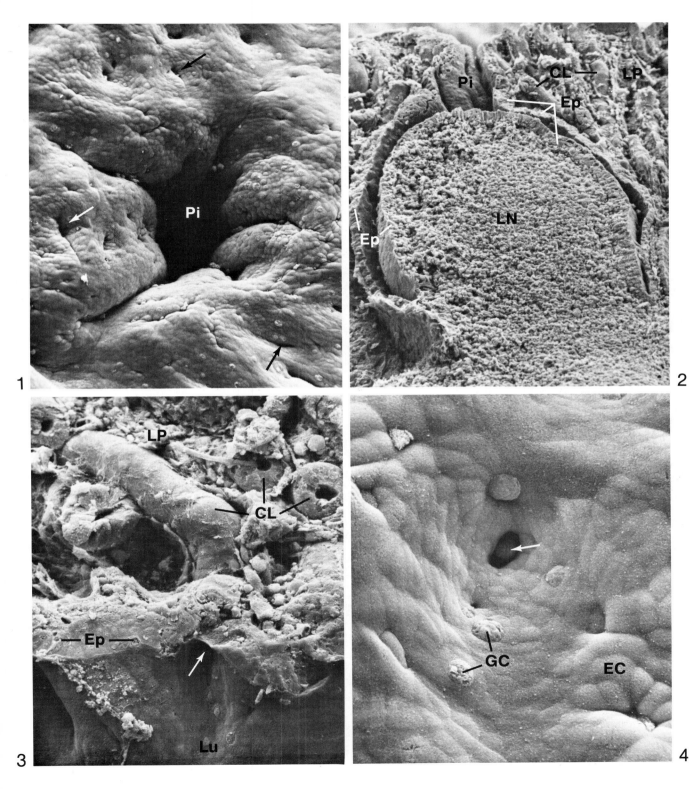

THE DIGESTIVE SYSTEM
Appendix

The surface view of the appendix mucosa (Fig. 1) reveals openings (**arrows**) into the tubular crypts of Lieberkuhn, as well as a larger opening that takes the form of a pit (**Pi**). These pits result from deep infoldings of the mucosal epithelium over lymphatic nodules, which in this case do not project as far into the mucosal surface as those in the two figures on the previous page. Fig. 2 is a cryofractured preparation of the mucosa. In such a preparation the infolded epithelial (**Ep**) layer can be followed as it extends inward along the pit (**Pi**) from the luminal surface to cover a lymphatic nodule (**LN**). Crypts of Lieberkuhn (**CL**) and lamina propria (**LP**) of the mucosa are also identified in Fig. 2. The size of the lymphatic nodule and the extent to which it projects toward the lumen produces the variations in the mucosal surface illustrated in the previous scanning electron micrographs. In sections of the appendix mucosa, such as illustrated in Fig. 3, it is possible to distinguish an opening (**arrow**) that leads from the appendix lumen (**Lu**) into a crypt. The tubular form of crypts of Lieberkuhn (**CL**) is observed both from their surface and in transverse section. The surrounding connective tissue lamina propria (**LP**) is also apparent. In the enlargement of the mucosal surface (Fig. 4), an opening (**arrow**) into a single tubular gland, or crypt of Lieberkuhn, is surrounded by surface epithelial cells (**EC**) and goblet cells (**GC**).

Fig. 1, ×275; Fig. 2, ×150;
Fig. 3, ×540; Fig. 4, ×1180

THE DIGESTIVE SYSTEM
Appendix

The surface epithelium of the appendix mucosa is continuous with the epithelial cells that line the crypts of Lieberkuhn. The epithelial cells are predominantly simple columnar in type and have a striated microvillus border. Goblet cells are interspersed among the simple columnar cells. The microvilli (**Mv**) covering the simple columnar cells are illustrated in these figures. The apical surface of the goblet cells projects (❋ in Fig. 1) above the surrounding columnar cells, and when the apical plasma membrane of the goblet cell is removed, it is possible to observe the many mucigen granules (**MG** in Figs. 1 and 4) that occupy this intracellular location. The distribution of microvilli on the apical surface of goblet cells is highly variable and may reflect different stages in the synthesis, accumulation, and secretion of mucigen granules. For example, in Fig. 2, the apical surfaces of two goblet cells possess a sparse number of microvilli, which may be related to the increased accumulation of underlying mucigen granules. It appears that the more mucigen granules a goblet cell contains, the further its apical surface projects above surrounding cells and the fewer the microvilli present. In sections of the epithelial sheet illustrated in Fig. 3, several goblet cells can be distinguished on the basis of the numerous mucigen granules (**MG**) that fill the apical cytoplasm. Some Paneth cells and argentaffin cells may also be distributed among the cells lining the crypts of Lieberkuhn.

Fig. 1, ×2400; Fig. 2, ×7510;
Fig. 3, ×2150; Fig. 4, ×8410

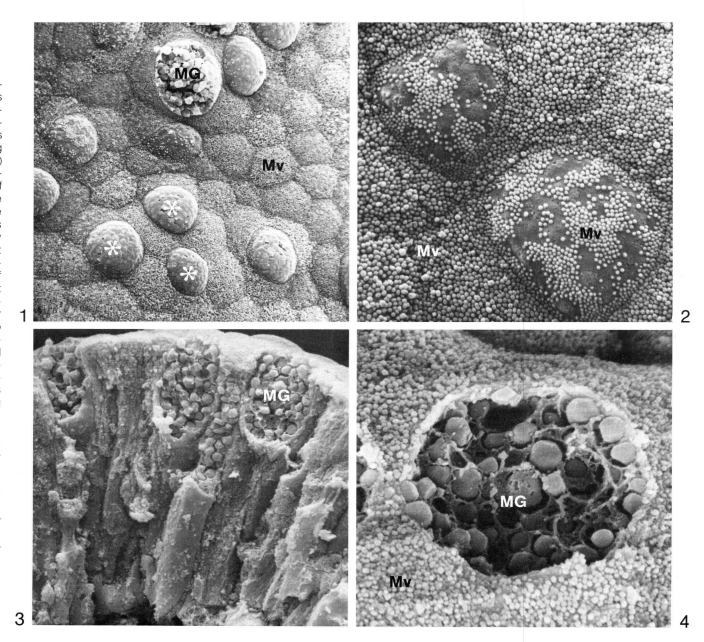

180

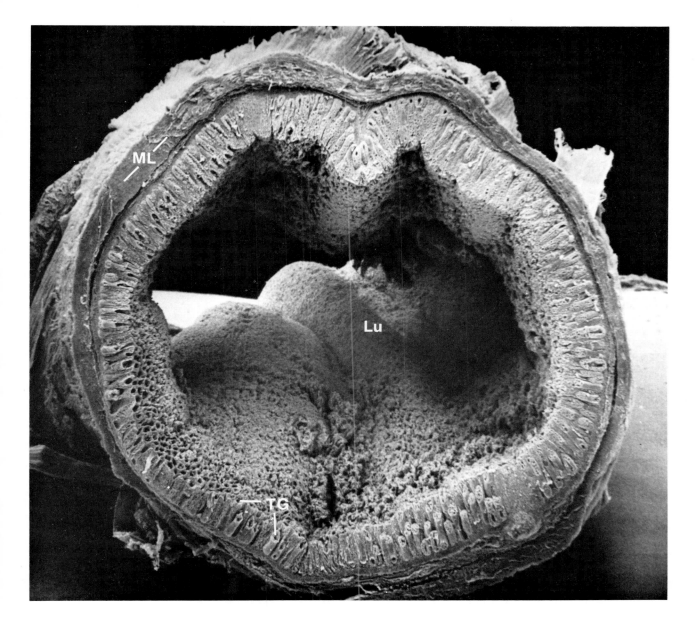

THE DIGESTIVE SYSTEM
Colon

The large intestine, also a tubular structure, is about 5 feet long and wider than the small intestine. It is divided into several regions, including the colon, appendix, rectum, and anal canal. The colon is divided into ascending, transverse, and descending segments, and is the longest portion of the large intestine. The colon is made up of the same basic layers as the small intestine, including a mucosa, submucosa, muscularis externa, and serosa. Villi, however, are absent in the large intestine. The absorptive capacity of the colonic mucosa is extensive, and includes the absorption of water, sodium, vitamins, and other minerals. Sodium is actively transported from the lumen of the colon, causing water to be moved along an osmotic gradient. As a result of the absorption of water, the watery chyme that enters the colon is converted into a semisolid feces. Many substances, such as sedatives, anesthetics, steroids, and tranquilizers can be rapidly absorbed in the colon. Some potassium and bicarbonate is secreted by the mucosal epithelium into the colon lumen. The figure shows a transverse section of a distended rat colon. Note the large central lumen (**Lu**), or cavity, through which contents of the tube are propelled. A very large portion of the inner layer of the colon, the mucosa, consists of many tubular glands (**TG**), or crypts of Lieberkuhn, which have been cut in various planes in this preparation. The tubular glands are oriented perpendicular to the long axis of the colon and are much longer than in the small intestine. The smooth muscle that forms the muscularis layer (**ML**) is also apparent. It is by means of the alternate contraction and relaxation of these muscle bands that the contents of the colon are transported.

×45

THE DIGESTIVE SYSTEM
Colon

A transverse section through a contracted portion of the colon wall is illustrated here. Layers of the colon surrounding the central lumen (**Lu**) are highly folded, as is typical of an empty organ. At this magnification the four basic layers of the colon can be observed. They include a mucous membrane, or mucosa (**Mu**), a submucosa (**Su**), the muscularis externa (**ME**), and the thin outer serosa (**Se**). The surface of the mucosa adjacent to the lumen has a relatively smooth contour, since villi are lacking in the colon. In addition to the lamina propria and the muscularis mucosa, another major constituent of the mucosa includes the numerous tubular intestinal glands (crypts of Lieberkuhn). The intestinal glands will be illustrated in subsequent figures. That portion of the colonic wall contained within the rectangle in this figure is enlarged in Fig. 1 on the next page.

×60

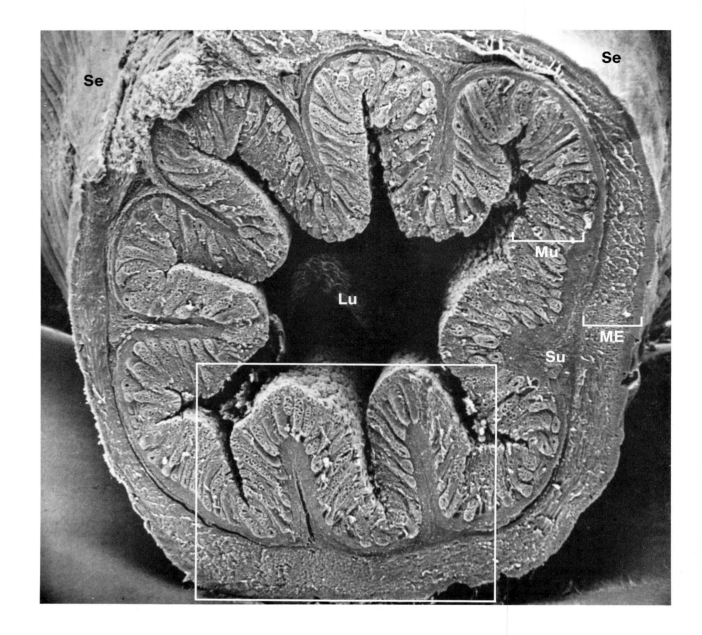

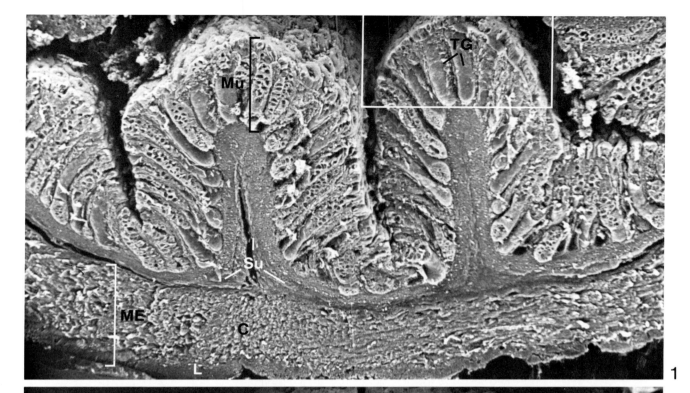

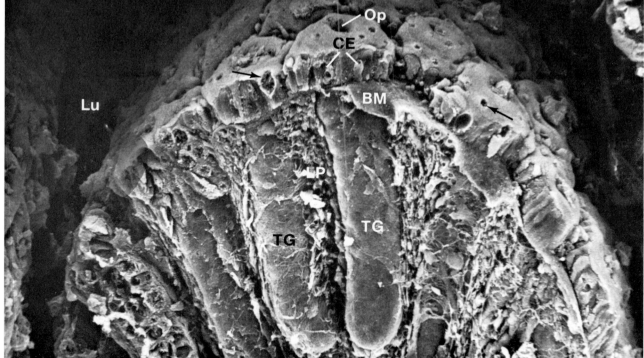

THE DIGESTIVE SYSTEM
Colon

The distribution of the numerous tubular intestinal glands (**TG**) of the mucosa (**Mu**) is shown in both figures. The region of the mucosa contained within the rectangle in Fig. 1 is enlarged in Fig. 2 to illustrate additional structural details. The surface epithelium consists of a single layer of absorptive columnar epithelial (**CE** in Fig. 2) cells and many mucous or goblet cells. The epithelial sheet is attached to a basal lamina (**BM** in Fig. 2). Goblet cells viewed both from the luminal surface and in sections of the epithelium are denoted by the **arrows** in Fig. 2. Also shown are the exterior of several simple tubular intestinal glands (**TG**), the opening (**Op**) into one of the glands from the colonic lumen (**Lu**) and the distribution of the connective tissue lamina propria (**LP**) around the tubular glands. Connective tissue cells and fibers make up the lamina propria in a manner similar to that described for the small intestine. Solitary lymphatic nodules may also be present. The muscularis mucosa portion of the mucous membrane, although not shown as a distinct layer in Fig. 1, usually consists of an inner circular and outer longitudinal layer of smooth muscle at the base of the intestinal glands. The submucosa (**Su** in Fig. 1) surrounds the mucous membrane and consists of irregular connective tissue with blood vessels, nerves, and lymphatics. The muscularis externa (**ME** in Fig. 1) is divided into an inner circular (**C**) and outer longitudinal (**L**) layer of smooth muscle. The outer portion of the muscularis externa is arranged into three longitudinal bands called the taeniae coli.

Fig. 1, ×110; Fig. 2, ×380

THE DIGESTIVE SYSTEM
Colon

The mucous membrane, or colonic mucosa, is shown in Fig. 1. All three layers that make up the mucosa are visible: the epithelial layer, which consists of surface columnar epithelium (**CE**) and tubular glands (**TG**), the surrounding connective tissue lamina propria (**LP**), and the muscularis mucosa (**MM**), which appears in the illustration at the base of the intestinal glands. The surface epithelium of the mucosa, consisting of both absorptive and goblet cells attached to a basement membrane (**BM**), is thus continuous with the single layer of cells that line the tubular glands. Figs. 2 and 3 show longitudinal and transverse sections of individual tubular intestinal glands (**TG** in Fig. 2). The lumen (**Lu**) of the gland is lined by goblet cells (**GC**), and the simple columnar absorptive cells possess microvilli (**Mv**) that project from the free cell surface into the lumen of the intestinal gland. In addition to absorptive cells and goblet cells, undifferentiated cells and some argentaffin cells are present in the epithelium. Large quantities of mucus are produced and released from goblet cells to lubricate and protect the mucosal surface of the colon. The undifferentiated cells are proliferative and serve to replace cells of the epithelial sheet, which only have a 2–3 day life span. Numerous bacteria reside in the large intestine. They produce a variety of substances, including vitamins, ammonia, and amines. The characteristic color of the stool is due to the pigments formed by bacterial action on bile pigments.

Fig. 1, ×380; Fig. 2, ×935; Fig. 3, ×1110

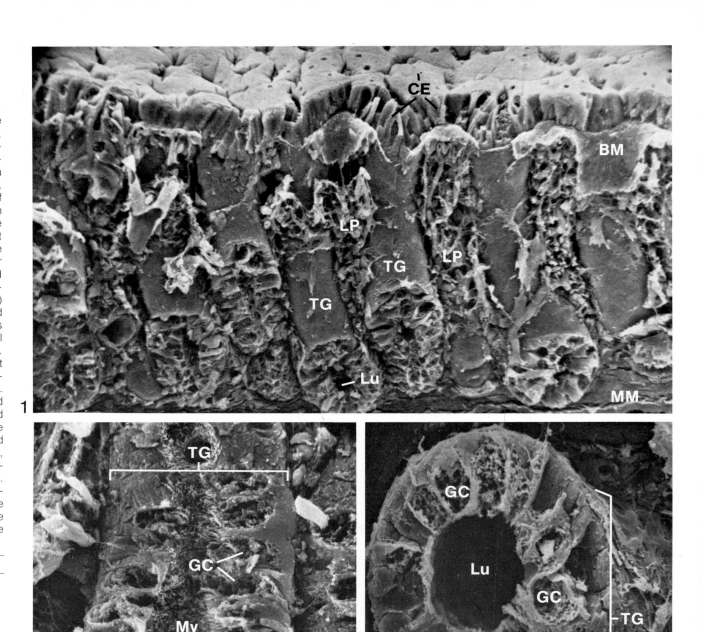

THE DIGESTIVE SYSTEM
Colon

This figure, a transverse section of the rat colon, illustrates the lumen (**Lu**) and mucosal surface. It also serves to illustrate the magnitude of the tubular intestinal glands, since each of the small black areas, a number of which are denoted by **arrows,** represents an opening from a tubular gland into the lumen of the colon.

×45

THE DIGESTIVE SYSTEM
Colon

Details of the colonic mucosa, as viewed from the lumen of the colon, are illustrated in Fig. 1. Also visible are the orifices (**O**) of numerous tubular intestinal glands. The surface of many goblet cells (**arrows**) are present in the epithelial layer surrounding the entrances into the intestinal glands. An extensive capillary plexus surrounds each of the tubular intestinal glands in the colon. A cast of this capillary plexus is illustrated in Fig. 2, as viewed from the luminal side of the mucosa. Although the contents of the mucosa were removed by maceration, the cast of the capillary plexus remains, and its distribution in relation to the tubular intestinal glands can be observed. The boundaries of individual epithelial cells of the colonic mucosa are visible in Fig. 3 as viewed from the lumen. Microvilli (**Mv**) associated with the free surface of the absorptive epithelial cells (**EC**) are illustrated in surface and lateral views in Fig. 4. Although goblet cells (**GC**) have a varied surface morphology, the microvilli on many of the cells shown in Fig. 3 are more sparsely distributed than usual. Large numbers of bacteria (**Ba** in Fig. 4) are present in the recesses of the folded colonic mucosa; Fig. 4 shows their spatial relationship to the microvilli (**Mv**) of surrounding epithelial cells.

Fig. 1, ×230; Fig. 2, ×170; Fig. 3, ×1410; Fig. 4, ×9985

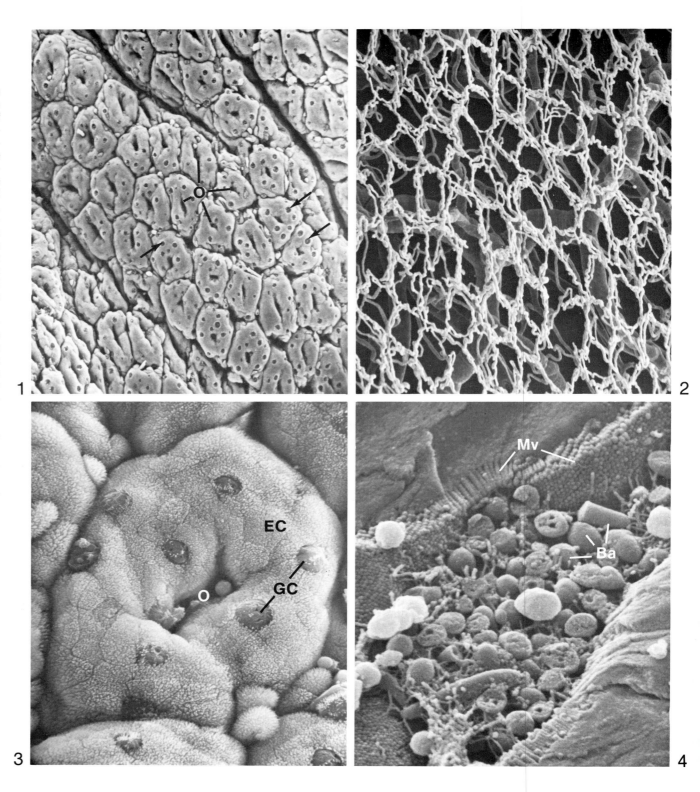

GLANDS OF DIGESTION

Introduction

The liver is the largest of all the organs in the body and the most diverse in function. Considered both an endocrine and an exocrine gland, it functions as a processing center and as a storehouse. As an endocrine gland, the liver releases lipids, glucose, proteins, glycoproteins, and lipoproteins into the bloodstream and in so doing determines much of the blood's composition. A few of the major blood proteins synthesized and liberated by the liver are albumin, fibrinogen, and complement. As an exocrine gland, the liver secretes bile into a system of ducts that convey their contents to the small intestine. Before reaching the intestine, bile may be stored and concentrated in the gallbladder. Coordinated muscle contractions of the gallbladder wall and sphincters within the bile duct promote emptying of bile into the duodenum. Within the intestine, bile emulsifies ingested fats into smaller aggregations which can then be absorbed. As a processing center, the liver transforms drugs and hormones into inactive and, in some cases, active or toxic metabolites. Because the liver metabolizes amino acids, it also releases substantial amounts of urea into the bloodstream. Additionally, the liver has the capacity to function as a storehouse for many raw materials of digestion.

These include sugars, fats, vitamins, and metal ions, all of which may eventually be used by the liver for its own functions or secreted into the blood to be used by other cells.

Much of the control over these hepatic functions is hormonal in origin, and therefore depends largely upon amount and source of the blood perfusing the liver cells at any given moment. This is why so much emphasis has been placed upon blood flow and its regulation within the liver. The liver, after all, has a special blood supply—special for three reasons. First, the liver has a *dual* blood supply, receiving both arterial blood and portal venous blood. Most of the blood comes from the portal vein, which drains the digestive tract, pancreas, gallbladder, and spleen; the remaining blood is supplied via the hepatic artery. This dual blood supply provides the liver with two sources of nutrients. The portal vein transports nutrients absorbed in the digestive tract, hormones from the pancreas, and breakdown products of red blood cells from the spleen. The arterial source supplies the liver with oxygenated blood containing hormones from other parts of the body. Thus the liver not only receives blood rich in resources and raw materials, but it also receives blood carrying hor-

mones, which influence the fate of these resources. The second reason why the liver's blood supply is so special has to do with the regulation of its flow and distribution. At any given time blood may be briskly percolating through liver tissue in one region while in adjacent areas it may be stagnant. Such regional variation in blood distribution and flow can have the effect of creating regions in which cells differ in their functional state. This further emphasizes the important relation between blood flow and function within the liver. The third reason is that the sinusoidal capillaries that supply liver cells have an unusually porous lining, which allows blood-borne substances to diffuse readily through the capillaries to the liver cell surface.

These three special characteristics of the liver vasculature—dual blood supply, regional regulation of blood flow, and high capillary permeability—allow the liver access to a wide variety of substances and are responsible for its diversity of function. Just how the liver is structurally organized to perform all of its functions will be presented in the first part of this chapter. The emphasis will not be placed on how an individual liver cell functions, but on how hepatocytes are organized to function as a unit.

GLANDS OF DIGESTION
Liver

The first impression of the liver is that it is an enormous mass of cells permeated by numerous blood vessels and bile ducts. Because the liver is a gland, its organization and function are best understood by dividing it into small repeating units. Each unit is commonly referred to as a liver lobule, an example of which is shown in the center of this figure. The classic liver lobule is polyhedral in shape, and consists of interconnecting plates of hepatocytes (**Hc**) that radiate toward a central vein (**CV**) located in the middle of the lobule. The classic lobule is bound on the periphery by groups of ducts and vessels, the most prominent of which is the conducting portal vein (**CPV**). Conducting portal veins branch at right angles to form distributing portal veins (**DPV**), which appear to delineate the periphery of the classic liver lobule. An arterial blood supply (**Ar**) with similar subdivisions also surrounds each lobule, and follows a path adjacent to that of larger portal vessels. Together, the portal venous blood and the arterial blood supply each lobule and mix upon entering a network of small sinusoidal capillaries within the lobule. Mixed blood, traveling via sinusoids, flows between intersecting plates of liver cells toward the middle of the lobule, where it empties into the central vein. From the central vein, blood flows into tributaries of the hepatic veins, which finally exit the liver to join the inferior vena cava. One of the tributaries of the hepatic vein, the sublobular vein (**SLV**), is sectioned so as to illustrate its continuity (**arrows**) with several longitudinally sectioned central veins.

×130

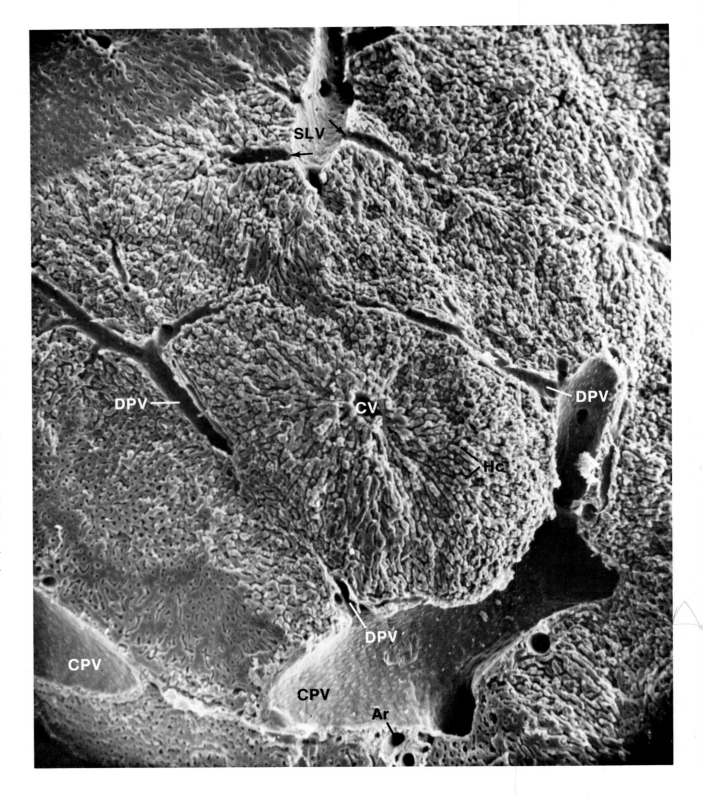

GLANDS OF DIGESTION
Liver

Fig. 1 diagrams the overall organization of liver tissue, including portions of adjacent lobules. The liver cells, termed hepatocytes, are organized into intersecting plates, usually one cell thick. The intersecting plates are collectively called a muralium (Elias and Sherrick, 1969), and tunnel-like spaces penetrating the muralium are called lacunae. Liver capillaries, known as sinusoids, travel within the lacunae and thus supply surrounding hepatocytes with blood. Divisions of blood vessels and bile ducts are also illustrated; details of their relationship with the cellular components will be the topic of subsequent pages. Fig. 2 shows how liver cells fit together to form interconnecting plates. The small bile canaliculi are extracellular channels formed by adjoining hepatocytes; they accumulate the bile secreted into them by liver cells. Canaliculi also reinforce the geometric stability of the cells. Although the plates are stable enough for the cells to hold their geometric relations with each other, they are easily flattened or stretched. Their configuration at any one moment depends largely upon regional blood flow, which tends to exert a directional force on the muralium. It is apparent from the organization of cell plates that hepatocytes can be divided into four major shapes, each containing a different volume. These four basic shapes (**A–D**), illustrated in Fig. 3, are the octahedron, pentahedron, decahedron, and dodecahedron. The shape of a hepatocyte is determined by its location within the muralium of liver plates (Elias and Sherrick, 1969). An example of the corresponding locations of each of the four major shapes is labeled in Fig. 2. [Fig. 1 redrawn from H. Elias and J. C. Sherrick, *Morphology of the Liver,* Academic Press, p. 135, 1969. Fig. 2 redrawn from H. Elias, The geometry of the cell shape and the adaptive evolution of the liver, *Journal of Morphology* 91:377 (1952). The Wistar Press. Fig. 3 redrawn from H. Elias, A re-examination of the structure of the mammalian liver. II. The hepatic lobule and its relation to the vascular and biliary system, *American Journal of Anatomy* 85(3):449, (1949). The Wistar Press.]

Sublobular vein

Perisinusoidal space (Disse)

Periportal arterial capillary

Lymph vessel

Limiting plate

Lacuna

Muralium of liver plates

Central vein

Sinusoid

Intralobular arterial capillary

Liver plate

Inlet venule

Periportal space (Mall)

Limiting plate

Portal vein

Bile duct

Artery

Intralobular ductule

Paraportal ductule (Hering)

Bile canaliculus

Central vein

1

Bile canaliculi

A B C D

2

A

B

C

D

3

GLANDS OF DIGESTION
Liver

Casts of the liver vasculature are useful tools for conveying important information about the distribution of blood vessels within the liver. This figure shows a low-magnification view of such a cast of liver circulation obtained after the tissue components were completely digested. The vascularization of a number of hepatic lobules is delineated by a bush-like distribution of sinusoids (**S**) that drain toward a central vein (**CV**). Many central veins are seen leading away from the lobules. Bordering each hepatic lobule are portal areas (**PA**) that consist of branches of the portal vein accompanied by thin, straight arterial divisions. The conducting branch of the portal vein (**CPV**) is shown dividing into the smaller distributing portal veins (**DPV**). The cast demonstrates that vessels within a portal area are shared by adjacent lobules. In studying the vascularization of the liver, Rappaport et al. (1954) and Rappaport (1958, 1963) proposed dividing the liver into functional units termed acini. Each acinus defines an area of liver tissue organized around the terminal branch of the portal vein in such a way that cells at the center of the acinus are the first to receive blood and cells located toward the periphery are the last to receive blood. In relation to the acinus, the central vein is no longer central in position; therefore, a more acceptable term would be "collecting vein." The concept of the liver acinus is frequently used when discussing the functional and pathological states of the liver. In considering the liver's function as an exocrine gland, Mall (1906) suggested dividing the liver into units termed portal lobules; each lobule is organized around a portal area, and its boundaries follow the distribution of bile ducts within the liver. It should be understood that no one concept, whether it be the classic lobule, the acinus, or the portal lobule, is either right or wrong. They should all be considered useful concepts for attempting to understand the liver's organization from a standpoint of both health and disease.

×60

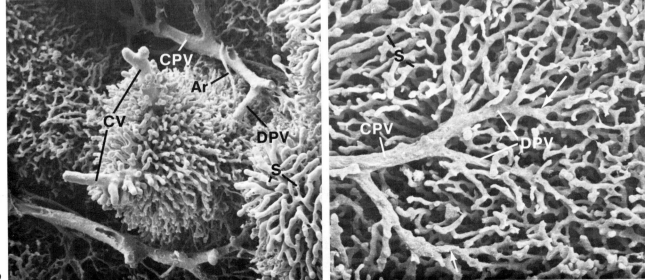

GLANDS OF DIGESTION
Liver

Further details of the liver vasculature are revealed by the casts shown in these figures. Portal areas containing the conducting portal vein (**CPV**), distributing portal vein (**DPV**), and arterial divisions (**Ar**) are apparent. At the periphery of each classic lobule, **arrows** denote where distributing portal veins communicate with sinusoids (**S**). The relationship between the distributing portal vein and sinusoids is particularly evident in Fig. 3. The terminal arterial division, defined as either an arterial capillary or an arteriole, is a straight, slender vessel that supplies sinusoids at either the periphery or toward the center of the lobule (✽), as shown in Fig. 1. In some of the larger portal areas, arterial branches may also give rise to a peribiliary capillary plexus that surrounds the wall of individual bile ductules located in these portal areas. The capillaries of the plexus may connect with sinusoids at the periphery of adjacent lobules. Connections have also been observed with divisions of the portal vein in the immediate area. The peribiliary plexus can be considered as a mini-portal system of the liver, since it is actually a plexus of capillaries that drain into another system of capillaries (the sinusoids). Another important feature of the hepatic circulation is the extensive anastomosing of sinusoids, both within and between adjacent lobules. This allows blood to reach sinusoids not directly supplied by either an arterial capillary or a distributing portal vein. Many of the anastomosing capillaries end blindly at the periphery of each lobule, but they all flow into a central vein (**CV**).

Fig. 1, ×115; Fig. 2, ×125; Fig. 3, ×245

GLANDS OF DIGESTION
Liver

As the portal vein penetrates the liver capsule and enters the substance of the liver, it divides extensively into many branches. Each branch is surrounded by a sheath of connective tissue that is a continuation of the liver's external capsule. Also contained within the connective tissue sheath are branches of the hepatic artery, bile ductules, lymphatic vessels, and nerves. Such regions are commonly referred to as portal areas, portal tracts, portal radicals, or portal triads. These regions are easily distinguished from areas surrounding central veins and other branches of the hepatic vein, which do not possess a connective tissue sheath. Specific components of the portal area, which supply two or more adjacent lobules, are identified in both figures. They include branches of the portal vein (**PV**), hepatic artery (**HA**), and bile ductules (**BD**) of different sizes. The bile ductules are distinguished from blood vessels at higher magnification by the presence of microvilli and cilia that project into their lumens (see Fig. 2 on p. 195). A single plate of hepatocytes, called the limiting plate (**LP**), always surrounds each portal area. Many anastomosing sinusoids (**S**) between plates of hepatocytes (**Hc**) are present within the lobule. In the living state, these sinusoids are packed with blood cells, but most were removed during the preparation of tissue. In Fig. 1, a central vein (**CV**) is cut longitudinally, illustrating areas of continuity with sinusoids (**arrows**). Fig. 1 is an example of a cut surface; the hepatocyte surface is rough, conforming to the external shape of each cell. Fig. 2 is an example of a surface exposed by cryofracturing at cold temperatures; the hepatocyte surface here appears smooth because the fracture passes through each cell's interior.

Figs. 1, and 2, ×120

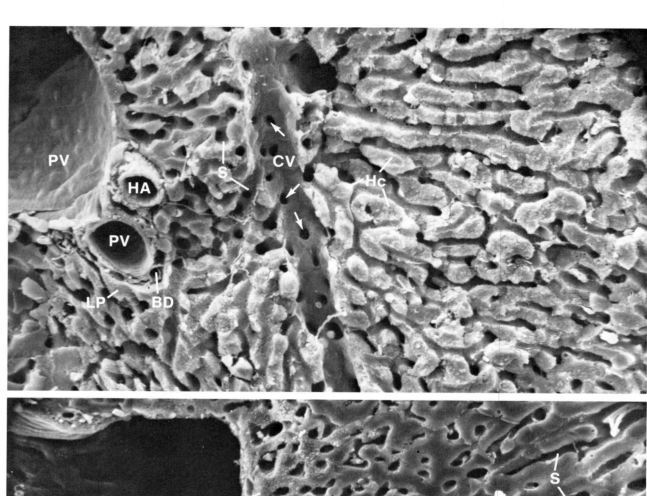

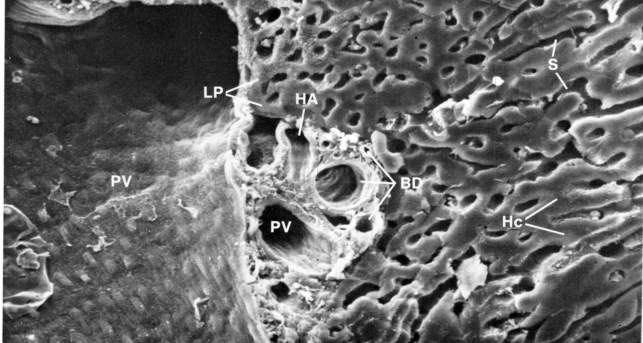

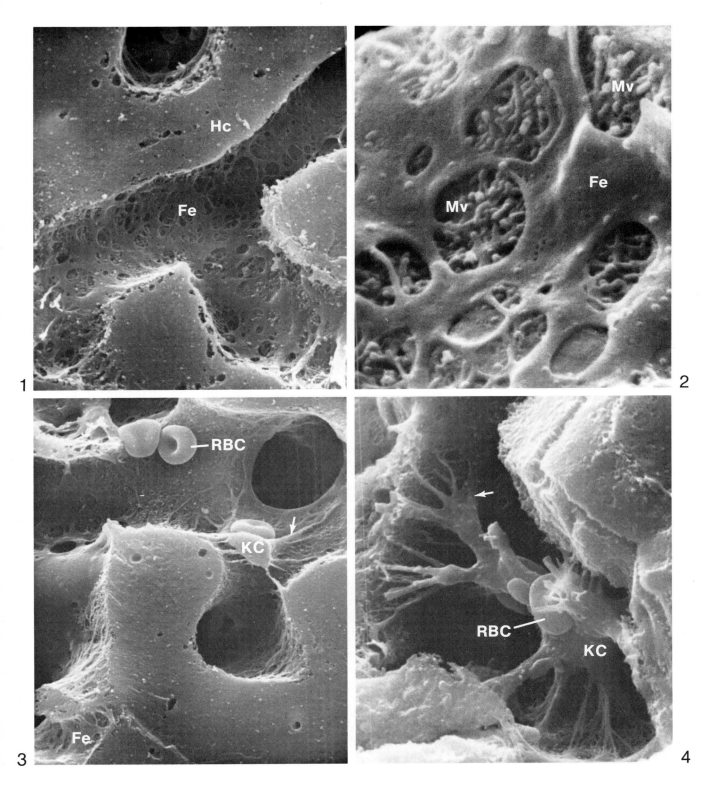

GLANDS OF DIGESTION
Liver

This series of figures illustrates some important features of the specialized liver capillaries that are referred to as sinusoids. The sinusoids are lined with a layer of porous endothelial cells. This fenestrated (**Fe**) lining possesses openings, or pores, that are highly variable in diameter. Since the sinusoids are incompletely lined, the surrounding hepatocytes (**Hc** in Fig. 1) have direct access to the fluid constituents of blood. Fig. 2, an enlargement of a portion of the fenestrated endothelium, shows how the microvilli (**Mv**) of the underlying hepatocyte surface abut on the fenestrated endothelium, allowing them to be seen through these pores. In addition to the endothelial cell, which forms the majority of the fenestrated lining, another cell type exists, as shown in Figs. 3 and 4. This cell, called the Kupffer cell (**KC**), is widely distributed within the hepatic sinusoids and is highly irregular in shape. Kupffer cell processes (**arrows**) may completely traverse the sinusoid. The cells are phagocytic (i.e., fixed macrophages) and are frequently located in regions where the sinusoids branch. Several red blood cells (**RBC**) still remain within the sinusoids in Figs. 3 and 4.

Fig. 1, ×3585; Fig. 2, ×12,725;
Fig. 3, ×2775; Fig. 4, ×4280

GLANDS OF DIGESTION
Liver

These figures illustrate how hepatocytes (**Hc**) are organized to form the small channels termed bile canaliculi (**BC**) into which they secrete bile. The manner in which hepatocytes fit together to form canaliculi was diagrammed in Fig. 2 on page 189. Branching of the canaliculi is denoted by **arrows** in all of these figures. The bile canaliculi, which are extracellular channels formed between apposing hepatocytes, are sealed off from the remainder of the extracellular space by tight junctions and are stabilized by desmosomes. Fig. 3 illustrates the location of these junctional complexes (**JC**). Microvilli (**Mv**), the finger-like projections of the hepatocyte surface, can be seen extending into the lumen of the canaliculus. Note that a small extracellular space, termed the space of Disse (**SD**), is present between the hepatocyte surface and the endothelial lining (**EL**) of the sinusoid (**S**). One of the pores (✱) of the endothelial lining is seen to open into the space of Disse. Exchanges between the hepatocyte surface and blood occur through the pores via the space of Disse. The hepatocyte surface projecting into the space of Disse also forms microvilli, increasing the surface area over which exchanges can occur. Also found within the space of Disse is a network of reticular fibers (**RF**) that form a supporting mesh of connective tissue between the hepatocytes and sinusoid. It is important to realize that the space of Disse is more of a potential space; under physiological conditions it is quite small, almost nonexistent. If tissue fluid does accumulate within it, the fluid can drain toward the portal area passing into a surrounding space called the space of Mall (see Fig. 1 on p. 189). From here it may diffuse into lymphatic vessels originating within the portal area and become lymph. [Fig. 2 courtesy of R. G. Kessel and C. Y. Shih, *Scanning Electron Microscopy in Biology,* Springer-Verlag, 1974.]

Fig. 1, ×1350; Fig. 2, ×2255; Fig. 3, ×14,170

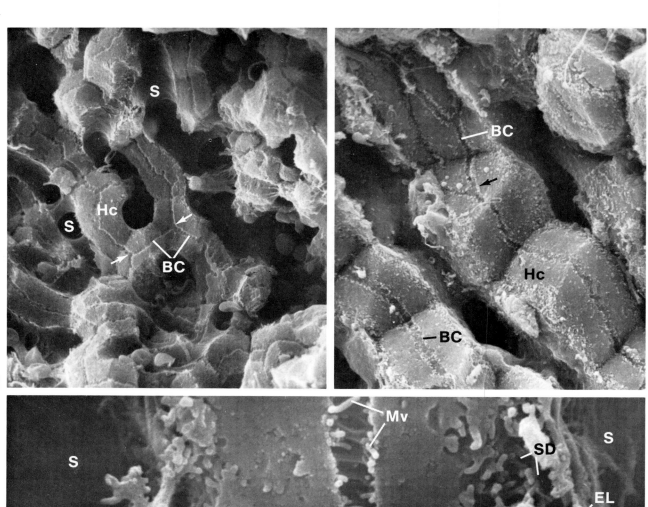

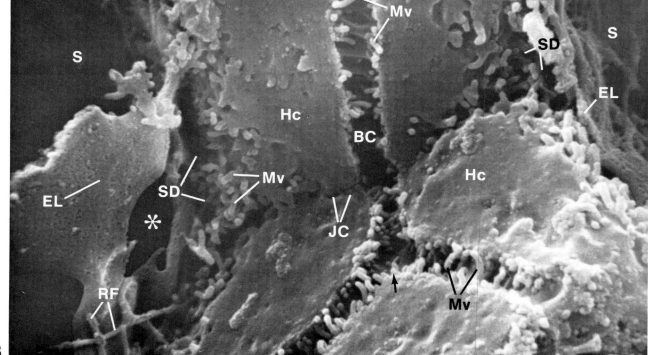

194

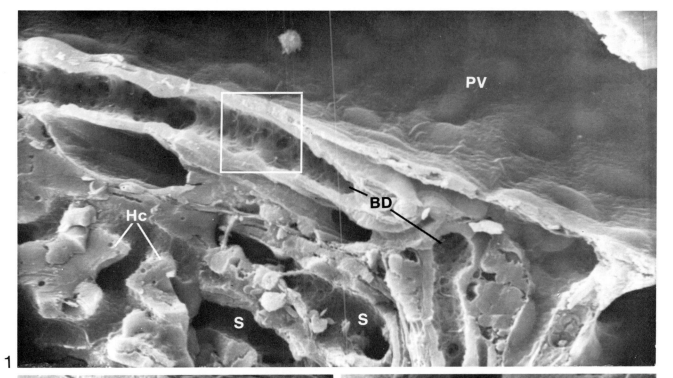

1

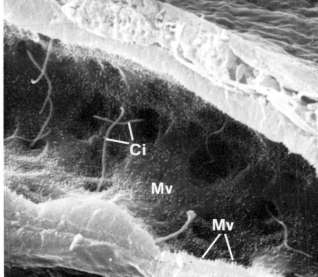

2

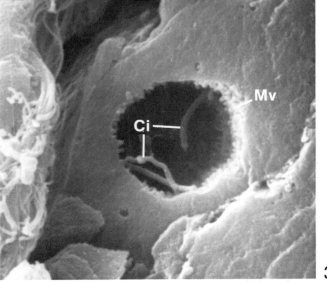

3

GLANDS OF DIGESTION
Liver

A portion of a portal area at the margin of a hepatic lobule is illustrated in Fig. 1. The portal vein (**PV**), bile ductule (**BD**), sinusoids (**S**), and plates of hepatocytes (**Hc**) are identified. Longitudinal and transverse sections of bile ductules are illustrated in Figs. 2 and 3, respectively. The longitudinal section in Fig. 2 is an enlargement of the square outlined in Fig. 1. Usually, the luminal surface of the simple epithelium of the ductules possesses numerous, but short microvilli (**Mv**). In addition, the cells appear to possess a single, long cilium (**Ci**) that projects into the lumen. These cilia may serve to assist in the movement of bile through the ductules. Bile is secreted by the hepatocytes into the canaliculi, and the movement of the bile within the canaliculi of the lobule is in a direction opposite the flow of blood within the sinusoids. Bile in the canaliculi drains into bile ductules of different sizes in the portal areas and eventually leaves the liver via the hepatic bile duct.

Fig. 1, ×975; Fig. 2, ×3905; Fig. 3, ×9460

GLANDS OF DIGESTION
Liver

The transmission electron micrograph shown in Fig. 1 illustrates many of the cellular constituents and intracellular contents of liver tissue. Aspects of the intracellular organization of hepatocytes (**Hc**) can also be observed in the scanning electron micrograph shown in Fig. 2. Most of the cells in Fig. 1 are hepatocytes (**Hc**) arranged in the form of interconnecting plates, as previously described. The boundaries of one hepatocyte are outlined, and many bile canaliculi (**BC**) formed by adjoining cells are illustrated in both transverse and longitudinal sections. Identified here are some major intracellular constituents of the hepatocyte: the nucleus (**N**), mitochondria (**M**), endoplasmic reticulum (**ER**), and accumulations of lipid (**L**). Although the nucleus can be positively identified in Fig. 2, the identity of the other constituents is less certain. Most of the small rounded structures may be mitochondria. Cells other than hepatocytes are identified in Fig. 1. Endothelial cells (**EC**) bulge into the sinusoids (**S**) in areas overlying the location of their nuclei; the remaining cytoplasm forms the thin fenestrated lining of the sinusoid. Kupffer cells (**KC**), which also form part of the sinusoidal lining, often contain lysosome-like inclusions (**arrows**). A third type of cell, termed the fat-storing cell (**FS**), does not participate in the actual lining of the sinusoid, but is associated with the space of Disse. Lipid droplets are present in the cytoplasm of fat-storing cells, but differ from those found within hepatocytes; they are enclosed by a membrane and are thought to contain vitamin A. Although blood was removed during tissue preparation, an occasional red blood cell (**RBC**) is present in the sinusoids. [Fig. 1 courtesy of E. Weibel. From A. Blouin, R. P. Bolender, and E. Weibel, Distribution of organelles and membranes between hepatocytes and nonhepatocytes in the rat liver parenchyma, *Journal of Cell Biology* 72(2):445, (1977). The Rockefeller University Press.]

Fig. 1, ×2000; Fig. 2, ×1515

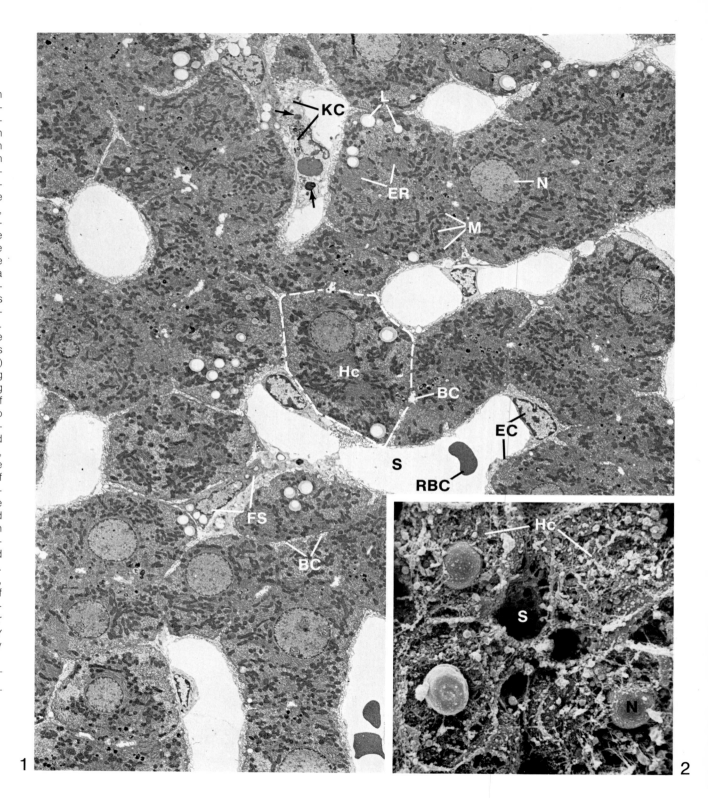

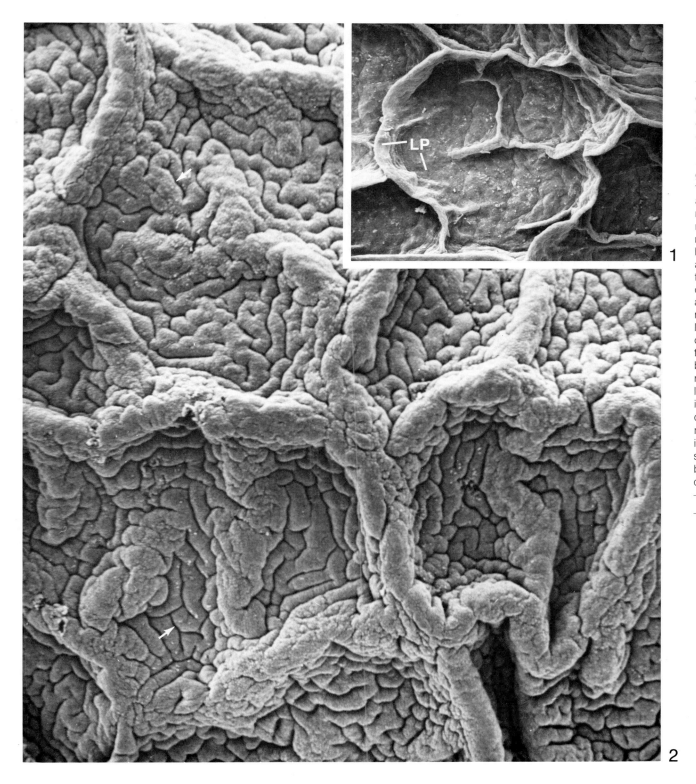

GLANDS OF DIGESTION
Gallbladder

The gallbladder is a pear-shaped sac located on the inferior surface of the liver. Bile produced by the liver may be shunted to the gallbladder, where it can be stored and concentrated before it is emptied into the duodenal portion of the small intestine. The topography of the luminal surface of the rabbit gallbladder is revealed in the low-magnification micrograph shown in Fig. 2. The most apparent feature is the folded gallbladder mucosa, which consists of a sheet of columnar epithelial cells anchored by a basement membrane to an underlying connective tissue layer. When the epithelial layer is removed, an underlying pattern of connective tissue folds is exposed, as shown in Fig. 1. This connective tissue, the lamina propria (**LP**), appears smooth because the basement membrane was not removed with the epithelium and still covers the connective tissue. By comparing Figs. 1 and 2, it becomes apparent that the folded surface of the mucosa is brought about mainly by a folding of the connective tissue layer. The epithelial layer follows these contours and, in addition, folds independently of the lamina propria to produce minor creases (**arrows** in Fig. 2). Beneath the folded mucosa of the gallbladder lie interlacing sheets of smooth muscle. When stimulated, the muscle sheets within the gallbladder wall contract, emptying bile into a duct system leading to the small intestine.

Fig. 1, ×100; Fig. 2, ×145

197

GLANDS OF DIGESTION
Gallbladder

This enlargement of the folded surface seen in Fig. 2 on the previous page illustrates the contours of the mucosal epithelial layer. The apical borders of the individual cells that form the layer are easily distinguished. Numerous creases (**arrows**) in the epithelial sheet are seen in the valleys among the interconnecting folds. The extensive folding probably functions to increase the absorptive surface area in contact with bile, thus increasing the efficiency by which the gallbladder can concentrate bile.

×350

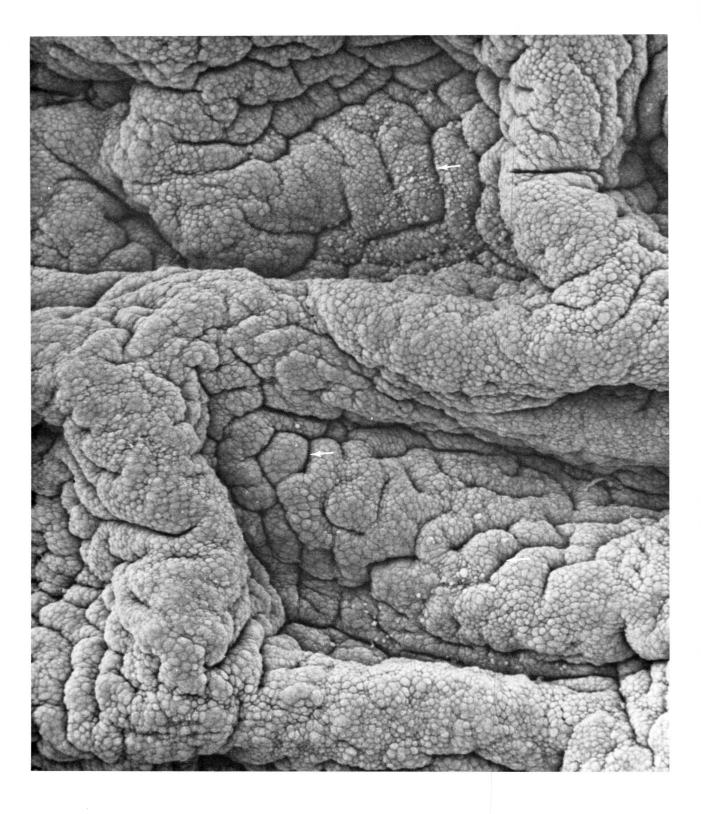

GLANDS OF DIGESTION
Gallbladder

The relationship of the epithelial sheet (**ES**) to its underlying tissue components is best illustrated when a portion of the epithelium is removed. Exposed in this figure is an underlying fold of the lamina propria (**LP**), covered with a smooth basement membrane. Also visible are the borders of each cell forming the epithelial sheets. From this view it is easy to see why this type of epithelial cell (**EC**) is termed columnar. It is important to realize that the connective tissue beneath the epithelium is richly vascularized, because it is here that exchange of materials takes place. The epithelial cells can transport material from the bile across their surface to underlying blood vessels. Conversely, vital nutrients and oxygen in the blood can diffuse outward to supply the overlying epithelium.

×695

199

GLANDS OF DIGESTION
Gallbladder

The columnar epithelial cells of the gallbladder are shown from a lateral view in Fig. 1. Many folds (**Fo**), or ridges, are associated with the plasma membrane of the lateral cell surface, whereas microvilli (**Mv**) are associated with the apical surface. Figs. 2 and 3 reveal details of these surface specializations at higher magnification. Viewed from the apical surface (Fig. 2), the borders (**arrows**) between epithelial cells are apparent. Many short microvilli populate the apical cell surface. Each cell is thought to pump ions through their lateral cell membranes and into the intercellular space. This creates an osmotic gradient that tends to draw water from the bile into the intercellular space, and in this way, bile within the lumen of the gallbladder becomes more concentrated. The epithelial cells are connected together at their apices by bands of junctional complexes. The approximate location of the junctional complex (**JC**) was exposed in Fig. 3 when adjacent cells were separated at their lateral borders. The banded area corresponds to the location of the zonula occludens and zonula adherens, two components of the junctional complex that can be identified in transmission electron micrographs. The zonula occludens serves as a seal, preventing the leakage of bile constituents from the lumen into the intercellular space.

Fig. 1, ×3895; Fig. 2, ×9970; Fig. 3, ×19,470

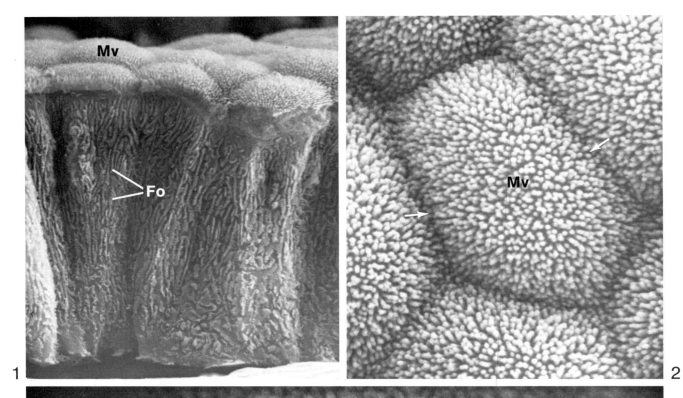

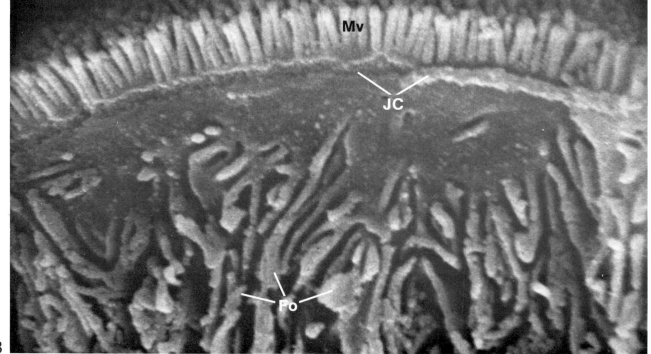

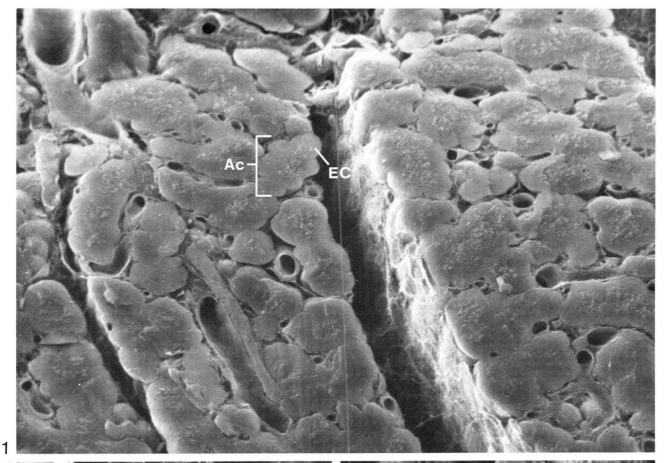

GLANDS OF DIGESTION
Pancreas

The pancreas can be divided into endocrine and exocrine portions. The endocrine portion consists of small aggregations of cells, the islets of Langerhans, which secrete hormones such as insulin, glucagon, and somatostatin into the portal venous system. The glandular epithelial cells (**EC**) of the exocrine pancreas (Figs. 1 to 3) are organized into acini (**Ac**), which convey their secretion into a branching system of ducts. The parenchyma of the gland is divided into lobules by thin connective tissue septa that contain ducts, blood vessels, nerves, and lymphatic vessels. Each acinus is also closely surrounded by a supportive meshwork of connective tissue (**CT**). Adjacent acinar cells are joined by junctional complexes, and their secretion is released at the apical pole of the cell. The secretory product of the acinar cells consists of proteolytic, lipolytic, and glycolytic enzymes that, in their activated form, aid in the digestion of food in the duodenum. Upon stimulus by the intestinal hormone pancreozymin, the cells that form the acinar unit express their secretory product into small intercalated ducts that are lined by a single layer of cells ranging in shape from squamous to cuboidal. The intercalated ducts communicate with larger ducts in the lobule (intralobular), and these lead into interlobular ducts located within the septa. The height of the cells lining these ducts varies with the diameter of the duct lumen. By way of the branching duct system, the secretion is eventually emptied into the duodenum. Secretin, a hormone released by endocrine cells along the gastrointestinal tract, causes the cells lining the ducts to increase the transport of electrolytes, bicarbonate, and water into the duct lumen.

Fig. 1, ×600; Fig. 2, ×515; Fig. 3, ×1110

1

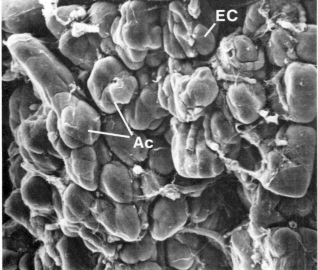

2

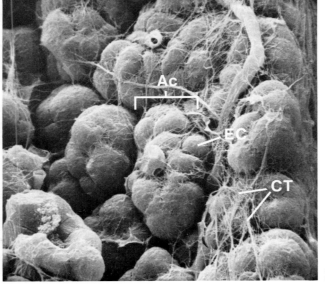

3

GLANDS OF DIGESTION
Pancreas

This transmission electron micrograph of acinar cells illustrates many of the intracellular constituents essential for the synthesis, packaging, and release of digestive enzymes. Two nuclei (**Nu**) of adjacent cells are characterized by areas in which electron-dense heterochromatin (**He**) is concentrated around the periphery of the nucleus. A prominent nucleolus (**Ne**) occupies an eccentric position within each nucleus. The majority of each cell's cytoplasm is occupied by stacked cisternae of rough-surfaced endoplasmic reticulum (**ER**), where the translation of messenger RNA into secretory proteins takes place. After nascent polypeptides have accumulated in the cisternae, they are transported to the supranuclear Golgi apparatus via small membranous transition vesicles. The major function of the Golgi apparatus in the pancreas is to package the proteinaceous secretory product into condensing vacuoles. During transformation of the condensing vacuoles into mature zymogen granules (**ZG**), water is removed. In those cells that produce a glycoprotein secretion, amino acid side chains of the protein are glycosylated in the Golgi apparatus in an orderly series of enzymatic steps. The electron-dense zymogen granules accumulate toward the apical cell membrane. Upon stimulation, the granule membrane fuses with the plasma membrane of the cell. Thus the secretion is discharged by exocytosis into the acinar lumen (**Lu**), which is continuous with the intercalated duct. This well-characterized secretory pathway is common to many forms of glandular epithelium. Most of the steps involved in the synthesis and release of protein are energy dependent, and most of the energy is provided by elongated mitochondria (**Mi**).

×9700

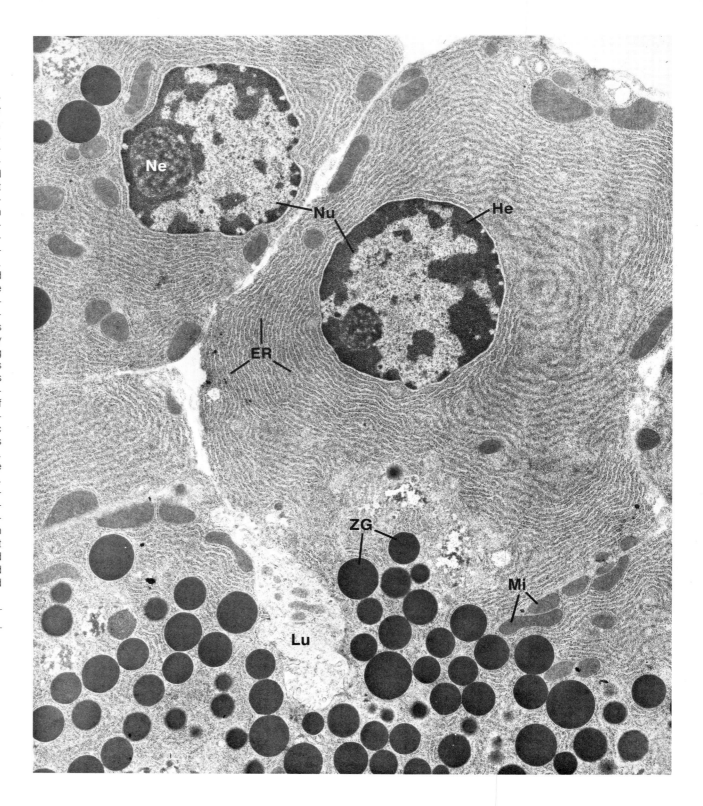

THE RESPIRATORY SYSTEM

Introduction

The respiratory system consists of three functional components: a conducting system of airways, a respiratory area in which exchange of gases takes place between blood and air, and a ventilating mechanism that provides for inspiratory and expiratory movements. The conducting portion consists, in succession, of the nasal cavities and their associated sinuses, the nasopharynx, the larynx, the trachea, and the bronchial airways, which collectively provide a conduit for the inhaled and exhaled air to enter and exit the respiratory area. The walls of the conducting passageways consist partly of cartilage and/or bone, which prevent their collapse during ventilation. The inner surface of the conducting airways is lined with a mucous membrane consisting of pseudostratified, ciliated columnar epithelium and an underlying connective tissue layer, the lamina propria. The epithelium also contains mucous-secreting cells, brush cells with afferent nerve innervation, and endocrine-like cells called small-granule cells. The widespread presence of ciliated epithelial cells, mucous cells, glands, and blood vessels in the walls of the conducting part of the respiratory system renders this portion capable of warming, humidifying, and filtering inspired air. Warming and humidifying take place mainly in the nasal cavities, which have large venous sinuses and numerous capillary loops that are distributed close to the mucosal surface. It is thought that the direction of air inflow across the mucosal surface is opposite that of the blood flow, so that countercurrent heat exchange can take place between air and blood. The venous plexus consists of thin-walled vessels of large diameter, resembling erectile tissue. Autonomic nerve stimulation causes constriction of the deeper veins and dilation of the arterioles that supply the capillaries leading into the venous plexus. The net effect is engorgement of the venous sinuses with blood, causing them to swell. For this reason, the venous sinuses have been collectively referred to as swell bodies. The vasoactivity of the vessels supplying and draining the venous sinuses is regulated so that hourly periods of swelling occur on alternate sides of the nasal cavity. On the side engorged with blood, the airflow is retarded and the air is preferentially deflected upward toward the olfactory mucosa, causing an increased perception of odoriferous molecules on that side. The olfactory regions consist of specialized epithelium, which is present on the roof of the nasal cavities and also covers a portion of the nasal septum and superior chonchae (see Chapter 6). The mucous secretion present on the surface of the respiratory mucosa serves to trap larger dust particles and clear the air of harmful gases, such as ozone and sulfur dioxide. Such pollutant gases are dissolved in the mucous secretion and become absorbed or moved to the pharynx by ciliary action.

The trachea constitutes a major portion of the conducting region of the respiratory system, and it branches to form right and left primary bronchi that enter the hilus of each lung. The hilus is the region where the primary bronchus, pulmonary artery, pulmonary vein, nerves, and lymphatics enter and leave the lung. Secondary divisions of the bronchi, connective tissue septae, and fissures on the surface of the lung divide it into lobes; the right lung has three lobes, whereas the left lung has two. The third- and fourth-order bronchial divisions further subdivide each lobe into bronchopulmonary segments, and successive branching gives rise to subsegments. Distally, the bronchi continue to become reduced in diameter, smooth muscle becomes a prominent constituent of the wall, and only a small amount of cartilage, if any, is present. At this point the conducting airways are lined with cuboidal epithelium and are classified as bronchioles, measuring 0.5–1.0 mm in diameter. Terminal bronchioles and the respiratory region of pulmonary tissues that they supply constitute a pulmonary lobule. As the terminal bronchioles ramify, they give rise to successive divisions, which constitute the respiratory portion of the lung. The divisions include the respiratory bronchioles, alveolar ducts, alveolar sacs, and alveoli, all of which possess an extremely thin epithelial lining that allows for the exchange of gas between air and blood.

Ventilation of the lungs involves the thoracic cage and its associated intercostal muscles, the muscular diaphragm, and the elastic connective tissue of the lungs. Contraction of the intercostal muscles and diaphragm increase the volume of the thoracic cage, causing expansion of the lungs and inspiration. During expiration, the muscles relax and the passive recoil of the formerly stretched elastic connective tissue fibers causes the lungs to decrease in volume.

THE RESPIRATORY SYSTEM
Nasal Septum

The nasal septum, illustrated in transverse section in Fig. 1, divides the nose into right and left nasal cavities. The septum consists of a central plate of cartilage (**Ca**) or bone, and it is covered by a mucous membrane (**MM**), which in this preparation was detached from the cartilage during sectioning. The mucous membrane consists of a respiratory epithelium (**Ep**) that resides on a highly vascular connective tissue lamina propria (**LP**). In addition to supplying the cellular elements of the mucous membrane with nutrients, blood contained within the large vessels (**BV**) serves to warm or cool the air passing over the epithelial surface. Three curved bony plates, termed the superior, middle, and inferior nasal conchae, project into the nasal cavities. These plates are covered either by a mucous membrane similar to that associated with the nasal septum or by an olfactory epithelium (see Chapter 6). The lamina propria of the respiratory epithelium covering the conchae also contains numerous blood vessels, as illustrated by the replica of the blood vessels (**BV**) from this region shown in Fig. 2. All cellular and fibrous elements of the mucous membrane were removed by digestion during preparation of the replica, but an underlying piece of bone (**Bo**) of the concha is present. The respiratory epithelium consists (Fig. 3) of pseudostratified columnar epithelium (**CE**), which has numerous cilia (**Ci**) associated with the free surfaces of the cells. Mucus (**Mu**), which is visible on some of the cell surfaces in Fig. 3, is synthesized and secreted by goblet cells (unicellular mucous glands), which are also present in the epithelial sheet. Secretion from the goblet cells and glands in the underlying lamina propria serves to humidfy the air, to entrap particles in the nasal passageways, and to inactivate certain bacteria.

Fig. 1, ×175; Fig. 2, ×30; Fig. 3, ×2420

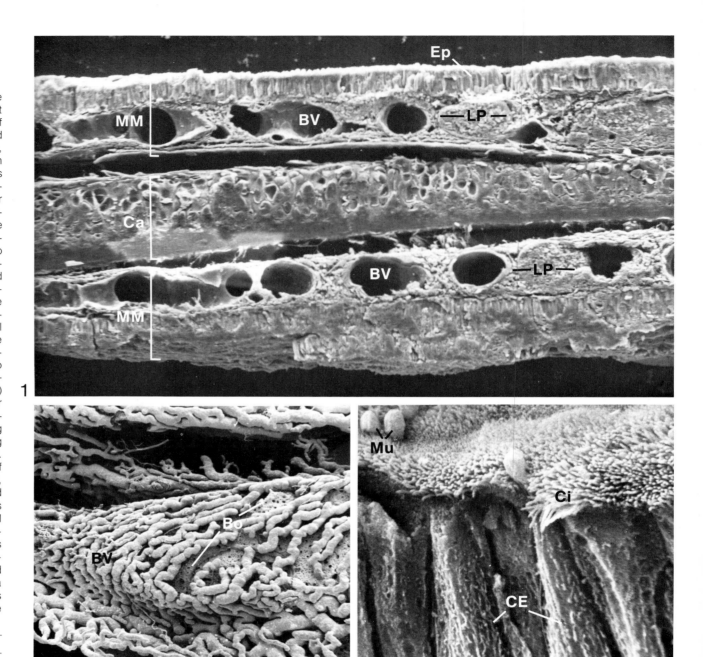

THE RESPIRATORY SYSTEM
Nasal Cavity

In a representative area of the surface of the rabbit nasal epithelium (Fig. 1), cells exhibiting three major morphological variations can be seen. Ciliated cells (**CC**) are widely distributed in the epithelium; their cilia range from 5 to 6 μm in length. Another cell type (Figs. 1 and 2), perhaps representing a progenitor cell (**PC**), possesses cilia (**Ci**) that are shorter but highly variable in length. Short microvilli (**arrows**) are also interposed between the cilia of these cells, and are similar in size and shape to the microvilli (**Mv**) of surrounding, nonciliated cells. The variation in ciliary length is suggestive of a progenitor cell in which the cilia are undergoing morphogenesis. Nonciliated cells (**NC**) of the respiratory epithelium are polyhedral in shape and variable in size (Fig. 1). In many cases, these cells are uniformly populated with numerous, short microvilli (**Mv**) like those illustrated in Figs. 2 and 3. In Fig. 3, however, the apical surface of some nonciliated cells exhibits variation in the distribution and length of microvilli (**Mv**). Such cells are often larger and project further into the nasal cavity than adjacent cells. This condition may reflect the intracellular accummulation of secretory product. The boundaries between the ciliated and nonciliated epithelial cells are clearly distinguished in all the scanning electron micrographs.

Fig. 1, ×6205; Fig. 2, ×7830; Fig. 3, ×7120

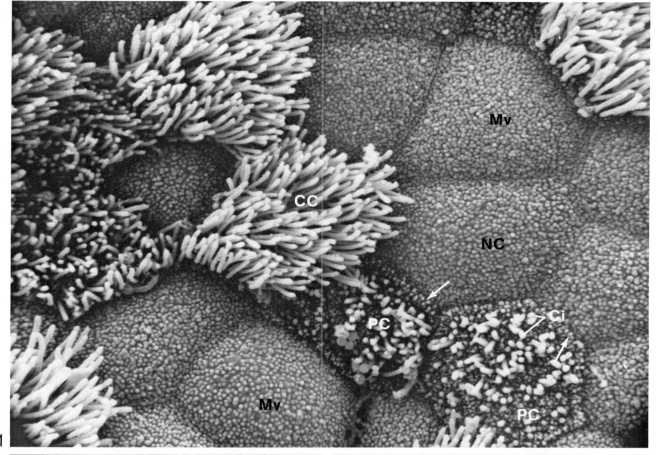

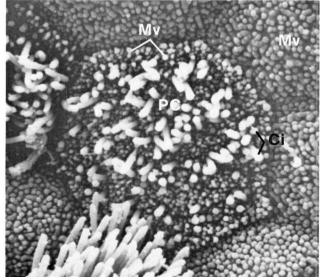

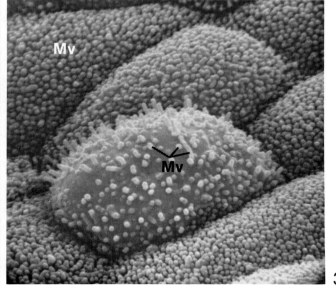

THE RESPIRATORY SYSTEM
Nasal Cavity

The nasal cavity is lined by a mucous membrane, or mucosa, that is functionally divided into two major areas: a respiratory region, and a region specialized for smell, termed the olfactory region. The surface of both the olfactory area and the adjacent respiratory region is illustrated in Fig. 1. The respiratory epithelial cells appear to be organized into alternating rows, as viewed from the surface. The surface of the olfactory epithelium contains numerous thin strands (✳) representing highly modified cilia. Further structural details of the olfactory epithelium are presented in Chapter 6. Secretory droplets (**SD** in Fig. 1) adhere to the surfaces of both the olfactory and the respiratory epithelium. Glands are present in the connective tissue beneath the epithelium of both the respiratory and the olfactory areas of the nasal cavities. The secretory products produced by these tubuloalveolar glands may be serous, mucous, or both types; they are conveyed by short ducts onto the epithelial surface. The orifices (**arrows**) of a number of ducts are identified in Fig. 1. The respiratory epithelium is illustrated in surface and lateral views in Figs. 2 and 3, respectively. The apical cilia (**Ci**) of epithelial cells form rows that alternate with rows of microvilli (**Mv**) from the nonciliated cells. Secretion (**Se**) from goblet cells protrudes above the epithelial surface, and the small openings in the epithelium may represent areas where such secretion has recently been liberated. A single goblet cell (**GC**) is shown in Fig. 3 from the exposed lateral surface of this region of respiratory epithelium. Cytoplasmic projections extend from the lateral margins of the goblet cell to the adjacent epithelial cells.

Fig. 1, ×865; Fig. 2, ×3585; Fig. 3, ×4040

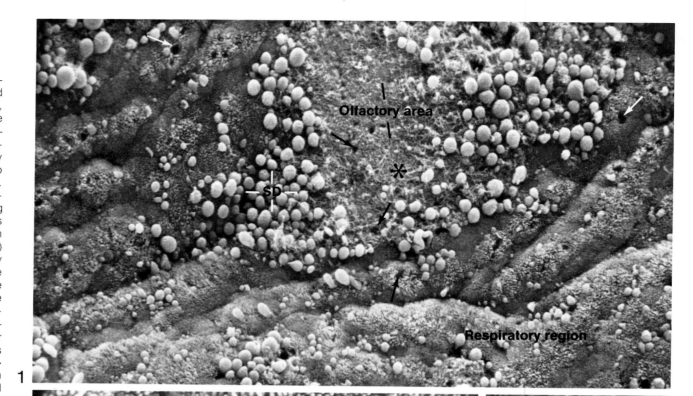

1

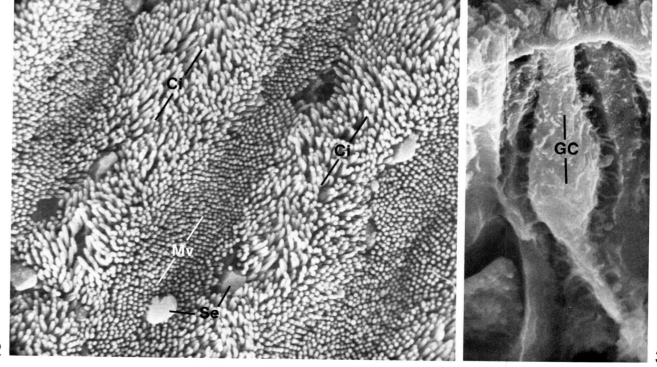

2

3

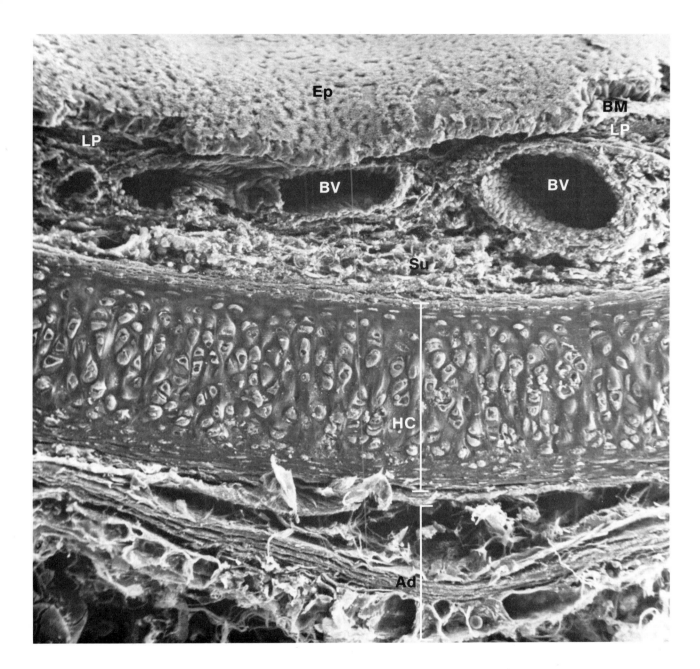

THE RESPIRATORY SYSTEM
Trachea

The trachea is a tubular structure that extends from the larynx and terminates by branching into two extrapulmonary bronchi. The trachea serves not only to conduct air to the lungs, but also to filter or strain the air of foreign particles. This figure illustrates a section through the wall of the trachea. The epithelial (**Ep**) layer and thin connective tissue lamina propria (**LP**) constitute the mucous membrane, or mucosa. A basement membrane (**BM**) between the epithelium and lamina propria is exposed in a small portion of the illustration. Numerous elastic fibers are present in the thin lamina propria. The mucosa is surrounded by a connective tissue submucosa (**Su**), which contains numerous blood vessels (**BV**) and mucous, or seromucous, glands. The outermost layer of connective tissue constitutes the adventitia (**Ad**). Located between the submucosa and the adventitia is a C-shaped plate of hyaline cartilage (**HC**), which is surrounded by a thin fibrocellular sheet called the perichondrium. The incomplete cartilage plates in the trachea are connected by smooth muscle.

×300

THE RESPIRATORY SYSTEM
Trachea

The figures on this page illustrate the epithelial surface of the trachea. The ciliated cells contain more than 200 cilia (**Ci**) per cell, each of which measures several micrometers in length. The ciliated cells also possess microvilli (**arrows** in Fig. 3), but they are usually obscured from view. The effective stroke of the ciliary beat cycle serves to move the mucous layer and any entrapped particulate material toward the larynx. The apical surfaces of goblet cells (**GC**) readily distinguish them from the ciliated cells. The goblet cell surface frequently has a peripheral area covered with short microvilli (**Mv**) and a central portion devoid of surface projections. Local environmental factors, such as temperature or humidity, may stimulate the goblet cells to release mucus onto the epithelial surface. Goblet cells that have released their contents are commonly referred to as brush cells. They contain little, if any, mucigen and possess a more prominent microvillous border than goblet cells, similar to that illustrated in Fig. 3. Although it might appear that brush cells are merely goblet cells in a particular state of secretory activity, it is known that a number of brush cells are innervated by afferent nerve endings. Consequently, an innervated brush cell may represent a type of sensory receptor. Another type of epithelial cell also exists within the conducting airways, but is less numerous than the other cell types. This cell, called the small-granule cell, is characterized by the presence of small granules within its cytoplasm and will be discussed in association with the bronchus.

Fig. 1, ×1310; Fig. 2, ×4670; Fig. 3, ×4110

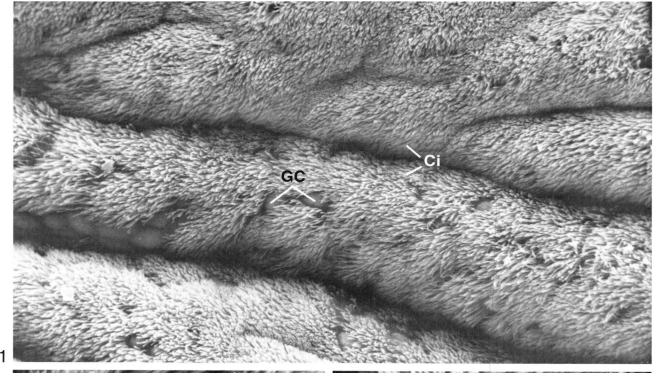

1

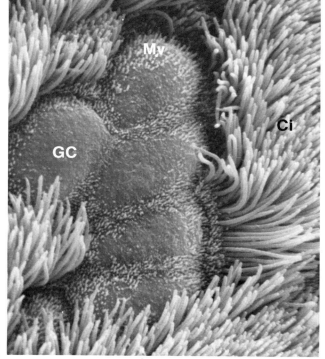

2

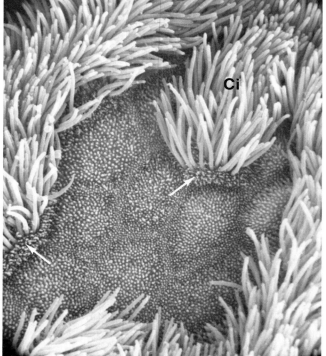

3

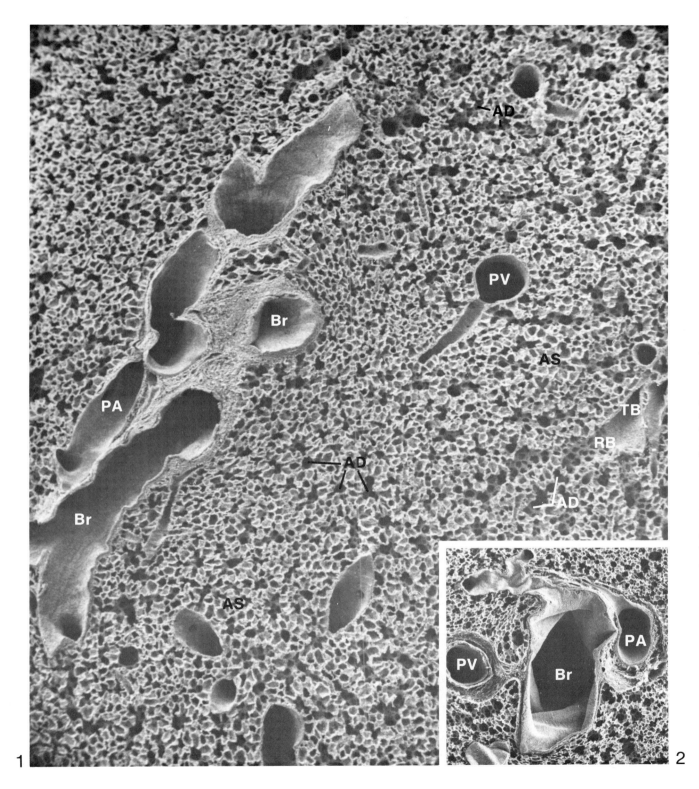

THE RESPIRATORY SYSTEM
Lung—Organization

As can be seen in Fig. 1, the spongy appearance of the lung is due to the presence of a multitude of thin-walled air sacs (**AS**) or alveoli. Alveoli share extremely thin partitions with other alveoli that are immediately adjacent to them. Exchange of gases occurs between air within the alveoli and blood in the surrounding extensive capillary networks. Air is conducted to the alveoli through a highly branched system of tubes that make up the bronchial tree. Divisions of the bronchial tree present in this section include bronchi (**Br**) as well as bronchioles and their branches. The continuity between a terminal bronchiole (**TB**), respiratory bronchiole (**RB**), and alveolar duct (**AD**) is denoted in Fig. 1. Alveolar ducts (**AD**) that are sectioned in different planes are identified in a number of areas in Fig. 1. Components of the circulatory system evident in this section include branches of the pulmonary artery (**PA**) and the pulmonary vein (**PV**). The pulmonary artery subdivides to accompany branches of the bronchi and bronchioles to the level of the respiratory bronchiole. As branches of the pulmonary vein originate from the capillaries surrounding the numerous alveoli, they initially pass through the substance of the lung at some distance from the branches of the bronchial tree and accompanying pulmonary artery. Larger branches of the pulmonary vein (**PV**), however, become closely associated with large bronchi (**Br**) and with the pulmonary artery (**PA**) near the hilus of the lung (Fig. 2). Unless otherwise indicated, all illustrations are of the rat lung.

Fig. 1, ×45; Fig. 2, ×20

THE RESPIRATORY SYSTEM
Lung—Bronchus

A transverse section of a bronchus (**Br**) and accompanying pulmonary artery (**PA**) is illustrated in Fig. 1. The epithelial layer (**EL**) and underlying lamina propria (**LP**) are identified. Contraction of the smooth muscle (**SM**) in the bronchial wall produces a scalloped appearance of the mucosa. The epithelium is usually pseudostratified ciliated columnar with mucous-secreting cells. Irregularly shaped cartilage plates are frequently embedded in the connective tissue lamina propria, but they are not clearly distinguished in the plane of section shown here. In addition, mucous or seromucous glands as well as numerous blood vessels (**BV**) may be present in the lamina propria. The bronchus in Fig. 1 is of the intrapulmonary type; many alveoli (**AI**) surround it. Lateral and surface views of the ciliated epithelium are shown in Figs. 2 and 3, respectively. The columnar cells reside basally on a basement membrane, and cilia (**Ci**) extend from the apical surface of the cells. A small amount of connective tissue belonging to the lamina propria (**LP**) is present in the lower portion of Fig. 2. The rounded apical surfaces of a number of goblet cells (**GC**) appear among the cilia, and short microvilli may be present on their free surface. Narrower bronchial divisions are lined with simple columnar epithelium. Small-granule cells are also present within the epithelium at various locations along the conducting airways. The small-granule cells occupy a basal position within the epithelium and include two general classes: neurosecretory cells and polypeptide-hormone-secreting cells. The neurosecretory cells are innervated and often occur in small clusters. The polypeptide-hormone-secreting cells occur singly and are presumably endocrine. Both classes of small-granule cells can take up and store amine percursors in cytoplasmic granules; the neurosecretory cell is thought to synthesize 5-hydroxytryptamine (serotonin). The small-granule cells are thought to discharge their contents upon receiving neuronal stimulation or perhaps in response to a changing oxygen tension, acting as chemoreceptors. The granules released at the cell base may enter the local circulation and possibly affect airway diameter or vascular resistance.

Fig. 1, ×240; Fig. 2, ×2360; Fig. 3, ×2865

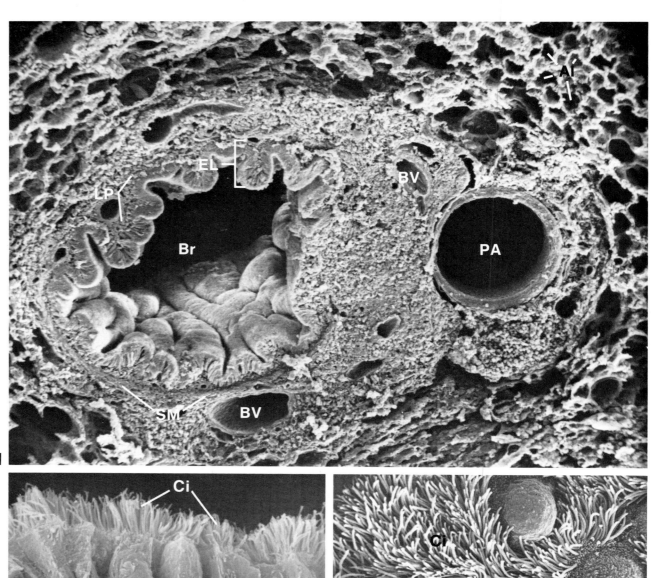

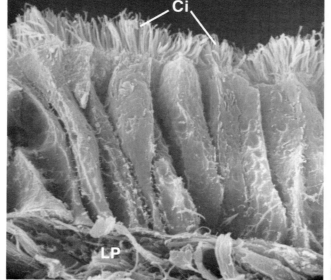

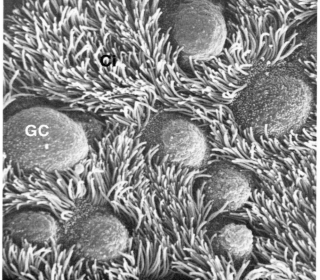

210

THE RESPIRATORY SYSTEM
Lung—Bronchiole

Bronchioles are extensions of bronchi, and they vary in diameter from approximately 0.5 to 1.0 mm. Bronchioles lack cartilage plates, and the smaller divisions are devoid of glands. Smooth muscle bands in the walls of bronchioles form a muscularis mucosa layer. There are reportedly some 20 generations of bronchioles, the smallest of which are called terminal bronchioles. Each terminal bronchiole may branch into several respiratory bronchioles; the latter are short and may be absent in man. Unlike terminal bronchioles, respiratory bronchioles may exhibit alveolar outpocketing along their entire length. Fig. 1 illustrates a region of transition between terminal bronchiole (**TB**), respiratory bronchioles (**RB**), and alveolar ducts (**AD**). A pulmonary arteriole (**PA**) is located close to the terminal bronchiole. Areas at which a terminal bronchiole branches into respiratory bronchioles mark the beginning of a primary pulmonary lobule. The primary pulmonary lobule, therefore, includes a respiratory bronchiole, alveolar duct, alveolar sacs, and alveoli (**AI**). The epithelium of bronchioles may vary from simple columnar to simple cuboidal, and has both ciliated and nonciliated cells. The apical epithelial cell surface of a bronchiole is illustrated in Fig. 2. The cilia (**Ci**) are visible, as are the dome-shaped protrusions of the nonciliated Clara cells (**CC**). It has been suggested that Clara cells may secrete a component (hypophase) of the surfactant material that covers the alveoli. An overview of the terminal divisions of the branched bronchial tree is provided in Fig. 3. Note the continuity between terminal bronchioles (**TB**), respiratory bronchioles (**RB**), and alveolar ducts (**AD**).

Fig. 1, ×265; Fig. 2, ×1245; Fig. 3, ×50

THE RESPIRATORY SYSTEM
Lung—Alveolar Ducts, Alveoli

Respiratory bronchioles divide into 2–11 alveolar ducts. When sectioned longitudinally, as in Fig. 1, the alveolar ducts (**AD**) appear as tubular conduits. Many alveolar sacs or alveoli (**AI**), arranged side by side, are continuous with the alveolar duct along its length. Around their circumference, alveolar ducts may open directly into single alveoli or into clusters of alveoli termed alveolar sacs. Alveolar sacs frequently occur at the terminations of the alveolar ducts. An alveolar sac (**AS**) with four apparent subsidiary alveoli is illustrated in Fig. 2. Since millions of alveoli are present in each lung, the thin walls of several alveoli (**AI**) may be closely apposed and share the intervening capillary bed and connective tissue elements (interstitium). An interalveolar septum thus includes portions of adjacent alveolar walls, together with the capillaries and connective tissue elements located between them. In Fig. 2, the interalveolar septa appear porous (**arrows**). These openings are sectioned capillaries from which blood was perfused during preparation. The extensive capillary bed associated with the respiratory alveoli is well illustrated in Fig. 3. This preparation illustrates a replica (cast) of the alveolar capillaries. All cellular elements of the lung and vessels were digested away in this preparation to illustrate in some depth the extensive capillary system associated with the alveoli. The major respiratory function of the lung is performed in this extensive capillary bed of the pulmonary alveoli.

Fig. 1, ×160; Fig. 2, ×615; Fig. 3, ×270

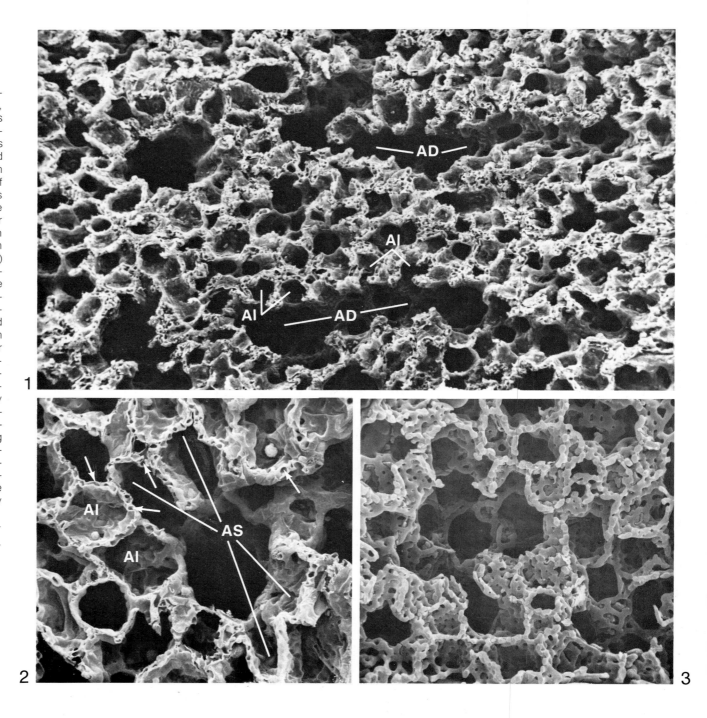

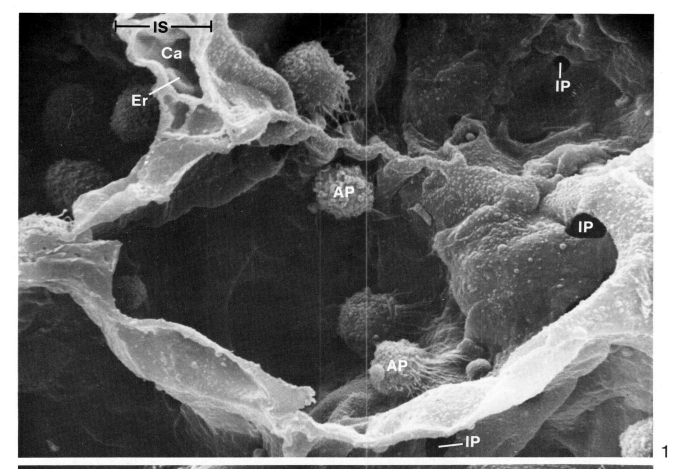

THE RESPIRATORY SYSTEM
Lung—Alveoli

Fig. 1 illustrates portions of adjacent alveoli. Small openings, called interalveolar pores (**IP**), provide for continuity between adjacent alveoli and traverse the interalveolar septum through interspaces of the alveolar capillary network. The interalveolar pores are variable in size, round or oval, and exhibit a smooth, even margin. Interalveolar pores could prevent collapse of those lobules that are supplied by an occluded bronchiole. Portions of alveolar phagocytes can be observed to lodge partially and temporarily within the pores. Pores thus serve to equalize air pressure within the lung, act as devices for collateral air circulation, and provide a means by which alveolar macrophages can migrate between alveoli. The interalveolar septum (**IS**) is sectioned at one region so as to expose a capillary (**Ca**) containing an erythrocyte (**Er**). Several alveolar phagocytes (**AP**), or dust cells, appear to be attached to the alveolar surface by thin tendrils. A small portion of the surface of an alveolus is illustrated in Fig. 2. The larger ridges denote the position of capillaries (**Ca**) on the underlying surface of the alveolar epithelium. The exposed surfaces of three alveolar type II (**II**) cells (granular pneumocyte, septal cell) can be distinguished from adjacent cells because of the presence of short microvilli. A second cell type called the alveolar type I cell (**I**), or membranous pneumocyte, is very flat, and has a smooth surface. The narrow linear ridges (**arrows**) most likely represent intercellular junctions between the cells that form the alveolar wall. Both epithelial cell types reside on an underlying basal lamina, which may be continuous with the basal lamina of underlying capillaries.

Fig. 1, ×2195; Fig. 2, ×3205

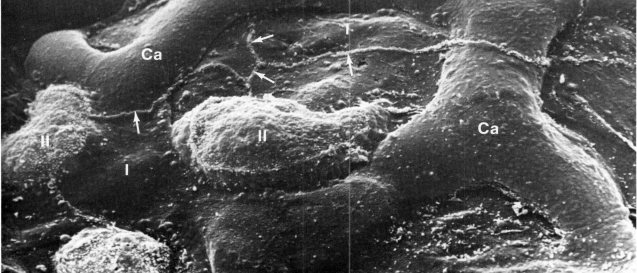

THE RESPIRATORY SYSTEM
Lung—Alveolus

The surface of an alveolus is illustrated in Fig. 1. The distinction between alveolar type I cells (**I**), alveolar type II cells (**II**), and an alveolar phagocyte (**AP**) is apparent. Short microvilli (**Mv**) found on the surface of alveolar type II cells are illustrated at higher magnification in Fig. 3. The alveolar type I cells have a relatively smooth surface. As previously mentioned, the narrow linear ridges (**arrows**) in Fig. 2 probably indicate regions of intercellular junctions between alveolar cells. An alveolar type I cell in Fig. 2 apparently was removed during tissue preparation, thus exposing the underlying basal lamina. The former position (✽) of the squamous alveolar type I cell is denoted. The alveolar type I cell is extremely large (perhaps 50–100 μm) but very attenuated. Its cytoplasm contains ribosomes, pinocytotic vesicles, and an occasional lysosome. Alveolar type II cells, as shown in Fig. 3, not only possess surface microvilli (**Mv**), but also exhibit rounded or irregularly shaped projections (**Pr**) on their surface. The latter structures may be related to the release of a special secretory product called surfactant. The surfactant includes surface active phospholipids that are rich in lecithin, which reduces the alveolar surface tension, thus preventing their collapse. The surfactant is thought to have a hypophase, perhaps produced by Clara cells, that includes proteins and polysaccharides. Current evidence suggests that the alveolar type II cells are capable of differentiating into type I cells, thus providing a potential source for their formation.

Fig. 1, ×4110; Fig. 2, ×2580; Fig. 3, ×13,885

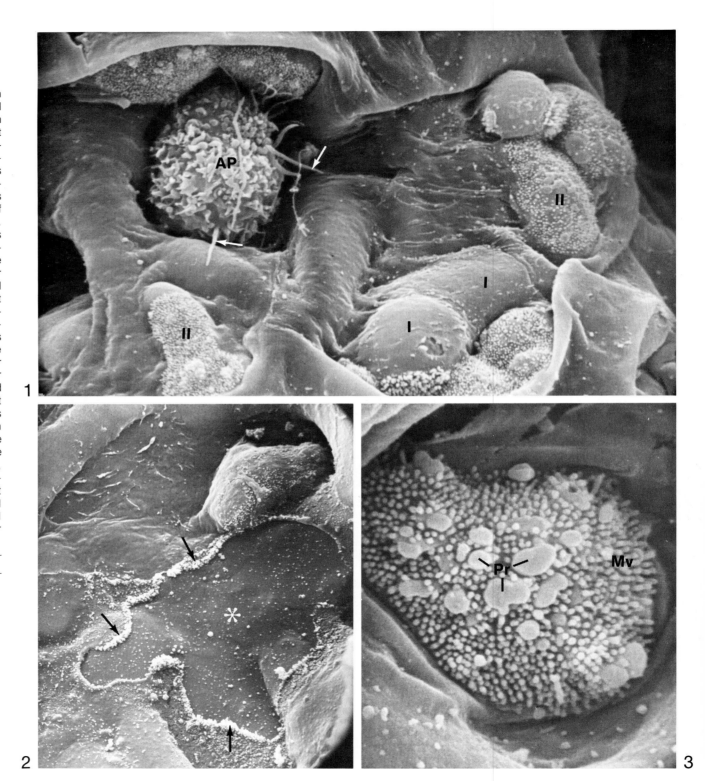

THE RESPIRATORY SYSTEM
Lung—Alveolar Macrophages

Phagocytic cells, called alveolar macrophages or phagocytes or dust cells, are widely distributed in the lung. They migrate over the surface of alveoli and function to ingest particulate material that might reach the lung with inspired air. The particulate material is digested intracellularly by lysosomes. The cells can migrate into the interalveolar septum and gain entrance into the lymphatic vessels of the lung. They may also migrate into bronchioles and be carried in the mucus by the beat of cilia toward the bronchi, trachea, and larynx to be expectorated or swallowed. Different configurations of alveolar macrophages from the rat lung are illustrated in Figs. 1 to 4. The surface of the rat alveolar macrophage is usually covered with numerous microfolds (**Fo**). Long slender tendrils (**arrows**) extend from the macrophage in Fig. 3 and attach the cell to the alveolar surface. The tendrils are commonly found on stationary macrophages, and they may continually be extended from the cell and withdrawn into it as the macrophage explores its immediate environment. In those alveolar macrophages illustrated in Figs. 1, 2, and 4, a pseudopod-like structure (**Ps**) is present at the base of what were probably migratory macrophages at the time of chemical fixation. In the fractured preparations shown in Fig. 5, the interior of two alveolar macrophages is exposed. The peripheral cytoplasm contains numerous vacuoles (**Va**) that probably represent the remains of dissolved lysosomal elements. The macrophage tendrils (**arrows**) and several interalveolar pores (**IP**) are identified. Fig. 6 illustrates several tendrils (**arrows**) of an alveolar phagocyte projecting through an interalveolar pore (**IP**). Alveolar phagocytes have previously been observed to become lodged within or to pass through interalveolar pores.

Fig. 1, ×4920; Fig. 2, ×4450; Fig. 3, ×5520
Fig. 4, ×5695; Fig. 5, ×3470; Fig. 6, ×9255

THE RESPIRATORY SYSTEM
Lung—Interalveolar Septum

Fig. 1 illustrates parts of two alveoli (**Al**) and the interalveolar septum (**IS**) from the rat lung. The septum was obliquely sectioned so as to expose collagenic and elastic fibers in the connective tissue (**CT**) of the interalveolar septum. In that portion of the micrograph illustrating the intact alveolar surface, it is possible to identify an alveolar type II (**II**) cell as well as an alveolar phagocyte, or macrophage (**AP**). The very thin interalveolar septum and many of its constituents in the rabbit lung are favorably displayed by cryofracture in Fig. 2. Portions of two alveoli (**Al**) are identified, and one contains an alveolar phagocyte (**AP**), or macrophage. The rabbit alveolar macrophage is uniformly populated with long, thin microvilli, and differs markedly from those in the rat lung, which possess many surface microfolds. The capillary (**Ca**) extending through the transversely sectioned interalveolar septum (**IS**) contains a few erythrocytes (**Er**). Alveolar type I cells (**I**) are extremely attenuated and are closely apposed to the capillary. The transmission electron micrograph (Fig. 3) shows the intracellular constituents of the interalveolar septum. The structures identified include: the attenuated cytoplasm of alveolar type I cells (✱), an alveolar type II cell (**II**) with secretory granules (**SG**), the fused basal laminae (**BL**) of the alveolar epithelium and capillary endothelium (**En**), and a longitudinally sectioned capillary (**Ca**) containing erythrocytes. Junctions between alveolar type I and type II cells are denoted by **arrows**. Short microvilli (**Mv**) are present on the free surface of alveolar type II cells. [Fig. 3 courtesy of E. R. Weibel.]

Fig. 1, ×1960; Fig. 2, ×3415; Fig. 3, ×8900

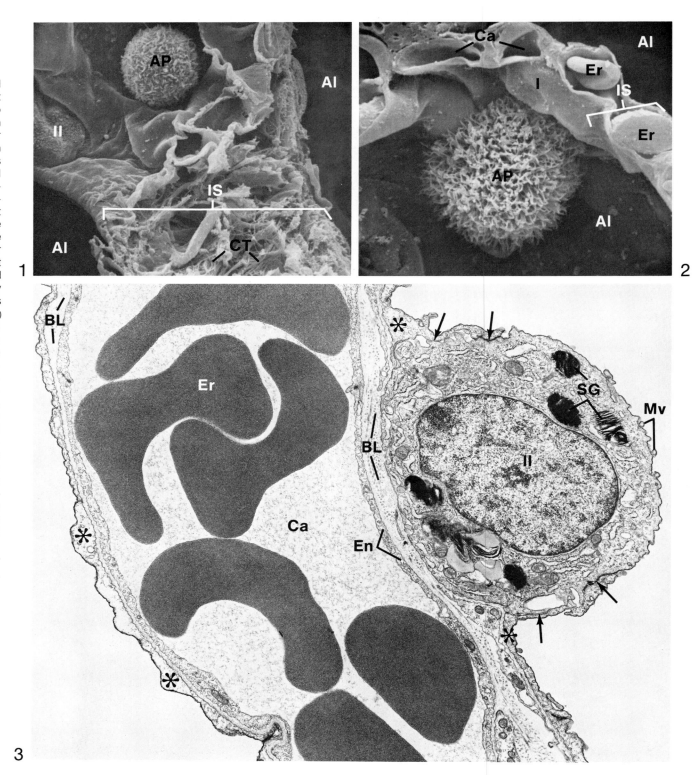

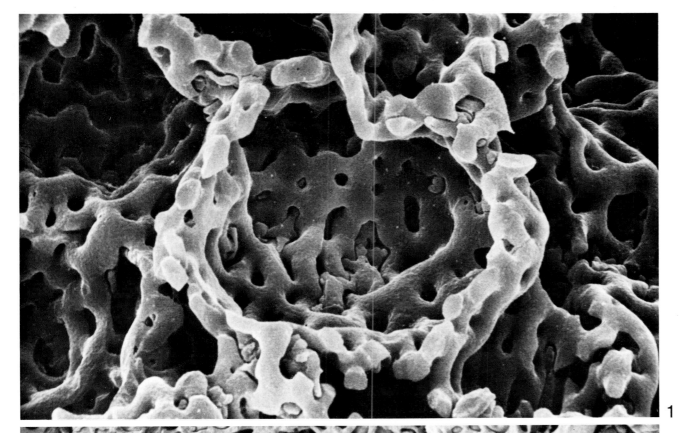

THE RESPIRATORY SYSTEM
Lung—Alveolar Circulation

Branches of the pulmonary artery form an extensive capillary network around each of the many air sacs in the lung. This network is functionally important in the diffusion of oxygen and carbon dioxide between the air sacs and capillaries. Scanning electron micrographs of the injected vascular system provide a clear, three-dimensional picture of extensive capillary networks around individual alveoli. Such casts are prepared by injecting a resin into the pulmonary artery so as to fill all branches of the vessel. After the resin polymerizes, the cellular and fibrous elements of the lung, including the blood vessels, are digested, leaving only casts of the circulatory system. Fig. 1 illustrates the capillary network surrounding a portion of one alveolar sac in the center of the figure. The capillary network surrounding a number of alveoli is shown in Fig. 2. In this preparation, the resin also filled the numerous alveoli so that their position and shape in relation to the surrounding capillary network can be seen. Larger bronchial divisions within the lung are supplied by branches of the aorta rather than the pulmonary artery. All venous blood, however, drains from the lung via the pulmonary vein.

Fig. 1, ×850; Fig. 2, ×415

1

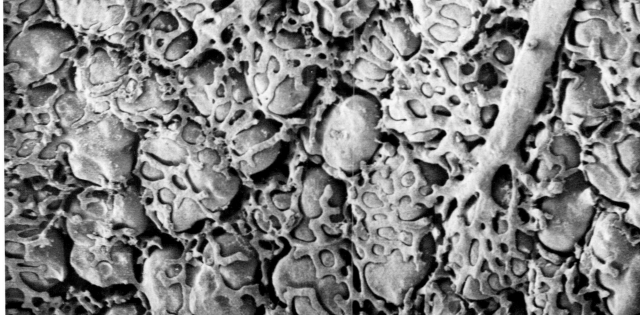

2

THE URINARY SYSTEM

Introduction

The urinary system consists of the kidneys, ureters, bladder, and urethra. The kidneys function as regulatory organs, helping to maintain the constancy of the internal environment in regard to both its volume and composition (i.e., water, electrolytes, metabolites). The ureters, which are continuous with each of the kidneys at the renal pelvis, transport urine down their length by peristaltic contractions of smooth muscle located within their tubular walls. In this manner, urine is conveyed to the bladder, which functions as a reservoir for urine until it is voided. The kidneys possess a specialized vascular supply that is uniquely associated with compound tubular glands called uriniferous tubules, which make up much of the kidney parenchyma. The arrangement of the tubular glands with respect to the vasculature is of utmost importance in understanding renal function. The most striking feature of the renal vasculature is the presence, in humans, of more than a million coiled glomerular capillary tufts, each of which is interposed between an afferent and an efferent arteriole. The evolution of the kidney as a filtering organ is attributed to the formation of these specialized vascular structures. It is the hydrostatic pressure generated between the two arterioles that forces water and crystalloids of blood through the wall of the intervening capillary tuft, resulting in the formation of an ultrafiltrate.

Almost one-quarter of the cardiac output perfuses the kidneys under normal conditions, far exceeding the oxygen requirement of renal tissue. In light of the evolution of the kidney as a filtering organ, however, it is not surprising that such a high rate of blood flow is maintained in relation to the kidney's size. This high rate of blood flow through more than one million glomerular filtering units produces almost 180 liters of filtrate per day. The quantity of filtrate produced is even more astonishing when one considers that the total amount of water (extracellular as well as intracellular) in the adult human is only 42 liters. The existing disparity between the total volume of available water and that which is filtered is accounted for by the fact that the kidneys are capable of resorbing more than 99% of the essential substances that they filter. The process of resorption is accomplished by the tubular portion of the kidney. Filtrate that has been forced through the wall of each glomerular capillary is collected in the expanded end of the uriniferous tubule that encapsulates each glomerulus. As the filtrate passes through the organized system of tubules, essential substances, such as water, glucose, amino acids, small proteins, and electrolytes, are resorbed by the tubule cells and returned to the surrounding vasculature. The process of resorption is selective, so that metabolic waste products are not resorbed but eliminated as part of the urine. In addition to resorption, the tubular cells are also capable of selectively secreting substances into the lumen of the tubule, thereby providing an additional mechanism (besides filtration) whereby metabolites or electrolytes may be eliminated from the body. Overall, the process of urine formation consists of two major steps. First, an ultrafiltrate is formed from blood by each of the glomerular capillaries; second, this filtrate is collected in uriniferous tubules, where it becomes modified through selective resorption and secretion by cells lining the tubules. The resulting modified ultrafiltrate is emptied from the tubules into the renal pelvis, where it continues as urine into the ureter.

From this brief description of urine formation it would at first appear that the kidney's main function is the elimination of waste products. Of equal, if not greater, importance is the kidney's ability to maintain the body's internal environment in the face of a changing intake of salt or water. The regulation of filtration and tubular resorption is controlled by hormonal and neuronal mechanisms. Vasodilators, vasoconstrictors, or neurogenic input may affect the smooth muscle tonus (and hence the resistance) in the walls of arterioles, causing an increase or decrease in the hydrostatic pressure developed across the intervening glomerular capillary. This provides effective control over the filtration rate, which in turn influences the electrolyte and water balance of the body. There is also regional specificity to this vascular control, because glomeruli in a certain area of the kidney may be preferentially influenced to produce more or less filtrate compared to those of adjacent areas. The kidney is also capable of autoregulating its glomerular filtration rate and renal blood flow in response to changes in renal blood pressure. The phenomenon of autoregulation is associated with the inherent ability of the kidney to maintain a constant glomerular filtration rate and renal blood flow in the face of a changing renal arterial pressure (over the pressure range between 80–180 mm Hg). It is thought that the compensatory change in resistance that must take place occurs at the level of the afferent arterioles.

The kidney also possesses an important homeostatic mechanism by which extracellular volume and sodium excretion can be regulated. Each nephron in the kidney possesses a specialized region, termed the juxtaglomerular apparatus (JGA), that is responsible for this regulatory process. The JGA is the region where the distal convoluted tubule of each nephron returns toward its respective glomerulus to form an attachment to the vascular pole of the glomerulus. The cells of the distal convoluted tubule, which form the attachment region, are collectively referred to as the macula densa. The basal ends of these cells are intimately associated with specialized myoepithelial cells located in the wall of the afferent arteriole that forms the vascular pole. The myoepithelial cells (often called juxtaglomerular cells) are in contact with the intima of the arteriole at their opposite ends. The juxtaglomerular cells contain intracellular granules that have been shown by histochemical techniques to contain the enzyme renin. Upon stimulus, these cells release renin into the blood, where it catalyzes the conversion of the plasma alpha globulin angiotensinogen into the decapeptide angiotensin I. A converting enzyme then cleaves two terminal amino acids from this peptide to form the bioactive octapeptide angiotensin II. At high concentrations, angiotensin II is one of the most potent vasoconstrictors known, acting to increase the resistance of blood vessels. Angiotensin II also stimulates the release of aldosterone from the adrenal glands and perhaps some antidiuretic hormone (ADH) from the posterior pituitary gland. These hormones fine tune the renal handling of sodium and water. Aldosterone increases the capacity of cells in the distal tubules of the kidney to reabsorb sodium. ADH causes the cells that make up the collecting ducts to become permeable to water, allowing it to be osmotically drawn out of the tubule lumen into the surrounding blood vessels, a process that conserves water. Angiotensin II is also present in the cerebrospinal fluid of the ventricle; it is not clear, however, how the hormone gets to this site. When angiotensin II is injected into the third ventricle of laboratory animals at very low concentrations, it stimulates drinking. This response has yet to be demonstrated in humans. Many of the physiological actions of angiotensin II have been postulated to result from the interaction with a receptor on the anterior surface of the third ventricle, which may stimulate nerves innervating this area.

The overall effect of renin release (and hence the formation of angiotensin II) appears to be the retention of water and sodium in the extracellular fluid. Conversely, when the release of renin is suppressed, diuresis and natriuresis ensue. What, then, exerts the control over renin secretion? At the present time there is evidence for at least three means of control over renin release: a baroreceptor, a sodium receptor, and neuronal input. The baroreceptor is thought to be located in the wall of the afferent arteriole of each juxtaglomerular apparatus. It appears that changes in the wall tension modulate renin release. A decrease in wall tension, either as a result of a drop in blood pressure or a decrease in extracellular volume, stimulates renin release, whereas an increase in wall tension has the converse effect. The sodium receptor, located in the macula densa of the distal tubule, also regulates the release of renin. A decreased sodium concentration at this site stimulates the release of renin, and an increased sodium concentration decreases its release. The movement of sodium into the cells of the macula densa appears to be required to produce an effect, since amiloride, an agent that blocks the entry of sodium into the cell, prevents renin release. Finally, sympathetic nerves innervating the JGA appear to play a role, since their stimulation or the infusion of beta-adrenergic agonists stimulates the liberation of renin. In summary, the JGA responds to an increased extracellular volume by causing excretion of sodium and water through inhibition of the renin-angiotensin system. Alternatively, when the JGA senses a volume depletion, or hyponatremia, the renin-angiotensin system is stimulated to restore the extracellular sodium and/or water through increased drinking and the release of ADH and aldosterone. The overall effect of the system is to keep the extracellular volume and sodium concentration at a constant level.

In addition to renin, two other hormones are synthesized and released from renal tissue. One is erythrogenin, which is released from the kidney in response to hypoxia. Erythrogenin appears to convert a precursor in the blood, called erythropoietinogen, to its biologically active form, erythropoietin. Erythropoietin stimulates an increased production of erythrocytes from precursor cells in the bone marrow. The distribution of erythrogenin in the kidney suggests that its source is tubular in origin. It is evident that the kidney plays an important role in regulating erythropoiesis, and clinical observations support this. For example, moderate to severe anemia often accompanies chronic renal insufficiency. Conversely, erythrocytosis may be associated with renal tumors. Another substance, prostaglandin (the E series) is also produced in the kidney. Recently it has been shown that many of the interstitial cells closely associated with the exterior of the tubules and vessels in the kidney medulla can synthesize and secrete prostaglandins, which may play a role in regulating medullary blood flow and transport of electrolytes. Other prostaglandins also appear to be released in the kidney, but their physiological role is still unclear.

By far one of the most widely investigated aspects of renal physiology is the ability of the kidney to conserve more than 99% of the water that is filtered at the glomerulus, which results in the excretion of a concentrated urine. From the onset, it is important to realize the necessity for such a process. It would at first seem that the entire process of urine concentration is not truly necessary, since the very first segment of the uriniferous tubule, the proximal portion, resorbs more than 80% of the water from the filtrate. It must be remembered, however, that the kidneys filter as much as 180 liters per day. Consequently, if only 80% of this volume were resorbed, the remaining 20% excreted would amount to a significant fluid loss (36 liters per day). Thus, because of an ex-

tremely high filtration rate, it is essential for the kidneys to conserve more than 99% of the volume that has been filtered, particularly during states of dehydration. In simplest terms, the mechanism of water conservation is accomplished by the production of a large gradient of osmolarity between the outer medulla (300 milliosmols) and the tip of the papilla in the inner medulla (1800 milliosmols). The high osmolarity at the papilla is partially attributed to the presence of urea at this location. At the distal portions of the uriniferous tubules (the collecting ducts), near the papilla, the high osmolarity of the surrounding interstitium draws water out of the collecting ducts, resulting in its conservation as well as the secretion of a concentrated (hyperosmotic) urine. As previously discussed, antidiuretic hormone (vasopressin) regulates the water permeability of the collecting duct wall, and hence the degree of diuresis. The production of the osmotic gradient, as well as its maintenance, is accomplished by two mechanisms, the single effect and the countercurrent multiplier system. In an energy-consuming step, solute must be separated from water. This is thought to occur in the thick ascending limb of the loop of Henle, the cells of which actively transport chloride from the tubule lumen into the interstitium. Sodium passively follows the chloride ion in order to preserve electroneutrality. These cells are also impermeable to water, preventing its passage out of the tubule lumen with the sodium and chloride. This process of separating solute from water has been termed the single effect. The single effect is "multiplied" by the flow of filtrate within the loop and distributed as an osmotic gradient between the cortex and medulla by means of the countercurrent multiplier system. Countercurrent mechanisms re-

quire that the straight portions of conduits be arranged in parallel, that they be quite close to one another, and that flow in the two adjacent conduits be in opposing directions. The exchange of fluids can take place between adjacent conduits by the process of diffusion, provided that their walls are sufficiently permeable. The countercurrent process is much more efficient than concurrent mechanisms, in which the flow in adjacent conduits is in the same direction. Countercurrent systems may take the form of an exchanger or a multiplier.

There are a number of possible countercurrent arrangements in the kidney medulla. Among them is a parallel arrangement of tubules, as between two segments of a loop (e.g., the loop of Henle) or, similarly, between two segments of a vascular bundle (the arterial and venous vasa rectae). Countercurrent arrangements between vascular segments and tubular segments are also possible. Because the interrelationship between tubule segments and blood vessels and their structural organization will determine the type of effect achieved, a knowledge of renal morphology is essential for understanding the physiology of this system. This is particularly true of renal countercurrent systems, which may produce the osmotic gradient by countercurrent multiplication and maintain the high gradient of osmolarity by a countercurrent exchange between the interstitium and the blood passing through the region. As a result of the unique organization of tubules and blood vessels, more than 99.7% of the filtered water is capable of being resorbed. For a more complete discussion of countercurrent systems, the reader is urged to consult the bibliography for this chapter at the end of the book.

THE URINARY SYSTEM
Kidney—Organization and Vasculature

As diagrammed in Fig. 1, the convex surface of the kidney is covered by a renal capsule (**RC**) composed of dense connective tissue. The kidney parenchyma is organized into an outer cortex (**Co**) and an inner medulla (**Me**) and contains more than one million functional tubular units called nephrons. The highly coiled portion of each nephron lies in the cortex, whereas the long straight portion descends into the medulla. Each nephron drains into a collecting duct, and together they are called a uriniferous tubule. Uriniferous tubules coalesce into larger ducts (the ducts of Bellini) within the medulla and collectively form a conically shaped mass of tissue called a pyramid (**Py**). Each pyramid and its overlying cortical tissue constitute a lobe. There are 8 to 18 such lobes in the human kidney, in contrast to the unilobar rat kidney, which possesses only one lobe. Urine within the collecting ducts passes through openings at the papilla (apex) of the pyramid and subsequently drains into the minor calyces (**MiC**), major calyces (**MaC**), renal pelvis (located at the hilus), and ureter. Fig. 2 illustrates a cast of the rat's renal vasculature, from which all of the cellular constituents have been digested with a strong alkaline solution. The renal artery (**RA**) and renal vein enter and leave the kidney at the hilus. The renal artery branches into a number of interlobar (**Ilo**) arteries, which in turn give rise to arcuate arteries (**AA**). Interlobular (**Ilu**) arteries that originate from the arcuate divisions traverse the cortex at right angles to the external connective tissue capsule. The interlobular arteries or its branches are continuous with afferent arterioles, and these become renal glomerular (**Gl**) capillary tufts. The efferent arteriole draining the glomerulus subsequently breaks down into an extensive capillary plexus (**CP**) that surrounds the tubular portions of the uriniferous tubule in the cortex and medulla. Blood in the capillary plexus enters the venous circulation (**VC**), which has divisions and names comparable to those described for the arterial circulation.

Fig. 2, ×6

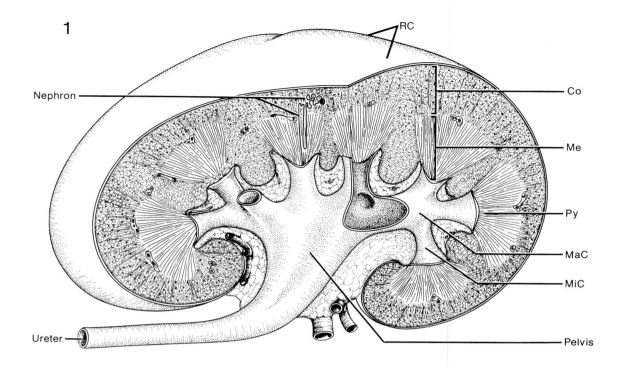

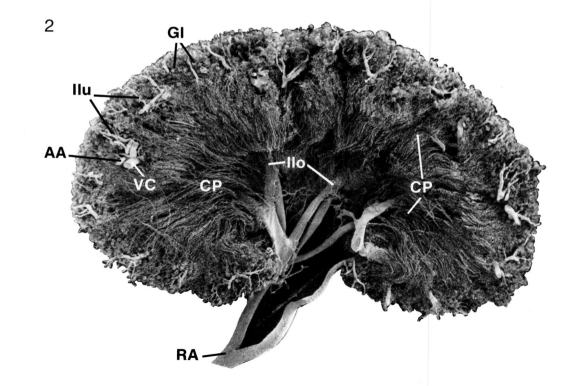

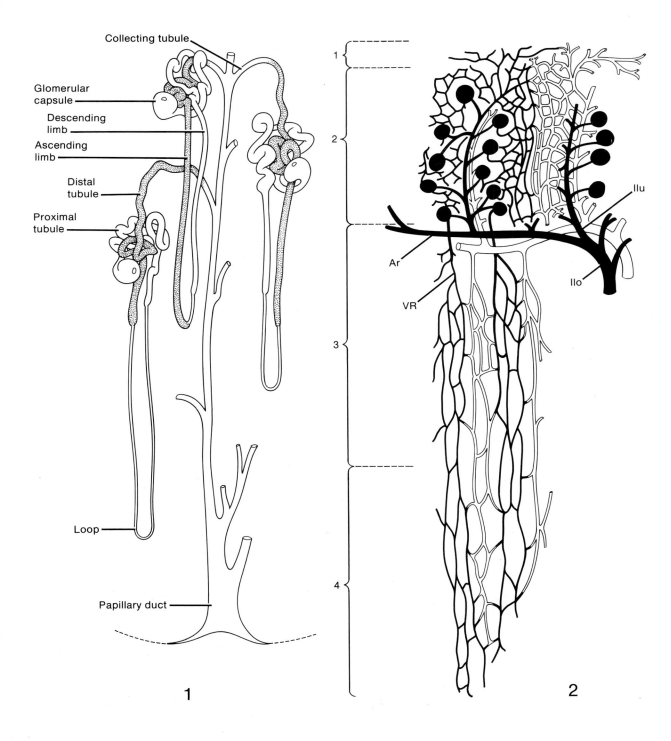

Collecting tubule

Glomerular capsule

Descending limb

Ascending limb

Distal tubule

Proximal tubule

Loop

Papillary duct

1

1

2

3

4

Ilu

Ar

VR

Ilo

2

THE URINARY SYSTEM
Kidney—Nephron, Vasculature

Three nephron units drained by collecting tubules are diagramed in Fig. 1. The collecting tubules converge at the apex of the pyramid into large papillary ducts, through which urine gains access to the calyces. Plasma filtrate, which is formed from each of the glomerular capillary tufts, is collected within a glomerular capsule of Bowman. Analogous to fluid passing through a funnel into a pipeline, the urine passes from the capsule into the proximal convoluted tubule, the loop of Henle (which consists of a descending limb, thin portion, and ascending limb), and subsequently into the distal convoluted tubule, which conveys urine into the collecting duct system. Those nephrons located deep in the cortex possess a much longer loop of Henle, often extending to the papilla of the medulla. Fig. 2 is a diagram of the vasculature that supplies corresponding regions of the nephron. Arteries and capillaries on the arterial side are black; veins and capillaries on the venous side are white. Tufts of capillaries forming the glomeruli appear as round blackened areas. The interlobar (**Ilo**) and arcuate (**Ar**) arterial divisions give rise to the interlobular (**Ilu**) arteries, from which afferent arterioles (supplying the glomeruli) extend like branches from a tree. The efferent arterioles leading from the glomeruli give rise to several types of capillary plexuses. Efferent arterioles extending from glomeruli located high in the cortex most often give rise to a capsular and subcapsular capillary plexus (Zone 1). Efferent arterioles originating in mid-cortical region (Zone 2) supply the peritubular capillary plexuses, which surround the proximal and distal convoluted tubules. The third major type of efferent arterioles, originating from the juxtamedullary glomeruli located in the lower cortex, may give rise to a peritubular capillary plexus and/or a descending arterial vasa rectae (**VR**). Vasa rectae extend between the long straight tubules in the outer medulla (Zone 3) and inner medulla (Zone 4). An extensive capillary plexus is interposed between the arterial vasa rectae and the ascending venous vasa rectae, which subsequently empty into the arcuate vein. [Fig.1 from W. F. Windle, *Textbook of Histology* (5th ed.), McGraw-Hill Book Company, 1976; redrawn from Carl Huber, *Special Cytology* (ed. E. V. Cowdry), Harper & Row, 1932. Fig. 2 from A. Maximow and W. Bloom, *Textbook of Histology* (6th ed.), W. B. Saunders Company, 1952.]

THE URINARY SYSTEM
Kidney—Cortex and Medulla

Fig. 1, a composite of two micrographs, provides an overall view of the kidney parenchyma in which the cortical (**Co**) and outer medullary (**Me**) regions can be distinguished. Further enlargements of the cortical and medullary regions are illustrated in Figs. 2 and 3, respectively. The cortex is characterized by the presence of renal corpuscles (**RC** in Figs. 1 and 2), each of which consists of a glomerular capillary tuft surrounded by a Bowman's capsule (**BC** in Figs. 1 and 2). Because many of the renal corpuscles were sectioned, the lumen of the glomerular capillary is exposed along its coiled length in a number of examples. In the process of sectioning, some of the glomeruli were also removed, so that only half of the Bowman's capsule remains in those cases. Portions of the uriniferous tubules are observed in both the cortex and the medulla. The highly coiled proximal and distal convoluted tubules (**CvT** in Figs. 1 and 2) are present only in the cortex, whereas the longer and straighter collecting ducts (**CD**) and loops of Henle (**LH**) predominate in the medulla. The collecting tubules, which drain the more superficial nephrons high in the cortex, descend through the cortical tissue in bundles along with the thick segments of the loop of Henle. These bundles of long, straight tubules extending through the cortex and into the medulla are known as the medullary rays (**MR** in Fig. 1). As observed in longitudinal section in Fig. 3, the collecting ducts converge (**arrows**) into the larger papillary ducts of Bellini located in the inner medulla.

Fig. 1, ×60; Fig. 2, ×105; Fig. 3, ×290

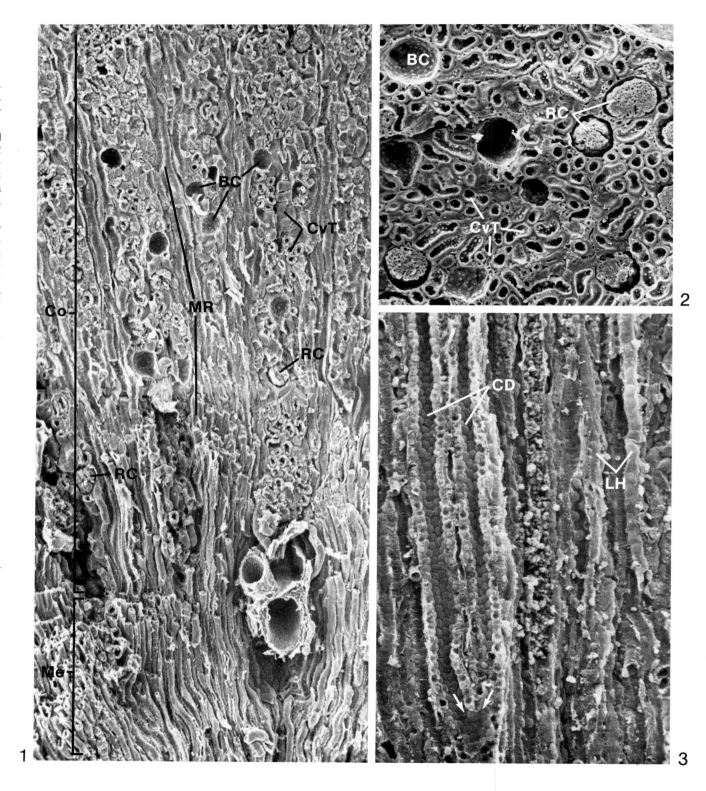

224

THE URINARY SYSTEM
Kidney—Circulation

The relationship between the cortical (**Co**) circulation and the medullary (**Me**) circulation is shown by a cast of the renal vasculature in Fig. 1. Whereas the cortical circulation consists primarily of glomeruli (**Gl**) with associated peritubular capillary plexuses (**Pt**), the medullary circulation contains long, straight vasa rectae (**VR**). The arcuate artery (**AA**), arcuate vein (**AV**), and interlobular arteries (**Ilu**) are also distinguished. Further details of the cortical circulation are illustrated in Figs. 2 to 4. Interlobular arteries (**Ilu**) branch extensively to form afferent arterioles (**Af**) which supply each glomerulus (**Gl**). As depicted in Fig. 3, the glomerulus is not simply a coiled capillary; it is a branched and anastamosing network of capillaries that is divided into lobules (**Lo**) by divisions (**arrows**) of the afferent arteriole. The efferent arterioles (**EF**) leading from glomeruli within the outer and middle cortex supply the extensive peritubular capillary plexus (**Pt** in Figs. 1, 2, 4), which surrounds portions of the proximal and distal convoluted tubules. The glomerulus is unique because it is interposed between two arterioles, a condition that generates a large hydrostatic pressure throughout the length of the capillary. The hydrostatic pressure can be increased or decreased by varying the tonus of muscular sphincters in the wall of both arterioles. Control over the muscle tonus may include hormonal or neuronal mechanisms, and the differences in arteriole diameter (**arrows** in Fig. 4) may result from differences in arteriole resistance. As a result, the hydrostatic pressure that is produced drives water, crystalloids, and small proteins across the capillary wall to be collected as an ultrafiltrate in the capsular space of Bowman. The ultrafiltrate subsequently drains into the proximal convoluted tubule, which is continuous with the parietal layer of Bowman's capsule at the urinary pole.

Fig. 1, ×20; Fig. 2, ×75;
Fig. 3, ×265; Fig. 4, ×260

THE URINARY SYSTEM
Kidney—Circulation

The replicated casts of kidney vasculature shown in these figures illustrate details of the long, straight arterial and venous divisions of the vasa rectae (**VR**) located in the kidney medulla. The vasa rectae are formed from the efferent arterioles (**Ef**) of glomeruli located at the junction between the cortex (**Co**) and the medulla (**Me**). The glomeruli located in the deep cortex are usually larger than those in the superficial cortex, and because of their location they are called juxtamedullary glomeruli (**JM** in Figs. 1 to 3). Under normal conditions, the glomerular filtration rate is higher for juxtamedullary glomeruli than for those located in the superficial cortex. The branches of efferent arterioles of the juxtamedullary glomeruli form a peritubular capillary plexus (**Pt**) in addition to the vasa rectae, as illustrated by the cast of a single isolated juxtamedullary glomerulus and its associated efferent arteriole (Fig. 3). The efferent arterioles of juxtamedullary glomeruli are greater in diameter than those of cortical glomeruli; in fact, many are almost the same size as the afferent arterioles. In a limited number of cases a double efferent arteriole may arise from a single glomerulus; one arteriole forms the peritubular capillary plexus and the other forms the vasa recta. The arterial vasa rectae also branch extensively (**arrows** in Figs. 2 and 3) in a fork-like manner to form vascular bundles. The vascular bundles often divide into extensive capillary plexuses (**CP** in Figs. 1 and 4) that surround the ascending loops of Henle and collecting ducts. The capillary plexuses drain into venous vasa rectae that ascend as bundles into the deep cortex, where they become continuous with arcuate veins.

Fig. 1, ×15; Fig. 2, ×80;
Fig. 3, ×70; Fig. 4, ×115

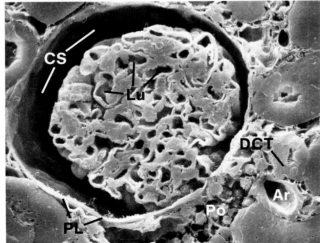

THE URINARY SYSTEM
Kidney—Cortex

The renal cortex (Fig. 1) is made up of many proximal and distal convoluted tubules (**Tu**), as shown in both transverse and longitudinal sections. It also includes numerous renal corpuscles (**RC**), which consist of highly coiled and branched glomerular capillary tufts with associated Bowman's capsules. Each Bowman's capsule consists of a parietal layer (**PL**) of simple squamous epithelium and a visceral layer (**VL**) that closely adheres to and completely covers the glomerular capillary. In most of these preparations, the parietal (capsular) layer has been cryofractured in half to reveal a glomerulus (**Gl**) as well as the adhering visceral layer. At the top of Fig. 1, the entire glomerulus was removed to reveal one-half of the remaining parietal layer (✱). Since the glomerulus invaginates Bowman's capsule at the vascular pole (**VP**), it is here that the visceral layer is continuous (**arrows**) with the parietal layer. Many of the glomeruli have been cryofractured, and at higher magnification (Fig. 2) the exposed lumen (**Lu**) of one such glomerular capillary is observed at various locations along its coiled length. The parietal layer (**PL**) of Bowman's capsule can be distinguished, as can the capsular space (**CS**), where ultrafiltrate produced by the capillary hydrostatic pressure is collected. One of the glomerular arterioles (**Ar**) has also been sectioned, and a portion of the distal convoluted tubule (**DCT**) appears to be closely associated with it. Details of this region are observed at higher magnification in Fig. 3. The cells in the wall (tunica media) of the afferent arteriole are thought to be specialized endocrine cells that form a portion of the juxtaglomerular apparatus; with appropriate stimulation, they secrete renin. Also in this area, the cells that constitute the wall of the distal convoluted tubule are taller and intimately associated with the arteriole wall, forming a structure known as the macula densa. A loose aggregation of spherical cells, collectively called the polkissen (**Po** in Figs. 2 and 3), populates a space surrounding the vascular pole.

Fig. 1, ×250; Fig. 2, ×650; Fig. 3, ×2140

THE URINARY SYSTEM
Kidney—Renal Corpuscle

The glomerulus and a portion of the parietal layer of Bowman's capsule were removed from the preparation in Fig. 1 to expose the half of the parietal layer that contains the vascular pole. It is at the vascular pole where the afferent and efferent arterioles (**Ar**) of each glomerulus invaginate the parietal epithelium. The protuberances (✱) on the surface of the capsular space are produced by the underlying nuclei of thin squamous epithelial cells in the parietal layer. The urinary pole (Fig. 2) of the parietal epithelium is located directly opposite the vascular pole; through this funnel-shaped structure, the collected ultrafiltrate drains into the beginning of the proximal convoluted tubule (**PT**). An example of a nearly intact Bowman's capsule (**BC**) is shown in Fig. 2. From the outer surface of the parietal layer, a portion of the underlying glomerulus (**Gl**) is visible through a small tear made in the capsule during specimen preparation. The specimen in Fig. 3 was cryofractured through the plane of the urinary pole, revealing the continuity between the capsular space and lumen of the proximal convoluted tubule. Details of the parietal epithelium surrounding the urinary pole is shown in Fig. 4. The nucleated portion (**NP**) of each squamous cell that projects into the capsular space possesses a single cilium (**Ci**) that extends from the cell surface into the capsular space. The surface of each parietal cell is also sparsely populated with short microvilli (**Mv**). They become denser, however, at the border between adjacent cells (**arrows**). Microvilli are also observed in the region where the parietal epithelium is continuous with the proximal convoluted tubule (**PT**); these longer microvilli constitute the brush border of the tubule.

Fig. 1, ×420; Fig. 2, ×635;
Fig. 3, ×535; Fig. 4, ×3890

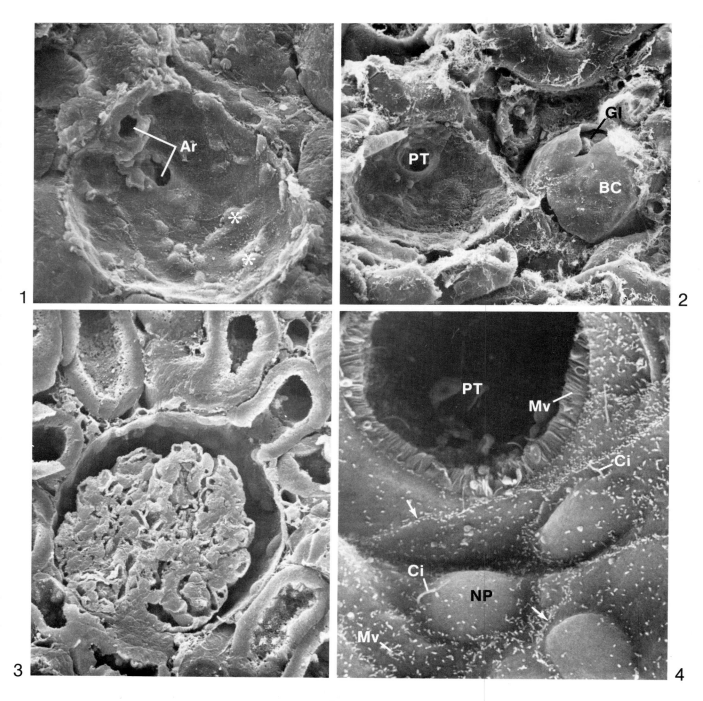

THE URINARY SYSTEM
Kidney—Renal Corpuscle

In Fig. 1, the parietal layer (**PL**) of Bowman's capsule has been partially removed so as to illustrate the capsular space (**CS**) of Bowman, in which the glomerulus resides. Although the parietal epithelial layer remains in the form of a sheet of simple squamous cells throughout embryonic development of the renal corpuscle, the morphology of the visceral layer changes appreciably, so that in the fully differentiated state the cells bear little resemblance to epithelial cells. The specialized cells that constitute the visceral layer, called podocytes (**Po**), are stellate in shape and possess cytoplasmic extensions (**arrows**) that originate from the nucleated portion of the cell body. Consecutive branching of the cytoplasm produces primary, secondary, and tertiary processes, which collectively enclose and surround the entire glomerular capillary. The coiled configuration of the underlying capillary can still be recognized. In Fig. 2, a portion of the capillary wall and its associated podocytes has been removed to expose the red blood cells (**RBC**) that flow through the lumen in the living state. Further enlargement of the exposed area (Fig. 3) reveals the close packing of the erythrocytes within the glomerular capillary.

Fig. 1, ×1595; Fig. 2, ×1270; Fig. 3, ×3200

THE URINARY SYSTEM
Kidney—Renal Corpuscle

This transmission electron micrograph presents a low-magnification survey of the renal corpuscle and the surrounding cortical area. The parietal epithelial cells (**PE**) in Bowman's capsule are very thin and attenuated except at the location of the cell nucleus, at which point the cell projects into the urinary space (**US**). Many of the podocytes (**Po**) in the visceral epithelium are sectioned through the cell nucleus, and their cytoplasmic processes (**arrows**) are seen to be closely associated with the basal lamina of the glomerular capillary. The capillary endothelium, basal lamina, and podocyte processes collectively constitute the filtration membrane (**FM**). Most of the blood cells in the capillaries have been removed by vascular perfusion. That portion of the endothelial cells containing the nucleus (**EN**) often bulges into the capillary lumen (**CL**). The basal lamina is continuous except at the region underlying the endothelial cell nuclei, at which point it breaks up to surround the mesangial cells (**MC**), which are thought to play a role in the phagocytosis and turnover of the basal lamina. The plane of section also includes portions of the proximal convoluted tubule (**PCT**), peritubular capillaries (**PCa**), and the afferent (**Af**) and efferent (**Ef**) arterioles. In the wall of the afferent arteriole, specialized endocrine cells called juxtaglomerular cells (**JG**) contain secretory granules that are thought to store renin. These cells are closely associated with the macula densa (**MD**) cells of the distal convoluted tubule (which are sectioned tangentially); together they form the juxtaglomerular apparatus. As was discussed in the introduction to this chapter, the overall function of the juxtaglomerular apparatus is to keep the plasma sodium concentration and extracellular volume at a constant level. [From J. A. G. Rhodin, *Histology, A Text and Atlas,* Oxford University Press, Inc. Copyright © 1974.]

×1280

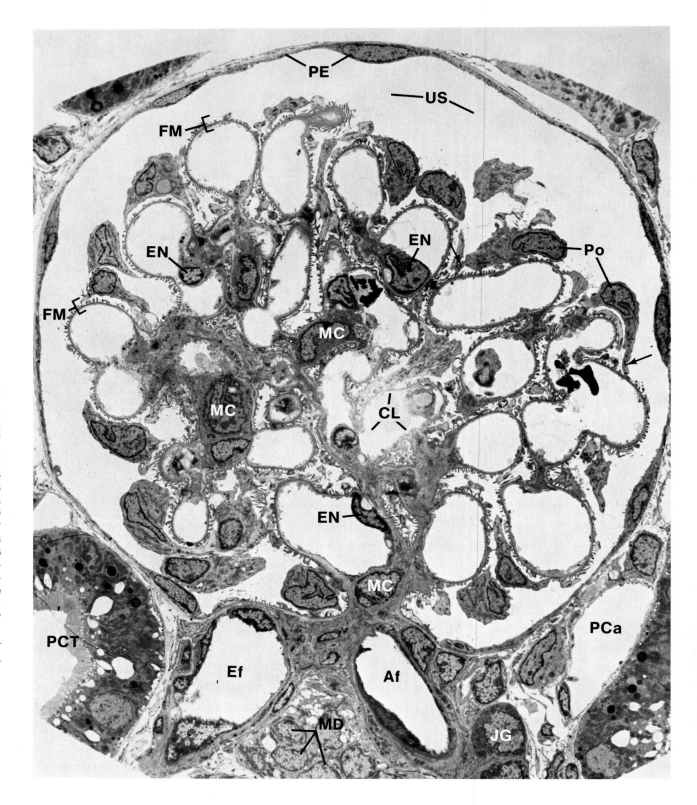

230

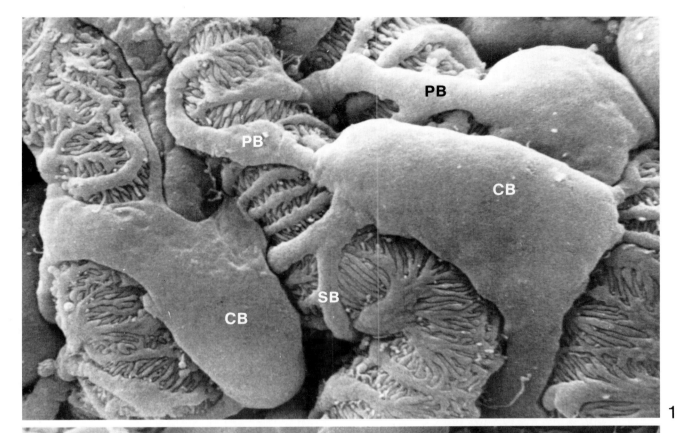

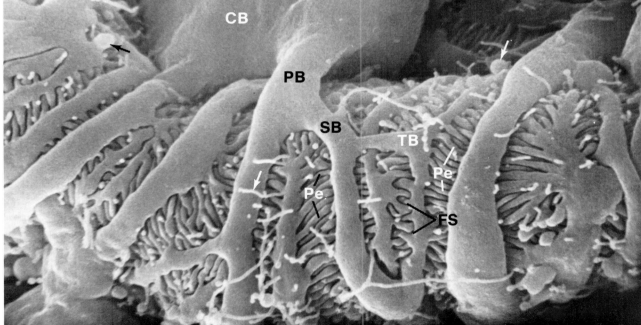

THE URINARY SYSTEM
Kidney—Podocytes

Figs. 1 and 2 illustrate the primary (**PB**), secondary (**SB**), and tertiary branches (**TB**) that extend from the nucleated portion of the podocyte cell body (**CB**). The terminal podocyte extensions, called pedicels (**Pe**), are finger-like in shape, and many of them extend at right angles from the larger podocyte branches. Although pedicels vary in their length and width, those of one cell always interdigitate with those of an adjacent cell. The underlying surface of the pedicels is in contact with the basal lamina of the glomerular capillary, and the specialized regions between interdigitating pedicels are called filtration slits (**FS**). The filtration slit is spanned by a diaphragm that represents the final barrier through which small molecules, electrolytes, and water must pass as they are forced through the wall of the capillary and into the capsular space. The filtration slit diaphragm restricts the passage of molecules with a molecular weight greater than 40,000, such as large proteins. In certain pathological states, the podocyte processes detach from the capillary basal lamina, allowing the unrestricted passage of large proteins into the tubules, a condition that results in high concentrations of protein in the urine (proteinuria). A variety of microprojections (**arrows**) are also commonly associated in varying numbers with the free surface of podocytes.

Fig. 1, ×5605; Fig. 2, ×9345

THE URINARY SYSTEM
Kidney—Glomerulus

The relationship between the investing podocyte processes (**Po**), basal lamina (**BL**), and the fenestrated endothelium (**FE**) of the glomerular capillary (**GC**) can be observed when a portion of the capillary is cryofractured (Figs. 1 and 2). The pedicels (**Pe**) are the smallest and most distal extensions of the podocyte and are continuous (**arrows**) with the basal lamina, which is also shared with the fenestrated endothelium. The filtration slits (**FS**) appear as clefts 300–400 Å wide between each podocyte. Spanning this slit is a thin diaphragm. In the living state, the lumen of the glomerular capillary is filled with red blood cells, but they have been removed in this preparation to reveal details of the capillary endothelial surface. As observed from the luminal surface (Figs. 3 and 4), the highly attenuated endothelium contains many pores, or fenestrations, (**Fe**), which are polygonal in shape and 500–1000 Å in diameter. Branched, thickened (**Th**) regions of the endothelial cells lack pores and may provide support for the thin and delicate fenestrated portion of the endothelium. Various forms of microprojections (**arrows** in Fig. 3) are associated with the thickened regions of the endothelial cell. It appears that the fenestrations are not simple perforations through the endothelial cell. Instead, it is likely that the microprojections extending from the thickened regions of the endothelial cell form an anastamosing network, or meshwork, in which the spaces of the netting appear as pores. Underlying the capillary endothelium is a specialized basal lamina (2500 Å in thickness) that consists of fine fibrils (which are similar to the unpolymerized form of collagen) embedded in an amorphous glycoprotein matrix.

Fig. 1, ×15,020; Fig. 2, ×31,820;
Fig. 3, ×13,015; Fig. 4, ×17,525

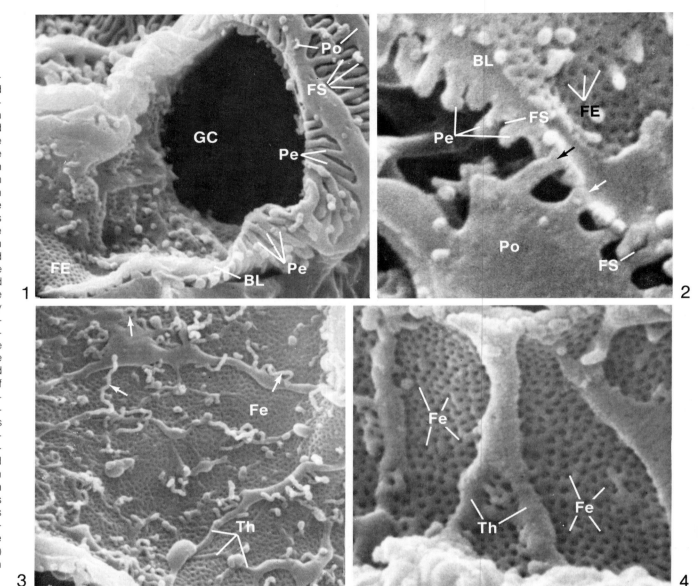

232

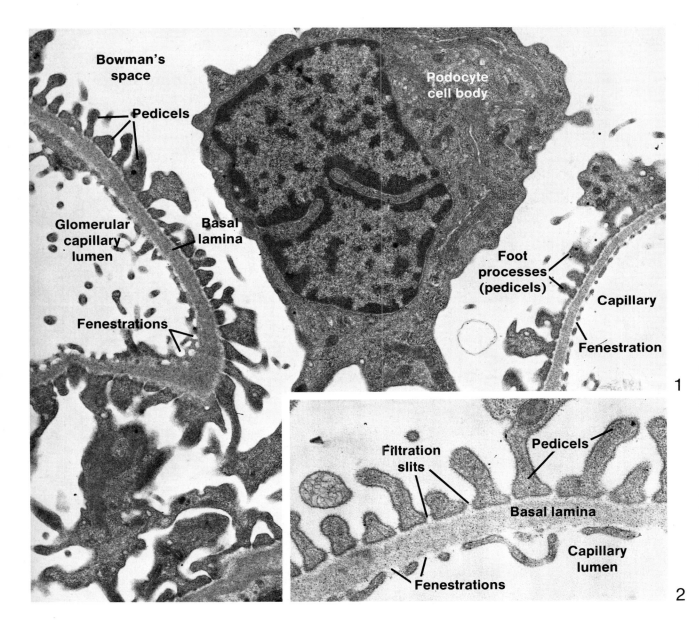

Fig. 1 — Bowman's space; Pedicels; Glomerular capillary lumen; Fenestrations; Basal lamina; Podocyte cell body; Foot processes (pedicels); Capillary; Fenestration

Fig. 2 — Filtration slits; Pedicels; Basal lamina; Fenestrations; Capillary lumen

THE URINARY SYSTEM
Kidney—Glomerulus

Further details of the glomerular capillary and investing podocytes can be seen in the transmission electron micrographs shown in these figures. The podocyte cell body, from which numerous cytoplasmic processes emerge, characteristically contains a nucleus of complex form, which often exhibits deep infoldings and clefts. Also within the cytoplasm, a well-developed Golgi apparatus, polyribosomes, and cisternae of rough-surfaced endoplasmic reticulum are observed. The podocyte processes contain numerous microfilaments and microtubules, and the terminal foot process of each pedicel is observed to be continuous with the thick basal lamina shared by the fenestrated endothelium of the capillary. The pedicels interdigitate closely (always with processes of neighboring cells and not with those of the same cell), so that 300–400 Å clefts remain between juxtaposed pedicels. Spanning each cleft is a thin, ribbon-like (70 Å thick) filtration-slit diaphragm, as seen in transverse section in Fig. 2. The diaphragm has a unique substructure consisting of a filamentous mesh that forms 60 × 140 Å pores. The fenestrated capillary endothelium, basal lamina, and slit diaphragm collectively form the complete filtration unit, which acts as a barrier to the passage of large proteins (those with a molecular weight greater than 40,000). Because blood plasma is continually forced through the basal lamina and filtration slit, a buildup of trapped proteins forms within the filter. "Unclogging" of the filter is apparently accomplished by mesangial cells that are in contact with the basal lamina. These cells most likely phagocytose the basal lamina and the accumulated debris within it and either return it to the circulation or degrade it. New basal lamina is thought to be synthesized by the surrounding podocytes.

Fig. 1, ×15,485; Fig. 2, ×42,720

THE URINARY SYSTEM
Kidney—Uriniferous Tubule

The schematic diagram on this page illustrates the major divisions of the uriniferous tubule, and corresponding regions are illustrated in the accompanying scanning electron micrographs. The divisions include the glomerulus (**GI**), which is supplied by the afferent (**Af**) and efferent (**Ef**) arterioles; Bowman's capsule (**BC**), which is continuous with the proximal convoluted tubule (**PCT**); thick descending (**TD**) segment; thin limb of the loop of Henle (**TLH**); thick ascending (**TA**) segment; distal convoluted tubule (**DCT**); collecting tubule (**CT**); and collecting duct (**CD**). The distal nephron and the collecting duct are **shaded**, and the direction of flow in the uriniferous tubule is denoted by **arrows**. The drawing is somewhat out of proportion; the tubular loop is much longer in relation to the glomerulus. If a renal corpuscle were the size of a penny, the loop of Henle would be about 7.5 m long. The juxtamedullary nephrons possess the longest loops, which extend toward the tip of the papilla. The presence of a loop of Henle is responsible for the kidney's ability to produce a hypertonic urine and to conserve water. The loop of Henle creates a hypertonic environment around the collecting ducts as they traverse this region, causing water to be osmotically withdrawn from the duct lumen. Besides the shortening of the uriniferous tubule in this diagram, it has also been spread out laterally. Many of the cells that make up different portions of the uriniferous tubule possess a single cilium (**Ci**). In a number of the micrographs, the nuclei (✲) of tubular cells have been exposed. [Diagram of nephron redrawn from R. Beeuwkes III and J. Bonventre, "Tubular organization and vascular-tubular relations in the dog kidney," *Am. J. Physiol.* 229(3):695–713 (1975). American Physiological Society.]

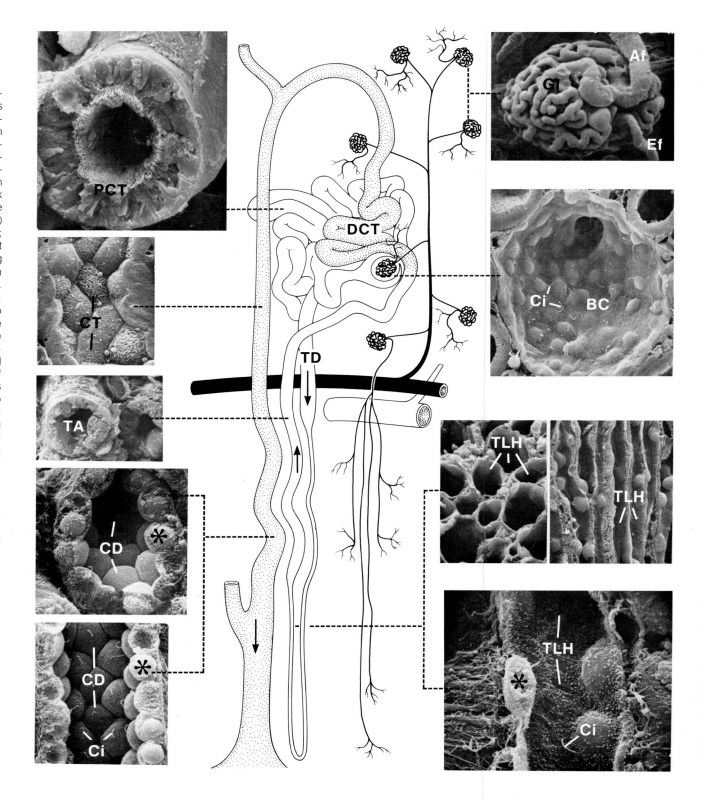

1

2

3

THE URINARY SYSTEM
Kidney—Proximal Convoluted Tubule

The proximal convoluted tubule, which arises from the urinary pole of Bowman's capsule, is divided along its length (about 18 mm) into a convoluted portion, the pars convoluta, and a straight portion, the pars recta. The pars convoluta usually turns back toward the Bowman's capsule and then becomes continuous with the pars recta as it forms the descending loop of Henle. The proximal convoluted tubule illustrated in Fig. 1 is sectioned transversely to reveal details of the organization and shape of its cells. The luminal border of the cells is populated with numerous long, slender microvilli (**Mv**). These structures constitute the brush border of the cells and are illustrated at higher magnification in Fig. 2. Apical projections (**AP** in Fig. 1) also commonly extend from tubular cells into the lumen. The lateral borders of the cells have irregularly shaped folds (**Fo**) that may reflect the interdigitation between adjacent cells in the tubule, and basal foldings often extend under neighboring cells. All of the cells reside on a basement membrane, and fibrous elements of connective tissue (**CT** in Fig. 1) surround the tubule. In some areas of the proximal convoluted tubule (Fig. 3), luminal regions lack microvilli. These regions, called microcraters (**MC**) or focal skip points, are usually round, measuring 1.5 μm in diameter, and are not uniformly distributed throughout the tubule. Because microcraters have been observed under a wide variety of fixation methods, it is generally agreed that they do not represent an artifact of preparation. It has been suggested that the microcraters are continuous with apical canaliculi (invaginations of the plasma membrane) of the proximal tubule cells, and that they may facilitate the uptake of proteins present in the filtrate. This suggestion is further supported by the fact that the number of microcraters increases during experimentally induced nephritis. [Fig. 1 courtesy of F. Feuchter.]

Fig. 1, ×5825; Fig. 2, ×8725; Fig. 3, ×125

THE URINARY SYSTEM
Kidney—Convoluted Tubules

A transverse section of proximal and distal convoluted tubules is illustrated in Fig. 1. The distal tubule can easily be distinguished by the paucity of secondary lysosomes and the short microvillous surface. The proximal tubule cell (Fig. 2) characteristically contains a large, rounded nucleus (**Nu**), primary and secondary lysosomes (**Ly**), apical canaliculi, apical vacuoles (**AV**), and numerous mitochondria (**Mi**). The height of the microvilli (**Mv**) that form the brush border may vary along the length of the proximal tubule. The cells are connected by junctional complexes (**JC**), and the tight junctions seal off the lumen from the intercellular space. A portion of the basal lamina is apparent, as is the fenestrated endothelium (**FE**) of an adjacent peritubular capillary. The proximal convoluted tubule resorbs much of the glomerular filtrate from its lumen into the circulation (more than 80% of the sodium chloride and water, 99% of the glucose). In addition, amino acids, potassium, phosphates, and sulfates may also be resorbed. Many of the small proteins in the filtrate are thought to be resorbed into apical canaliculi, which detach from the luminal surface and subsequently fuse to form apical vacuoles. Much of the protein is digested intracellularly by lysosomes, a process that allows reutilization of the amino acid breakdown products. Sodium passively enters each cell at the luminal border, but is actively pumped into the intracellular space by enzymes in the lateral cell membrane. This causes chloride ions to follow, preserving electroneutrality, and water is subsequently drawn into this space by osmosis. Fig. 3 diagrams many of the intracellular constituents of the proximal convoluted tubule as well as the complex arrangement of the interdigitating lateral cell processes. [Fig. 1 courtesy of A. B. Maunsbach. From "The influence of different fixatives and fixation methods on the ultrastructure of rat proximal tubule cells." *J. Ultrastruct. Res.* 15:242–282 (1966). Academic Press. Fig. 2 courtesy of J. A. G. Rhodin. From, *Histology, A Text and Atlas,* Oxford University Press, Inc. Copyright © 1974. Fig. 3 courtesy of R. E. Bulger. From "The shape of rat kidney tubular cells." *Am. J. Anat.* 116:237–256 (1965). The Wistar Press.]

Fig. 1, ×1780; Fig. 2, ×6765

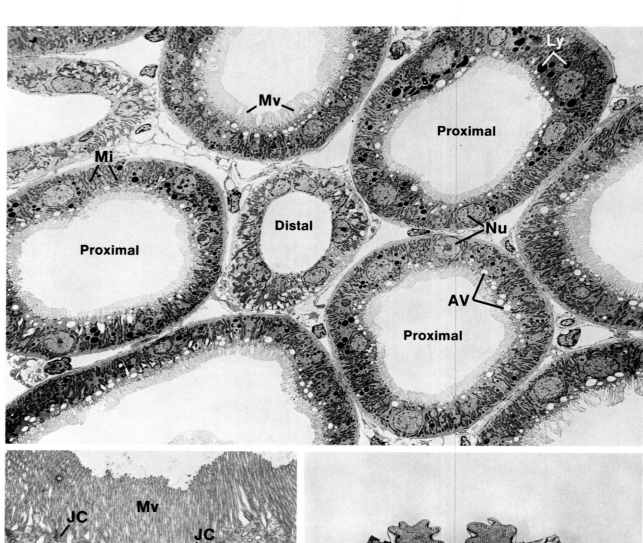

1

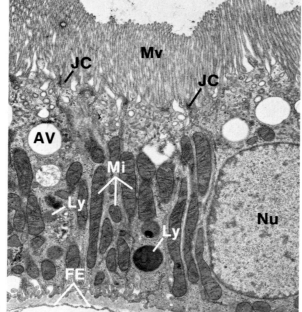

2

3

Ascending thick limb

Descending thin limb

DC

Mi

Vasa recta

Collecting tubule

LC

1

VR

TA

TA

CD

TLH

TLH

CD

2

VR

TD

3

THE URINARY SYSTEM
Kidney—Medulla

The transmission electron micrograph shown in Fig. 1 gives an overall survey of the rat kidney medulla. A transverse section through many of the vascular and tubule constituents reveals the ultrastructure of their lining cells. The wall of the collecting tubule consists of two types of cells, called the light cell (**LC**) and the dark cell (**DC**). The dark cell is characterized by the large number of mitochondria (**Mi**) and polyribosomes within its cell cytoplasm. Mitochondria are even more numerous in the cells of the adjacent ascending thick limb of the loop of Henle. The cells lining the descending thin limb are thin and flattened (squamous) and possess short microvilli on the luminal surface. Many of the vasa rectae are transversely sectioned to reveal the thin attenuated nature of the endothelial cells. A corresponding region of the medulla is shown in transverse section by scanning electron microscopy in Fig. 2 and in longitudinal section in Fig. 3. The collecting ducts (**CD**), thick descending (**TD**) and thick ascending (**TA**) portions of the loop of Henle, and thin segments of the loop of Henle (**TLH**) can be identified on the basis of cell height, diameter of the tubule, and surface specializations, such as cilia (one or two per cell) and microvilli (the height of which varies in different segments of the uriniferous tubule). The vasa rectae (**VR** in Figs. 2 and 3) often contain red blood cells, and the endothelial cells do not possess cilia, as is the case with the thin segments of the loop of Henle. The vasa rectae and the thin loops of Henle, however, are very similar in cross-sectional diameter. Further details of the tubular segments will be shown in subsequent micrographs, in which more definite criteria for differentiating portions of the uriniferous tubule will become apparent. [Fig. 1 courtesy of A. B. Maunsbach. From "The influence of different fixatives and fixation methods on the ultrastructure of rat proximal tubule cells." *J. Ultrastruct. Res.* 15:242–282 (1966). Academic Press.]

Fig. 1, ×3740; Fig. 2, ×825; Fig. 3, ×855

THE URINARY SYSTEM
Kidney—Loop of Henle

Fig. 1, a composite of three micrographs, illustrates the region of continuity (labeled **B**) between the thick descending (**TD**) segment and the thin segment of the loop of Henle (**TLH**). The thick descending segment, also termed the pars recta of the proximal tubule, originates from the convoluted portion of the proximal tubule in the cortex, and can be followed along its straight course. The thick segment is longer in the more superficial nephrons located higher in the cortex. The thick ascending (**TA**) segment is also observed adjacent to the thick descending portion; its cells possess fewer and shorter microvilli than those of the thick descending tubule. The lateral surface of the cells that make up the pars recta of the proximal tubule is shown in Fig. 2. The cells are shorter in height than those of the pars convoluta segment and possess a prominent brush border as well as basal foot processes (**FP**) that reside on a basement membrane (**BM**). There are, however, fewer lateral folds of the membrane in comparison to the convoluted segment. Details of the regions labeled **A** and **B** in Fig. 1 are shown in Figs. 3 and 4, respectively. At region **A** (Fig. 3) the cells of the thick descending segment possess numerous microvilli (**Mv**) that form a brush border, but the length of the microvilli become shorter as the tubule descends further into the medulla. A single cilium commonly projects from each cell into the lumen. At the transition between the thick and thin tubules (region **B** in Fig. 3), the cells lining the thin limb are simple squamous in contrast to the cuboidal cells of the thick segment. The squamous cells also possess a single cilium (**Ci**), but the microvilli are considerably shorter and sparsely distributed.

Fig. 1, ×980; Fig. 2, ×3295;
Fig. 3, ×3220; Fig. 4, ×2725

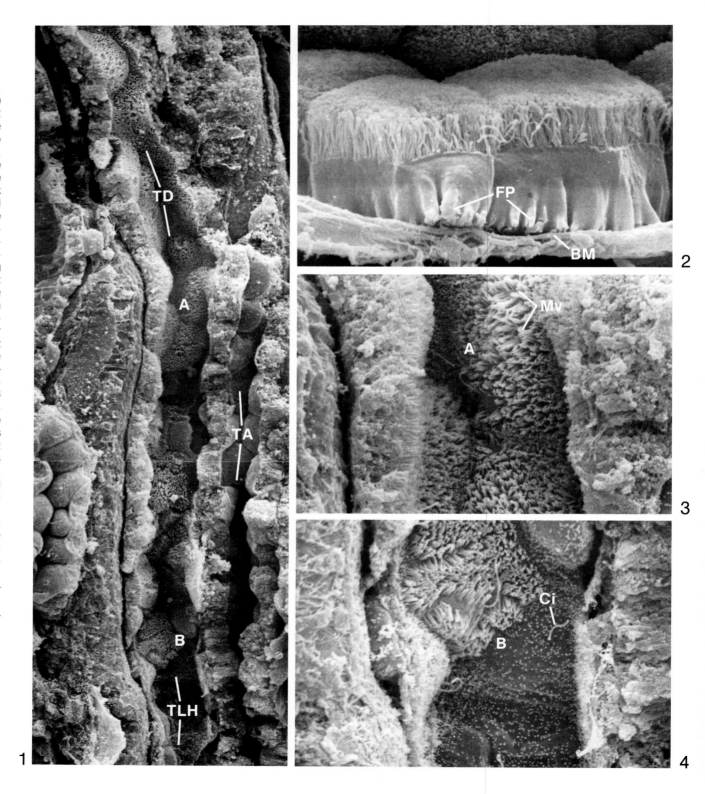

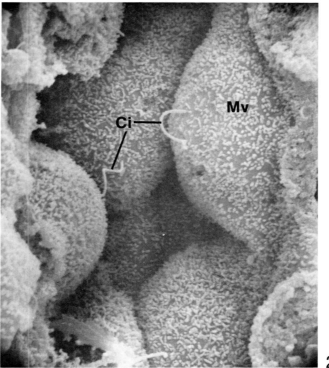

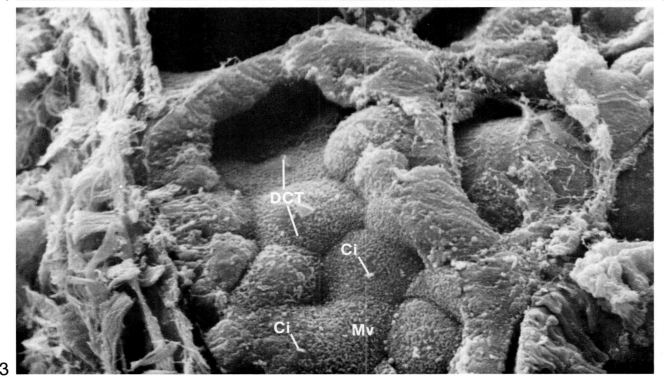

THE URINARY SYSTEM
Kidney—Thin Loop of Henle, Vasa Rectae, Distal Convoluted Tubule

The close association between a portion of the vasa rectae (**VR**) and a thin segment of the loop of Henle (**TLH**) is shown in longitudinal section in Fig. 1. The endothelial cells that line the vasa rectae possess a smooth surface, and the portion of each cell that contains the nucleus (✶) projects far into the lumen, almost occluding it. Although the thin segment of the loop is similar in diameter and is also lined by squamous epithelium, the surface morphology of these cells differs markedly from that of the vasa rectae. As observed at higher magnification in Fig. 2, cells of the thin loop possess stubby microvilli (**Mv**) that are sparsely distributed on the cell surface. In addition, a single cilium (**Ci**) commonly projects from each cell into the tubule lumen. The vasa rectae are associated with the thin limb, so that flow is in opposing directions between adjacent tubules and vessels. This promotes countercurrent exchange of electrolytes and water, which helps to maintain the osmolarity gradient produced in the medulla by the countercurrent multiplier system. Consequently, as the arterial vasa rectae descend deep into the medulla, their contents become hyperosmotic after equilibration with the surrounding interstitium. Conversely, as the venous vasa rectae ascend back toward the cortex, their contents gradually become more isosmotic as they equilibrate with the surrounding interstitium in the outer medulla and cortex. The distal convoluted tubule (**DCT**), which is continuous with the thick ascending portion of the loop of Henle, is shown in Fig. 3. In this segment of the nephron, the cells are cuboidal and possess microvilli (**Mv**) that are much shorter and less densely distributed compared to those of the proximal convoluted tubule cells. As in other regions of a nephron, the cells often contain a single cilium (**Ci**) that projects into the lumen of the tubule. During its tortuous course, the distal convoluted tubule returns toward the parent renal corpuscle in close association with the vascular pole of the glomerulus to form the macula densa.

Fig. 1, ×1860; Fig. 2, ×4095; Fig. 3, ×2180

THE URINARY SYSTEM
Kidney—Collecting Tubules

The collecting tubules begin in the cortex as arched collecting tubules and before entering the medulla merge into straight collecting tubules, one of which is observed in longitudinal section in Fig. 2. Two cell types, each exhibiting a different morphology, make up the wall of the collecting tubule. On the basis of their appearance in transmission electron micrographs, they have been classified as light and dark cells. In scanning electron micrographs (Figs. 2 and 3), the luminal surface of the more numerous light cell (**LC**) is characterized by the presence of a single (occasionally two) central cilium (**Ci**) measuring approximately 2.5 μm in length. In addition, microprojections in the form of short microvilli (**Mv**) are sparsely distributed upon the cell surface. The apical surface of the dark cell (**DC**) does not usually possess a cilium, but it does possess thin folds (**Fo**) called microplicae. Cytoplasmic projections (**CP**) are also occasionally observed on the surface of this cell. It has been shown that the number of dark cells increases during respiratory acidosis, suggesting that, in conjunction with the cells of the distal convoluted tubule, they have a role in the regulation of acid-base balance. The lumen of the collecting tubule (Fig. 1) may sometimes contain casts (**Ca**) that appear spherical in shape and variable in number and size. Casts may represent tubular cells that have been sloughed into the lumen, leukocytes or erythrocytes that have migrated into the lumen by diapedesis, or mucoprotein that is secreted by the distal tubule. These constituents often become dehydrated and precipitate during passage along the tubule, resulting in the formation of a hyaline cast. Fatty casts containing cholesterol esters may also be formed in some cases. The content, number, and appearance of tubular casts are sometimes useful in the diagnosis of a pathological state.

Fig. 1, ×2000; Fig. 2, ×1050; Fig. 3, ×3660

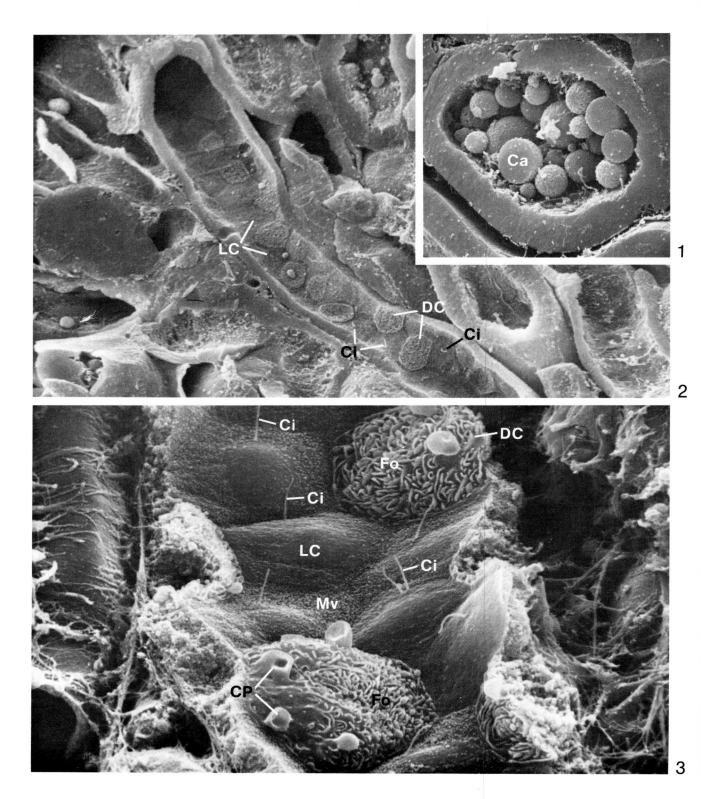

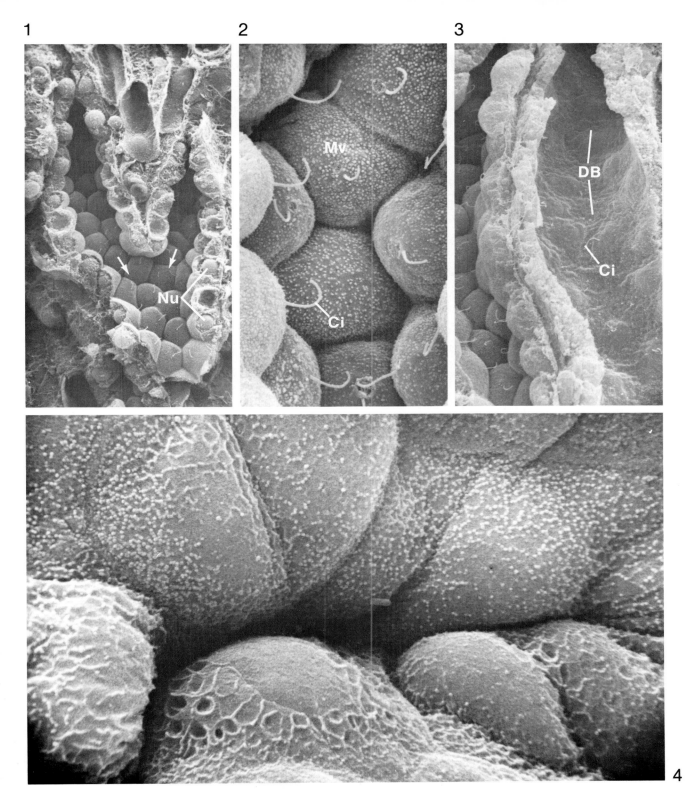

THE URINARY SYSTEM
Kidney—Collecting Ducts

As the collecting tubules descend deep into the medulla, they coalesce (**arrows**) to form larger collecting ducts, as shown in Fig. 1. Cells lining the collecting ducts are cuboidal in smaller divisions and columnar in the larger ducts. In the medulla, the cells lining the collecting ducts are primarily light cells; few, if any, dark cells are present. The cuboidal light cells (Fig. 2) possess a single cilium (**Ci**) and are sparsely populated with short microvilli (**Mv**). The cell nucleus (**Nu** in Fig. 1) has been exposed in a number of instances. Large collecting ducts, known as the ducts of Bellini (**DB**) contain few, if any, microvilli (Fig. 3). Occasionally, a cilium (**Ci**) may project from these cells, but they are rare. Toward the apex of the papilla, the collecting duct epithelium becomes continuous with the epithelium that lines the papilla. It is here that the luminal contents of the ducts are emptied into the calyces. An opening of a single collecting duct at the apex of the papilla is shown in Fig. 4, as viewed from inside the calyx. Duct cells that are continuous with the papillary epithelium bear short microvilli, which often form variable patterns upon the cell surface. Although one or two cilia per cell may be present in almost every segment of the uriniferous tubule, their functional significance in the mammalian kidney is unclear. Cilia are, however, more densely distributed in the kidney tubules of cold-blooded animals. Since blood pressure in cold-blooded animals is relatively low, and filtration pressure is low in the kidney tubules, the large number of motile cilia might function in the movement of filtrate through the tubules. The decrease in number of cilia in the mammalian uriniferous tubule suggests that they may merely be vestigial evolutionary remnants. They could create turbulence, however, which would help prevent laminar flow through the tubules and enhance mixing of the luminal contents. It has also been hypothesized that the single cilium on each cell may play a sensory role, detecting changes in flow and composition of the tubular contents.

Fig. 1, ×865; Fig. 2, ×2795;
Fig. 3, ×1290; Fig. 4, ×5695

THE URINARY SYSTEM
Kidney—Interstitium

In the renal medulla the arterial vasa rectae possess a continuous endothelial lining; the venous vasa rectae, however, possess an endothelium that is fenestrated (**arrows**), as can be seen in the longitudinal plane of section in Fig. 1. The venous vasa rectae are also larger in diameter, and connective tissue fibers (**CT**) are commonly observed in the vicinity of both arterial and venous divisions. In addition to the cells that form the walls of vessels and tubules in the cortex and medulla, cells within the interstitium are commonly observed, as illustrated in Figs. 1 to 3. These cells, termed interstitial cells (**IC**), are intimately associated with the outer walls of tubules and vessels. Two major types of interstitial cells have been described. In the inner cortex, a flattened fibroblast-like cell predominates (Fig. 2) that possesses highly attenuated cytoplasmic processes associated with the basal lamina of both proximal and distal convoluted tubules. The second type of interstitial cell is found primarily in the medulla (Figs. 3 and 4) between the ascending and descending loops of Henle or between the vasa rectae and collecting tubules. This type of cell is not as flattened as the first type, and often appears rounded with a few stellate processes. In transmission electron micrographs the cells resemble "modified" fibroblasts and are identified by their content of lipid droplets. Recent evidence indicates that the interstitial cells produce prostaglandin E, and possibly regulate blood flow within the renal medulla. In addition, they may also play a role in mediating the transport of electrolytes between the tubules and interstitium. In Fig. 4, the more rounded cells (**arrows**) are most likely blood leukocytes that have migrated into the extracellular space. They may leave the interstitium by passing into the lymphatic vessels that drain the region.

Fig. 1, ×8675; Fig. 2, ×1735;
Fig. 3, ×1425; Fig. 4, ×1480

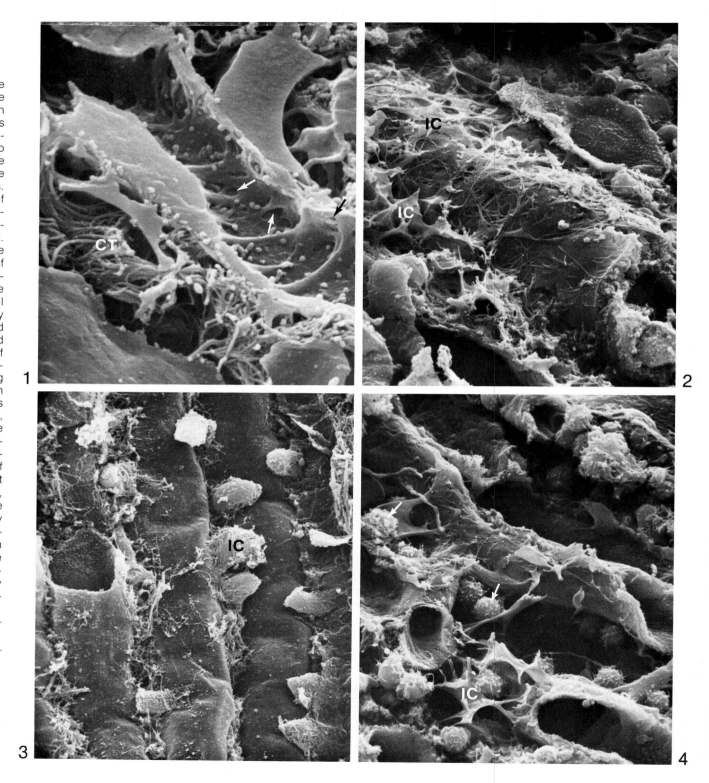

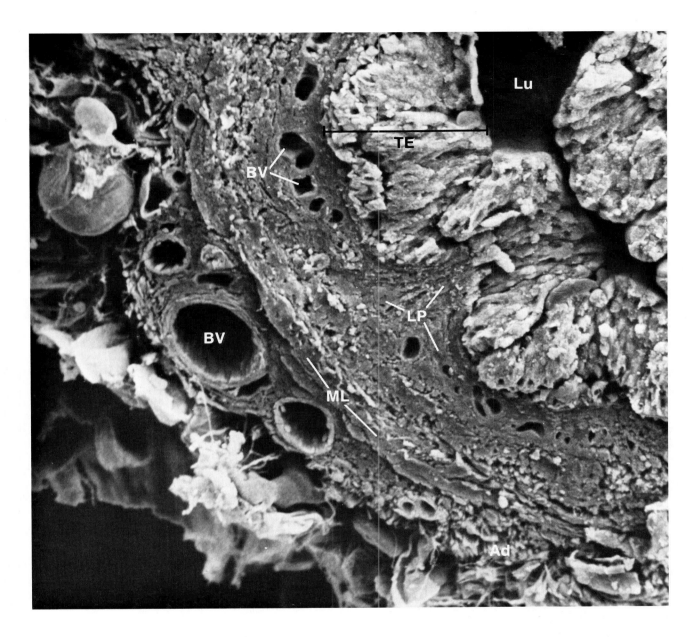

THE URINARY SYSTEM
The Ureters

The ureters are a pair of mucosal-lined tubes that transport urine down their length by peristaltic contractions of the smooth muscle located within the tubular wall. Each ureter begins as a continuation of the renal pelvis and is retroperitoneal, ending as an insertion into the posterior wall of the bladder. This micrograph shows a transverse section through the ureter. The lumen (**Lu**) is surrounded by a mucosa (mucous membrane) made up of transitional epithelium (**TE**) and an underlying lamina propria (**LP**) that consists of connective tissue. A number of blood vessels (**BV**) are transversely sectioned in the lamina propria. The lamina propria blends with a rather ill-defined circular and longitudinal arrangement of smooth muscle that constitutes the muscularis layer (**ML**). Externally, the ureter is covered by a connective tissue layer called the adventitia (**Ad**), which contains larger divisions of blood vessels (**BV**) that supply the ureter.

×400

THE URINARY SYSTEM
The Ureters

As Fig. 1 illustrates, the mucosa of the ureter becomes folded when the ureter wall is contracted and the lumen is empty. The transitional epithelium (**TE**) consists of six to eight layers of cells, and the superficial epithelial cells (**SE**) have a rounded or dome-shaped appearance. In the distended ureter, the epithelium is thinner, the superficial cells becoming more flattened and stretched. The superficial layer of epithelial cells, as viewed from the lumen of the ureter, is illustrated in Figs. 2 and 3. The cell boundaries are distinct, and a number of the epithelial cells are highly flattened (✱), whereas other nearby epithelial cells possess a more rounded contour (**arrow**). The epithelial cells in Fig. 3 possess an extensive branching network of short microfolds (**Mf**) on their surface.

Fig. 1, ×935; Fig. 2, ×1200; Fig. 3, ×2200

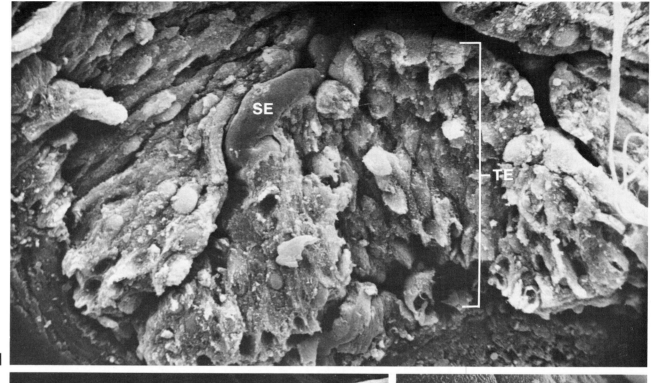

1

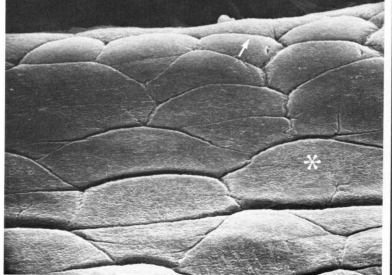

2

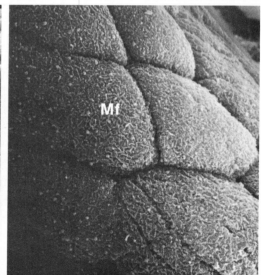

3

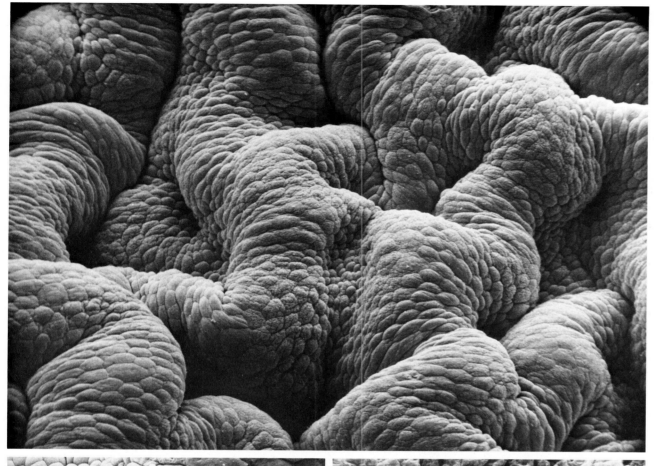

1

2

3

THE URINARY SYSTEM
The Urinary Bladder

The mucosa of the urinary bladder consists of transitional epithelium in association with an underlying lamina propria. As can be seen from the luminal view shown in Fig. 1, the bladder mucosa is folded; this folding is a result of contraction of the underlying layer of smooth muscle. The folding of the mucosa is a mechanism by which the lumen of the bladder can accommodate volume losses as urine is voided. The superficial layer of cells in the transitional epithelium, which are in contact with urine in the bladder lumen, are illustrated in greater detail in Figs. 2 and 3. These superficial epithelial cells are unique; they respond to forces of tension by changing their surface configuration. Depending upon the particular force exerted, these cells may have a smooth surface (**arrows**) or a folded (**Fo**), puckered surface. Extensive folding frequently produces deep, intersecting creases (**Cr** in Fig. 3) in the apical plasma membrane of the cells. The apical plasma membrane and the junctional complexes that anchor the cells together at their lateral borders (**LB**) are highly specialized. Both are very impermeable to fluid, restricting the diffusion of water from underlying capillaries into the bladder lumen and preventing the dilution of the concentrated urine within the bladder.

Fig. 1, ×150; Fig. 2, ×310; Fig. 3, ×1245

245

THE URINARY SYSTEM
The Urinary Bladder

A structural peculiarity associated with the apical plasma membrane of the most superficial transitional epithelial cells appears to play a special role in the functional properties of these cells. In the plasma membrane (Figs. 1 to 5) are arrays of rigid polygonal plaques (**PL**) that give the surface a cobblestone appearance. The technique of freeze-etching (Figs. 2 and 3) reveals that the plaques are planar aggregations of hexagonal particles in the membrane. The organization of the plaques and their relationship to the cytoplasmic microfilaments in the cell are diagrammed in Figs. 4 and 5. The hexagonal organization of the small particles (most likely proteins) that make up the plaques is diagrammed in the inset shown in Fig. 4. Bundles of filaments (**F**) within the cytoplasm (**C**) lie directly below the plaques. Filaments are anchored onto the plaques in such a way that contraction and shortening of the filaments exerts tension on the membrane. Flexible interplaque regions (**IN**) allow the membrane to fold between plaques when tension occurs (part B of Fig. 5). The resulting folds appear as deep creases (**Cr**) in the most superficial cells. The functional significance of the highly ordered membrane of transitional epithelial cells is not yet fully understood, but it would seem likely that interplaque regions act as hinges, enabling the membrane to accommodate large changes in surface area, which occur during distention or contraction of the urinary bladder. The rigid plaque regions might stabilize and strengthen the membrane. Since the underlying cytoplasmic filaments are anchored to the plaques, the plaques are probably the site at which the filaments exert a pulling force as they contract. The plaque structure of the membrane is also thought to play a role in the impermeable nature of the cell's plasma membrane. [Figs. 2 to 5 from L. A. Staehelin, F. J. Chlapowski, and M. A. Bonneville, "Luminal plasma membrane of the urinary bladder. I. Three-dimensional reconstruction from freeze-etch images," *J. Cell Biol.* 53(1):73–91 (1972). The Rockefeller University Press.]

Fig. 1, 21,025; Fig. 2, 40,940; Fig. 3, 112,140

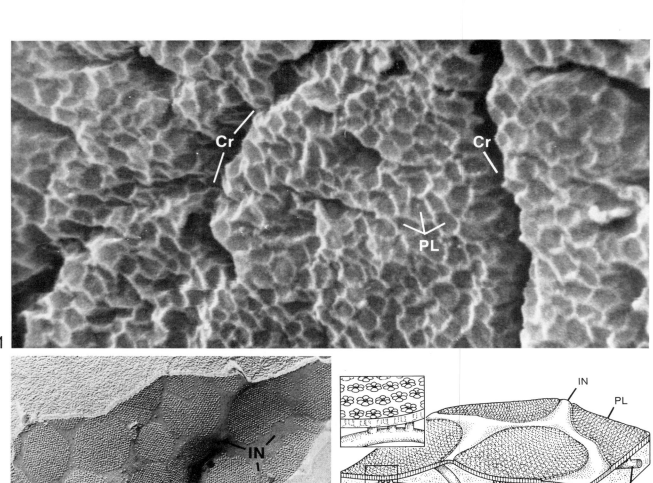

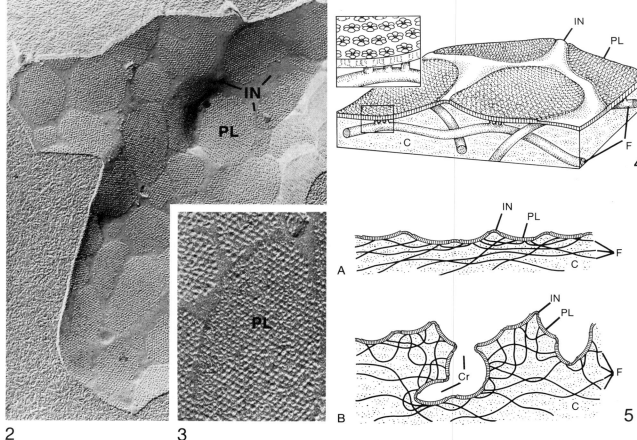

246

THYROID GLAND

Introduction

The thyroid gland consists of two lateral lobes connected by an isthmus that crosses the trachea just inferior to the cricoid cartilage. The parenchyma of the thyroid consists of irregularly shaped follicles that are lined by a single layer of epithelial cells. Each follicle contains a gelatinous colloid that represents the stored secretory product of the follicular cells. The capacity of the thyroid to store a precursor form of its hormone in an extracellular compartment makes it unusual among the endocrine glands. Two forms of thyroid hormone are secreted from the gland, thyroxine (tetraiodothyronine) and triiodothyronine. Each is an iodinated form of two tyrosine amino acid derivatives that exist in an O-linkage; thyroxine contains four iodines per molecule, and triiodothyronine contains three. The production of thyroid hormone is a two-step process. The first is the synthesis of thyroglobulin, the form from which thyroid hormone is made. Thyroglobulin is secreted into the follicle lumen from the apical pole of the cell. Since thyroglobulin is a glycoprotein (molecular weight 660,000), its formation involves the synthesis of a protein moiety, the synthesis and coadunation of carbohydrate to the protein, and the iodination of the tyrosyl amino acids that constitute much of the protein. The second step involves the reuptake of the stored thyroglobulin from the follicle lumen into the cell. The thyroglobulin molecules are then hydrolyzed by lysosomes within the follicle cells, causing the subsequent liberation of iodothyronines into the extracellular space at the base of the cell. The synthesis of thyroglobulin requires an adequate intracellular pool of iodine. The thyroid gland is unique in its capability of concentrating iodine against an electrochemical gradient; the cells are thought to accomplish this by the active transport of iodine into the cell. The mechanism appears to take place in the basal plasma membrane and is in some way linked to Na^+-K^+ ATPase. Negatively charged ions with a molecular volume similar to that of I^- (i.e., perchlorate, thiocyanate) can compete for the carrier site on the membrane and inhibit the uptake of iodine.

The amino acids necessary for the synthesis of thyroglobulin (particularly tyrosine) are also actively transported into the cell. Before iodine can be covalently attached to the tyrosine amino acids of thyroglobulin, it must be oxidized—a process that is thought to be performed by a peroxidase enzyme within the cell. Agents that interfere with this oxidation cause a deficiency of thyroid hormone. The thioamides are the most common group of such agents and include thiourea, 6-propylthiouracil, and methimazole. Another agent is progoitrin, a naturally occurring thioglycoside found in turnips. Progoitrin is converted by intestinal bacteria into the isothiocyanate form, goitrin, which mimics the effect of the thioamides.

Much of the control over the production of thyroid hormone comes from another hormone, thyroid-stimulating hormone (TSH), which is, in turn, liberated from the anterior pituitary gland in response to thyroid releasing hormone (TRH). TSH stimulates a multitude of steps that lead to the increased production of thyroid hormone. The appropriate stimulation for the release of TSH and TRH is a decrease in the concentration of unbound thyroid hormone in the blood (negative feedback effect).

The action of thyroid hormone has been widely studied. If the thyroid gland is removed, a decrease in the basal metabolic rate ensues, implying a regulative role in oxidative metabolism. Recently it has been demonstrated that thyroid hormone stimulates the synthesis of many enzymes, possibly including those involved in oxidative metabolism. The importance of thyroid hormone in the development of the brain and skeleton in children has also received widespread attention. The mechanism by which thyroid hormone acts on target tissues is still uncertain; it is known, however, that triiodothyronine is the more potent of the two forms and binds more specifically to the nuclei of target tissues. This has led some investigators to believe that thyroxine is converted to the more potent triiodothyronine at its site of action.

Failure of the thyroid gland to secrete adequate

amounts of thyroid hormone (hypothyroidism) can result from a variety of disorders. In primary thyroid deficiency (known as myxedema in humans), the disorder exists in the gland itself, and stimulation with TSH has a negligible effect. The results are general sluggishness, fatty deposits within the subcutaneous tissue, a lowered basal metabolic rate, dry hair, high plasma cholesterol level, slow cerebration, and constipation. Secondary thyroid deficiency occurs when the pituitary gland fails to secrete TSH or if the hypothalamus does not produce TRH in sufficient amounts. This condition is less severe than primary deficiency and can be alleviated by the administration of TSH. The presence of a thyroid deficiency in the developing fetus leads to cretinism, a condition characterized by irreversible mental retardation and stunted growth.

The abnormal ingestion of iodine can also result in hypothyroidism. When iodine is deficient in the diet, as occurs in certain geographical locations, there is an increased production and storage of thyroglobulin, but insufficient amounts of the iodinated hormone. Conversely, chronic ingestion of excess amounts of iodine may lead to the development of hypothyroidism. This phenomenon, known as the Wolff-Chaikoff effect, is due to the blocking effect that a high blood level of iodine has on its own transport into the thyroid. Many of the thyroid deficiencies result in the enlargement of the thyroid gland, a condition known as goiter.

Decreased production of thyroid hormone may result in a proliferation and hypertrophy of follicle cells (parenchymatous goiter)—a compensatory mechanism that serves to restore the synthesis of the hormone to normal levels. In the case of primary thyroid deficiency, or abnormal iodine intake, the increased secretion of TSH (due to negative feedback) may cause cell proliferation and hypertrophy. Alleviation of the condition often results in the storage of excess colloid synthesized by the hypertrophied cells. The overabundance of the stored thyroglobulin causes enlargement of the gland (colloid goiter).

Goiters may also be produced in hyperthyroid states caused by tumors of the thyroid gland (adenomas), which are manifested by hyperexcitability, high metabolic rate, tremor, excessive sweating, diarrhea, and cardiac abnormalities. Another form of hyperthyroidism is Grave's disease (exophthalmic goiter), which is characterized by forward displacement of the eyeballs. The sera of many of the patients contain an antibody (IgG) that appears to be produced against a thyroid antigen. The antibody, although chemically unrelated to TSH, stimulates the thyroid to produce excessive amounts of thyroid hormone. For this reason the antibody has been referred to as the long-acting thyroid stimulator (LATS). The apparent autoimmune response against a thyroid antigen also involves cross-reactivity with the extraocular muscles. The immune response causes

inflammation of the tissue surrounding the orbit, which results in the forward displacement of the eyeball. The exact cause of Grave's disease is not known; it is possible, however, that incidental damage to the thyroid (i.e., irradiation or thyroidectomy) may cause the leakage of a formerly secluded thyroid antigen into the blood stream, and this might initiate an immune response. It is also of interest that the lymphatic vessels draining the thyroid gland are continous with those draining the orbits of the eyes. Thus, during sleep, thyroidal lymph that is rich in antibodies and lymphocytes directed against the thyroid may reach the orbits by gravity flow and enhance the cross-reactivity against the extraocular muscles. The curious fact that some patients only experience exophthalmos in one eye might be explained by their sleeping habits; that is, sleeping on one side may predispose that eye to lymphatic drainage from the thyroid.

In addition to thyroid hormone, the thyroid gland also produces the hormone calcitonin. The cells responsible for the synthesis and release of calcitonin are called parafollicular cells, or C-cells. These cells are not a part of the follicle, but share the basal lamina with the follicle cells. C-cells are derived embryologically from the ultimobranchial body, which is formed from a portion of the pharyngeal pouches. Calcitonin produces a decrease in the level of blood calcium.

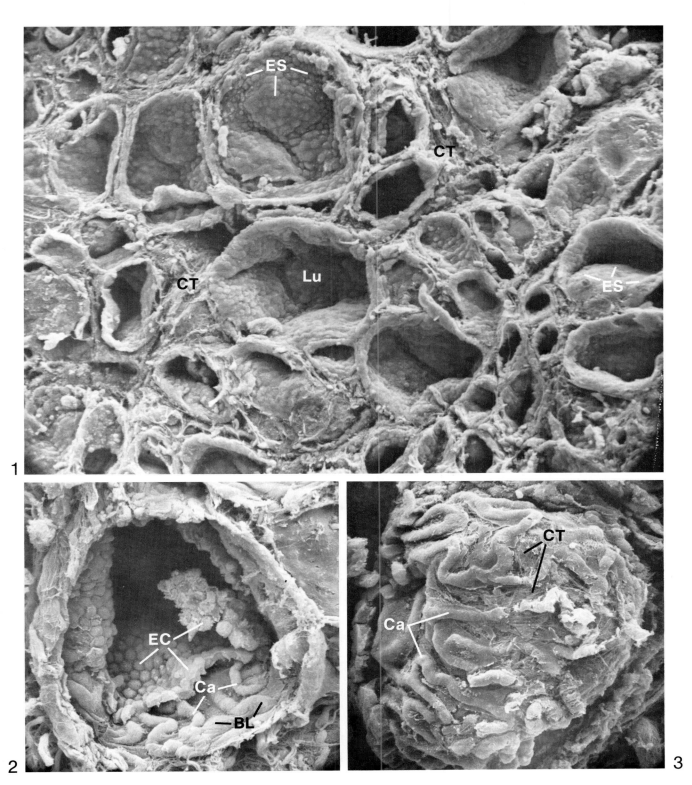

1

2

3

THYROID GLAND
Follicles

The thyroid gland is composed of many follicles, lined by epithelium, that appear roughly spherical in shape and highly variable in size (0.02–0.9 mm in diameter, the smaller follicles being more numerous). Each follicle consists of a single layer of epithelial cells that synthesize thyroglobulin and secrete it into the follicle lumen, where it is stored. Thyroglobulin is the precursor glycoprotein from which thyroid hormone is formed. The thyroid follicles shown in Fig. 1 were sectioned transversely, and the thyroglobulin that normally occupies the lumen (**Lu**) of each follicle was removed by repeated washings before the tissue was preserved. The removal of the secretory product allows the epithelial surface (**ES**) to be observed. Connective tissue (**CT**) occupies the interstices among the closely packed follicles and contains blood vessels, lymphatic vessels, and nerves. In Fig. 2 a single follicle is illustrated at higher magnification, and the sheet-like arrangement of the simple cuboidal epithelial cells (**EC**) is apparent. The height of the cells may vary from squamous to columnar, depending on their state of activity. In some areas, the epithelial sheet has been partially removed and detached from the smooth surface of the thin basal lamina (**BL**). An extensive capillary plexus is closely applied to the exterior surface of each follicle. The form and distribution of the capillaries (**Ca**) underlying the basal lamina can easily be observed. The external surface of a single follicle is illustrated in Fig. 3, and the associated capillaries (**Ca**) branch and anastomose to form an extensive vascular plexus supported by connective tissue (**CT**) that is rich in its content of reticular fibers.

Fig. 1, ×340; Fig. 2, ×595; Fig. 3, ×625

THYROID GLAND
Vascular Cast

The vascular casts shown in these micrographs illustrate the distribution of vessels comprising the thyroid circulation. The arterial (**Ar**) and venous (**Ve**) divisions can be distinguished, as well as the smaller arterioles (**Al**) that supply the capillary plexus (**CP**) surrounding each follicle. Amino acids, iodine and other substrates are brought to the epithelial cells via the capillary plexus, and thyroid hormone secreted by the follicle cells also enters the capillary plexus, thereby gaining entrance to the peripheral circulation.

Fig. 1, ×130; Fig. 2, ×140; Fig. 3, ×490

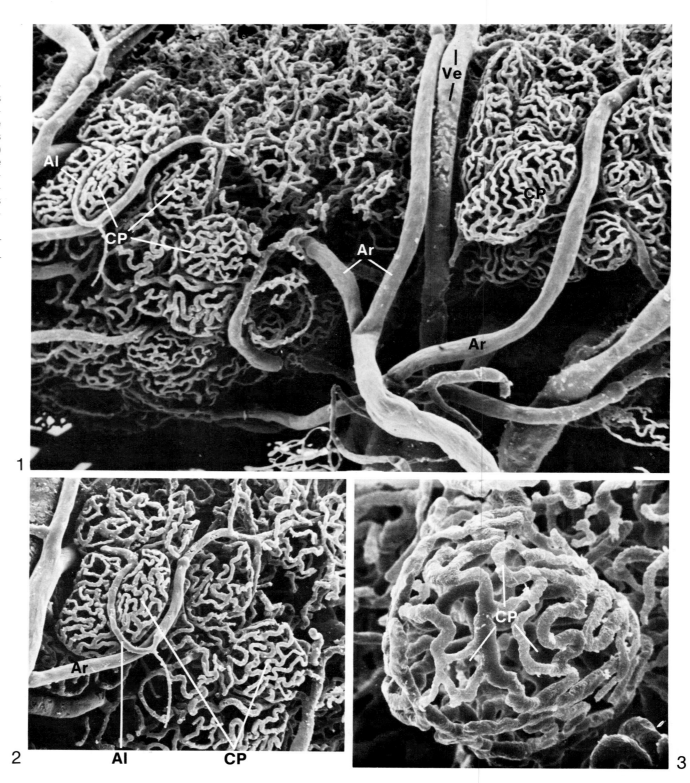

THYROID GLAND
Follicle Cells

Details of the structure of follicle cells are shown in these illustrations. In Fig. 1, the lateral surface (**LS**) of the cuboidal epithelial cells is observed in an area where adjacent cells were removed. On the apical cell surface (Figs. 1, 3, and 4), portions of the thyroglobulin secretory product (**SP**) that were not removed during specimen preparation are apparent. Short microvilli (**Mv**) populate the free surface of the cells, and in a number of cases a single cilium (**Ci**) projects from the cell surface into the follicle lumen. The transmission electron micrograph illustrates the internal organization of the follicle cell. Basally, the cell is closely associated with a capillary (**Ca**) network. The capillaries are of the fenestrated type, and a few erythrocytes (**Er**) are present within the lumen of the capillary shown. In addition to the nucleus (**Nu**), the cytoplasm of the follicular cell contains rough-surfaced endoplasmic reticulum (**RER**) and many primary and secondary lysosomes (**Ly**). The thyroid follicle cell is unique because it is perhaps the only endocrine cell that stores its product extracellularly before secreting it into the blood. The cell synthesizes the protein portion of thyroglobulin in the rough-surfaced endoplasmic reticulum and then transports the protein to the Golgi apparatus, where it is glycosylated. Tyrosyl amino acids of thyroglobulin are thought to be iodinated at the apical cell membrane as the glycoprotein is secreted into the follicle lumen. When the cells are stimulated to produce thyroid hormone, apical projections of the cytoplasm, called lamellipodia, phagocytose the thyroglobulin into membrane-bound vacuoles. Lysosomes in the cytoplasm then fuse with the vacuoles to form secondary lysosomes (phagosomes). Enzymes from the lysosomes hydrolyze the thyroglobulin, liberating thyroxine and triiodothyronine, which are secreted into the extracellular space, where they may enter into the capillaries and lymphatic vessels that drain the area.

Fig. 1, ×2660; Fig. 2, ×9290;
Fig. 3, ×7880; Fig. 4, ×8720

ADRENAL GLANDS

Introduction

The adrenal glands are paired endocrine glands located near the superior poles of the kidneys in close association with adipose tissue. In the human, each adrenal gland is divided into three areas: a head, a body, and a tail. The body forms a vertical crest and two horizontal wings, and the tail is similar in shape to that of the body. Each adrenal gland initially consists of two separate glands that become united during embryonic development to form an outer cortex and an inner medulla, each possessing a distinct morphology and function. The cortex is of mesodermal origin, whereas the medulla is a derivative of the neural crest, from which sympathetic ganglia also originate. In fact, the medulla has often been considered a modified sympathetic ganglion, in itself, because of its innervation by preganglionic cholinergic fibers and the capacity of its cells (called pheochromocytes) to synthesize catecholamines.

Morphological and histochemical studies have demonstrated that there are two cell types in the medulla; one synthesizes epinephrine, the other norepinephrine. The basic morphological difference is in the structure of the intracellular storage granules. The secretion of these catecholamines appears to be controlled largely by neuronal stimulation from the preganglionic fibers, particularly during states of emotional and physical stress. The major effects of catecholamines are glycogenolysis, mobilization of free fatty acids from adipose tissue, increased blood pressure, increased heart rate, and greater mental alterness. It is of interest that aggressive and predatory animals selectively secrete more norepinephrine than epinephrine, whereas timid and placid animals secrete little norepinephrine. The human secretes ten times more epinephrine than norepinephrine from the adrenal glands. The cortex is subdivided into three concentric layers on the basis of the disposition and appearance of its cells. The outer layer, or zona glomerulosa, is thin relative to the other layers, and consists of irregular clusters of cells that produce mineralocorticoids, the most prominent of which are aldosterone and deoxycorticosterone. Mineralocorticoids act on the distal tubules of the kidney, promoting sodium reabsorption and potassium excretion, and are thus involved in the homeostasis of electrolyte concentrations within the body. The primary stimulus for the release of mineralocorticoids is angiotensin II. Adrenocorticotropic hormone (ACTH) released from the anterior pituitary gland does have a minor stimulatory effect on the glomerulosa cells, but acts mainly to stimulate the synthesis and release of glucocorticoids produced by the remaining two layers of the cortex.

The middle layer, or zona fasciculata, is the most prominent, and its cells are arranged into cords that radiate inward. At the inner cortex the parallel arrangement of cords gives way to an anastomosing network of cells that form the final cortical layer, called the zona reticularis. In response to ACTH the cells in the zona fasciculata and zona reticularis synthesize and secrete glucocorticoids. The major glucocorticoids released are cortisone and cortisol or, in some species, corticosterone. Glucocorticoids are concerned with the regulation of protein, carbohydrate, and lipid metabolism. An increased concentration of glucocorticoids in the blood may result in protein catabolism (gluconeogenesis), glycogenesis, increased blood glucose, and mobilization of lipids from adipose tissue. Glucocorticoids, particularly cortisone, also have a supressor effect upon the immune system and are used in the treatment of certain autoimmune and allergic disorders. The glucocorticoids also exhibit weak mineralocorticoid activity because of similarities in chemical structure.

Only small amounts of the sex steroid dehydroepiandrosterone are normally produced in the cortex, and although not nearly as potent as the testicular androgens, they do produce masculinizing and anabolic effects. When specific congenital enzyme defects are present in the adrenal gland, the biochemical pathway for the synthesis of dehydroepiandrosterone may become favored. In such cases, the amount produced reaches significant proportions, resulting in precocious puberty in males or a virilizing effect in females.

It is apparent that the adrenal cortex is concerned with a wide variety of functions that are essential to life, for its removal or destruction in man ultimately leads to death unless exogenous adrenal cortical hormones are administered. In adrenocortical insufficiency, a profound electrolyte imbalance ensues, causing hypovolemia, shock, and eventually death unless treatment with mineralocorticoids or salt replacement is undertaken. The absence of glucocorticoids is almost as life-threatening as the lack of mineralocorticoids. In such cases, blood glucose is difficult to maintain, and the response of the body to stress is inadequate, often resulting in death. In contrast to the cortex, the adrenal medulla is not essential to life, most likely because there are other sources of catecholamines in the body (i.e., sympathetic ganglia).

Although the cortex and medulla appear to function separately, there is a unique dependency of the medulla upon the cortex. Within the chromaffin cells, the enzyme responsible for the conversion of norepinephrine to epinephrine apparently requires the presence of glucocorticoid hormones for its induction and maintenance. This enzyme, phenylethanolamine-*N*-methyltransferase, catalyzes the transfer of a methyl group onto norepinephrine, yielding epinephrine. In this respect, the pattern of vascularization of the medulla has important physiological consequences. The adrenal medulla has a dual blood supply. Arterial divisions may pierce the connective tissue capsule and pass directly through the cortex until they reach the medulla, where they branch to form a rich capillary plexus around the clumps of chromaffin cells. Alternatively, some of the arterial vessels form a plexus within the capsule, and arteries arising from this capsular plexus form the long anastomosing sinusoidal capillaries that traverse the cortex and supply the cords of cortical parenchyma cells. At the zona reticularis the sinusoids converge into a plexus of capillaries that anastomose across the corticomedullary junction with the capillary bed in the medulla. The latter route provides the medulla with blood from the cortex, rich in cortical steroid hormones. The extent to which a medullary cell can convert norepinephrine to epinephrine may thus depend upon its location with respect to steroid-rich blood originating from the cortex.

The large central vein, which drains the majority of the medulla, plays a significant role in the delivery of catecholamines into the general circulation. Because of its large capacity, the central vein is capable of storing a large amount of catecholamines, which normally have a half-life of less than a minute in peripheral blood. The store of catecholamines can be released in spurts into the bloodstream as the smooth muscle in the wall of the central vein contracts or relaxes. Shortening of the longitudinal smooth muscle in its wall may also obstruct inflow into the central vein, in which case blood may exit the adrenal gland by an alternative pathway provided by alar and emissary veins.

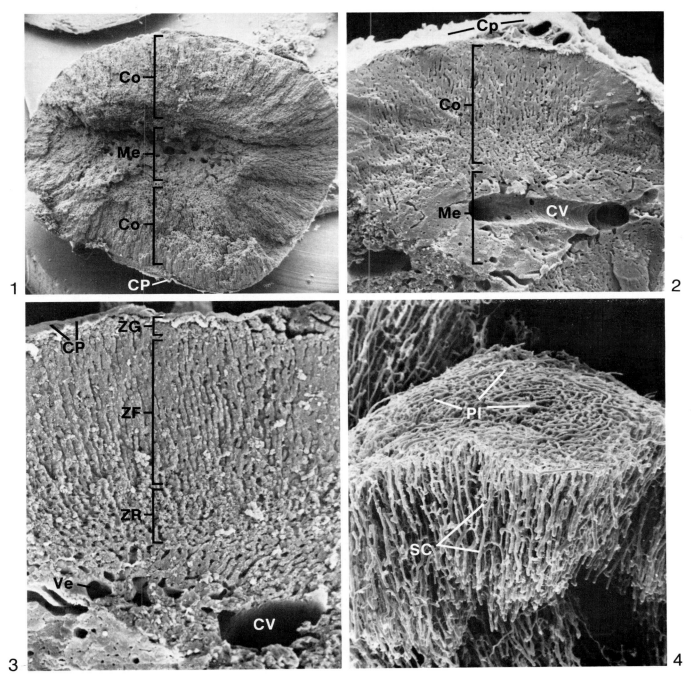

1

2

3

4

ADRENAL GLANDS
Organization

As shown in tranverse section in Figs. 1 to 3, the adrenal gland is surrounded on its exterior surface by a connective tissue capsule (**Cp**). The gland is embryologically derived from two sources: the cortex (**Co**) originates from embryonic mesoderm, and the medulla (**Me**) is derived from neural ectoderm (neural crest cells). The cortex is divided into three concentric layers, or zones: an outer zona glomerulosa (**ZG**); a middle and more prominent zona fasciculata (**ZF**), which consists of long cords of cells; and an inner zona reticularis (**ZR**). Mineralocorticoids are secreted by the cells of the glomerulosa; the glucocorticoids are released primarily from the remaining two zones. Sex steroids are secreted in large amounts from a ''provisional'' cortex during embryonic development, but not to any great extent in the adult. As demonstrated by the vascular cast in Fig. 4, the suprarenal arteries that supply the adrenal gland form a plexus (**Pl**) in the capsule. The plexus is continuous with a network of long anastomosing sinusoidal capillaries (**SC**) that surround the cords of cells in the cortex. Deep in the cortex the sinusoids form a reticular plexus that drains into the medulla to form venules (**Ve**). The venules coalesce to form a large central vein (**CV**), which is observed in both transverse (Figs. 1 and 3) and longitudinal (Fig. 2) sections. Alternatively, the blood may bypass the central vein, leaving the adrenal medulla by alar veins or emissary veins.

Fig. 1, ×25; Fig. 2, ×60;
Fig.3, ×125; Fig. 4, ×180

ADRENAL GLANDS
Zona Glomerulosa, Zona Fasciculata

The junction between the zona glomerulosa (**ZG**) and the cords of cells in the zona fasciculata (**ZF**) is illustrated in Fig. 1. A cryofracture through the zona glomerulosa is shown in Fig. 2. The zona glomerulosa is surrounded by a connective tissue capsule (**Cp**) and consists of clusters of closely packed cells that are associated with a network of capillaries (**Ca**). The capillaries extend into the zona fasciculata as straight sinusoids (**Si**), as observed in longitudinal section. When cryofractured, the cells appear vacuolated (**Va**) as a result of extraction of lipid from the cytoplasm during tissue preparation. The degree of lipid content may depend on the relative activity of the cells, and there is also wide variability in lipid content among different species. The glomerulosa cells secrete mineralocorticoids in response to decreased levels of sodium in the blood, presumably from angiotensin II stimulation.

Fig. 1, ×530; Fig. 2, ×2055

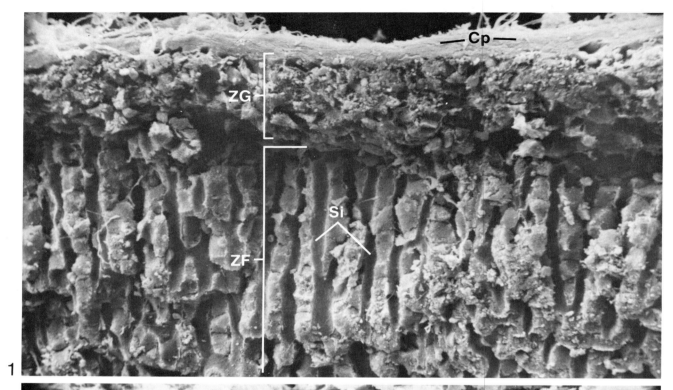

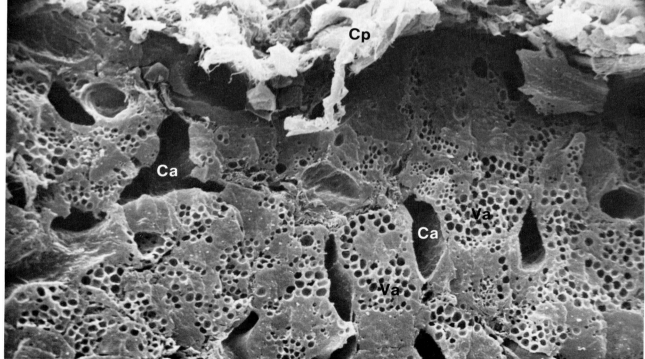

ADRENAL GLANDS
Zona Fasciculata, Zona Reticularis

Individual cells of the zona fasciculata, called spongiocytes (**Sp**), are characteristically arranged into cords of cells (Figs. 1 and 2) that radiate toward the adrenal medulla. The long, straight sinusoidal capillaries (**SC**) intervene between the cords, and are closely associated with the surfaces of the spongiocytes. A number of erythrocytes (**Er** in Fig. 1) are still present in some of the capillaries, but most of them were removed after perfusion of the gland. As in most endocrine glands, the capillaries are of the fenestrated type, which allow for rapid exchange of materials (particularly hormones) between the cells and blood. The endothelial surface (Figs. 3 and 4) consists of smooth, thickened (**Th**) areas that intervene between areas containing fenestrae (**Fe**). In the zona reticularis (Fig. 5), the cords of cells are arranged into anastomosing networks, and are surrounded by sinusoids that form a reticular plexus (**RP**). The cells of the reticularis contain many lysosomes and lipofuscin pigment granules. The cells are thought to be capable of secreting both glucocorticoids (cortisol) and androgens (dehydroepiandrosterone).

Fig. 1, ×65; Fig. 2, ×2545; Fig. 3, ×1075; Fig. 4, ×11,640; Fig. 5, ×950

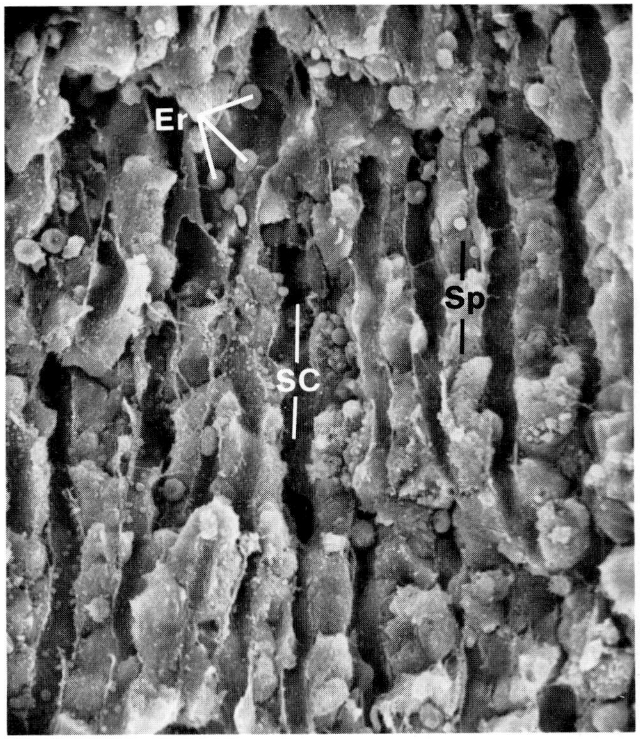

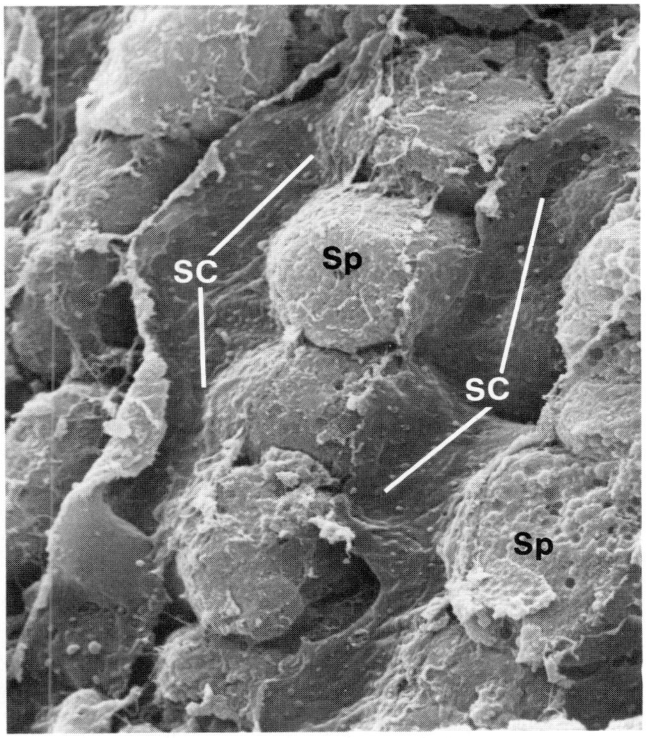

257

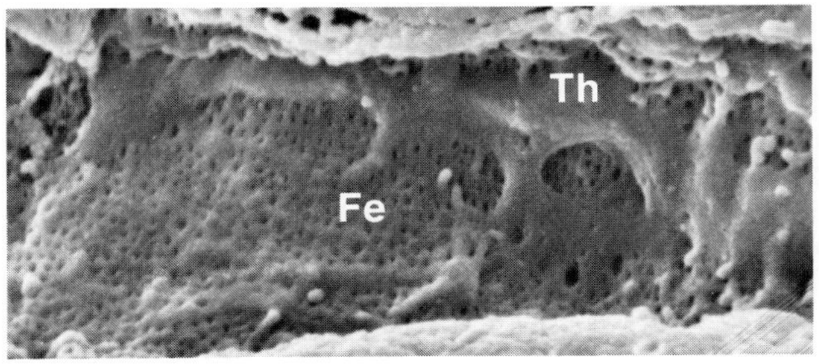

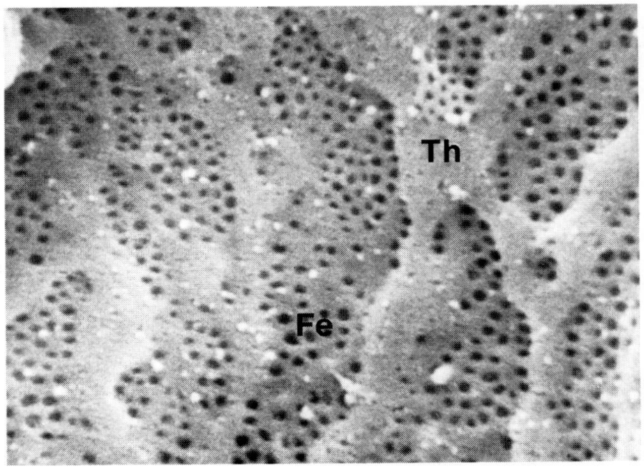

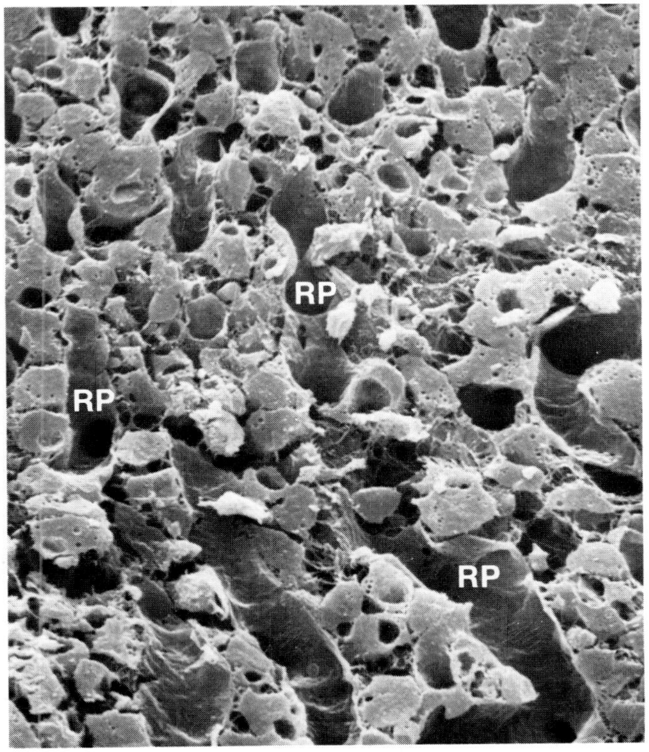

MALE REPRODUCTIVE SYSTEM

Introduction

The male reproductive tract consists of the testes, a system of genital ducts, accessory glands, and the penis. The testes are contained within the scrotum and are invested by three tunicae: the tunica vaginalis, the tunica albuginea, and the tunica vasculosa. The tunica vaginalis is a pouch of serous membrane that originates during embryonic development as a downgrowth of the peritoneal cavity. After its descent into the scrotum, that part of the tunica vaginalis which was in contact with the peritoneal cavity becomes obliterated, forming a closed sac. The subsequent descent of the testes into the scrotum results in an invagination of the closed sac, such that parietal and visceral layers of the tunica vaginalis are formed. The visceral layer, which covers the outside of each testis, becomes continuous with the parietal layer at the posterior of the testis, where blood vessels and nerves enter and exit. The intervening space between the parietal and visceral layers constitutes the cavity of the tunica vaginalis. The tunica vaginalis allows each testis to glide freely within its surrounding envelope without being subjected to damaging pressure and friction. When the connection between the cavity of the tunica vaginalis and the peritoneal cavity fails to become obliterated, a patency may persist through which the contents of the abdominal cavity (i.e., intestines) may pass. This congenital anomaly is known as oblique inguinal hernia. The tunica albuginea is a fibrous covering of the testis consisting of interlacing fibers of dense connective tissue. Beneath the tunica albuginea lies a dense layer of blood vessels, the tunica vasculosa, which is suspended by loose areolar connective tissue. At its posterior border, the tunica albuginea is reflected into the substance of the testis, forming an incomplete vertical septum of connective tissue called the mediastinum testes. Numerous septae of connective tissue originate from the mediastinum and radiate toward the periphery of the testis, eventually becoming continuous with the tunica albuginea. The thin fibrous septae, called septulae testis, divide the testis into approximately 250 conically shaped lobules. Each lobule of the testis contains 1 to 4 coiled seminiferous tubules, which may branch and anastomose freely. The seminiferous tubules are lined by stratified epithelium consisting of supportive nurse cells (Sertoli cells) and germinal epithelial cells. The germinal cells undergo differentiation within the tubular wall to become mature spermatozoa, a process known as spermatogenesis. The loose areolar connective tissue that supports the tunica vasculosa at the periphery continues into the substance of the testis to fill all of the interstices among the seminiferous tubules. The connective tissue contains fibroblasts, macrophages, mast cells, mesenchymal cells, and aggregations of special endocrine cells, known as the interstitial cells of Leydig. The interstitial cells synthesize and secrete testosterone in response to interstitial-cell-stimulating hormone (ICSH), which appears to be chemically identical to the luteinizing hormone (LH) present in females. The secreted testosterone may gain access to the general circulation by passing into the surrounding capillaries or by diffusing into the numerous lymphatic vessels that surround the seminiferous tubules. In addition to ICSH, another glycoprotein hormone, follicle-stimulating hormone (FSH), exerts its influence on the testes. FSH stimulates the differentiation of immature germ cells into spermatozoa, causing an increased production of sperm. It is not well understood how this is accomplished, but it is known that FSH causes Sertoli cells, which are intimately associated with the germ cells during the entire process of spermatogenesis, to produce an androgen-binding protein (ABP). Because of its affinity for androgens, ABP creates a concentration of testosterone in the vicinity of the developing germ cells. It is of interest that testosterone appears to be necessary for the completion of spermatogenesis.

In simple terms, the process of spermatogenesis is the development of germ cells in accord with a specific pattern of differentiation. Mitotic and meiotic divisions result in the production of fertile spermatozoa, each with a haploid number of chromosomes. The process is complex, and may proceed synchronously in various portions of the seminiferous tubules. Since spermatogenesis cannot take place at body temperature in most mammals, the testes are suspended from the body in the scrotum, where a temperature a few degrees lower than the rest of the body is maintained, enabling spermatogenesis to take place. The specialized circulation of the testes also functions to maintain the lower temperature by a mechanism of countercurrent heat exchange between adjacent arteries and veins. Failure of the testes to descend into the scrotum during embryonic development (cryptorchidism) causes infertility because the testes are subjected to body temperature. Similarly, congenital maldistribution of vessels or growth of fibrous tissue between the vessels may cause infertility by decreasing the efficiency of countercurrent heat exchange between testicular arteries and veins. The Sertoli cells and endocrine cells of the testes are not, however, so adversely affected by fluctuations in temperature.

Under normal conditions, fully differentiated spermatozoa are released into the lumen of the seminiferous tubule and continue toward the apex of the lobule, where they abruptly empty into the tubuli recti, the first segment of the system of excretory ducts. The tubuli recti are lined by Sertoli cells and are confluent with the rete testis, a plexus of spaces lined with epithelium and contained within the connective tissue of the mediastinum. From the posterior aspect of the testes, approximately 12 ductuli efferentes arise from the rete testis and form 5 to 10 conically shaped structures, called the coni vasculosi, which are produced by the spiral windings of the ducts. The ductuli efferentes are lined by groups of alternating tall and short cells. The short cells possess microvilli, but the tall cells are ciliated. The rhyth-

mic beating of the cilia provides a mechanism by which the nonmotile spermatozoa are moved along the length of the duct. At their distal ends, the coni vasculosi fuse to form the single ductus epididymidis. The epididymis is extremely long, but because it is highly coiled, it is a very compact structure. The epididymis is divided into three regions: head (caput), body (corpus), and tail (cauda). The epididymis is the site of accummulation and storage of spermatozoa; upon leaving the epididymis, the spermatozoa become motile, perhaps by a maturational process that occurs in this excretory duct. The epithelial cells lining the epididymis are reabsorptive (more than 90% of the fluid leaving the testes is reabsorbed here) and also contribute a secretory product. The epididymis gradually straightens and becomes continuous with the thick-walled ductus deferens (vas deferens); spirally arranged layers of smooth muscle in its wall help to move sperm down its length by peristaltic contractions. Each of the ductus deferens passes through the inguinal canal into the abdominal cavity and, after crossing the ureter, forms an enlargement called the ampulla. The seminal vesicles, which originate as evaginations of the ductus deferens during embryonic development, empty their secretion into the duct at this location. As the ductus deferens continues toward the base of the urinary bladder, it pierces the prostate gland and finally empties into the prostatic urethra. The epithelium of both the seminal vesicles and prostate gland responds to hormonal stimulus. Testosterone causes the

epithelium to become highly proliferative, whereas estrogen causes degeneration of the epithelium of these accessory glands. The small bulbo-urethral glands (Cowper's glands) also contribute a secretory product. These compound tubuloalveolar glands empty their secretion into the membranous urethra, and their secretory product is thought to aid in the gelation of seminal fluid after ejaculation.

The final component of the male reproductive tract is the penis, which consists primarily of three cylindrical columns of erectile tissue, each of which is surrounded by a sheath of dense connective tissue. Two of the columns, the corpora cavernosa penis, are paired and form a deep groove; the third column, the corpus cavernosum urethrae (corpus spongiosum) lies along this groove. The corpus cavernosum urethrae contains the urethra and terminates distally in an enlargement known as the glans penis. The exterior surface of the penis is covered by stratified squamous epithelium. The erectile tissue in the three corpora is a vast network of sponge-like vascular spaces that are supplied by afferent arteries and drained by efferent veins. In the flaccid state, little blood occupies the vascular spaces, and they are collapsed. In the erectile state, the inflow of blood to the spaces exceeds the outflow, resulting in a filling of the spaces. The increased volume of blood that is present causes enlargement and stiffening of the erectile tissue.

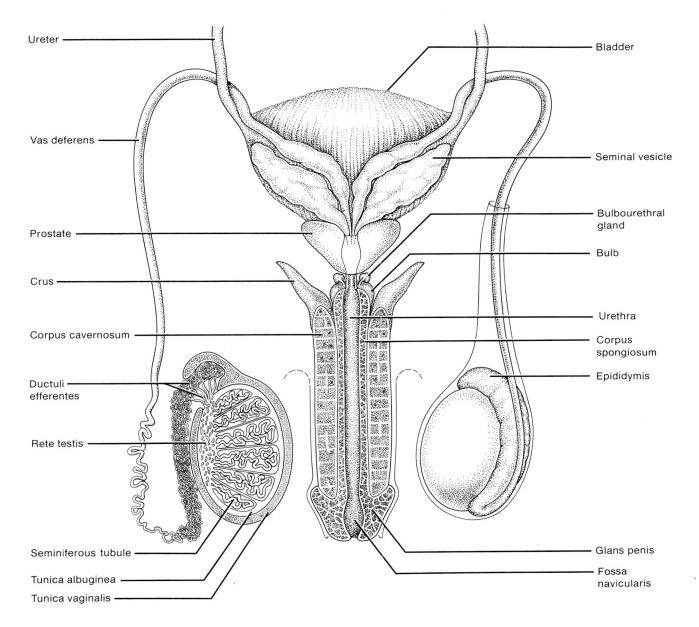

Ureter

Vas deferens

Prostate

Crus

Corpus cavernosum

Ductuli efferentes

Rete testis

Seminiferous tubule

Tunica albuginea

Tunica vaginalis

Bladder

Seminal vesicle

Bulbourethral gland

Bulb

Urethra

Corpus spongiosum

Epididymis

Glans penis

Fossa navicularis

MALE REPRODUCTIVE SYSTEM
Organization

The interrelationship between the various components of the urogenital tract of the male is diagrammed in this figure. The spermatozoa formed in the seminiferous tubules pass through a system of intratesticular genital ducts. Initially, the spermatozoa enter the tubuli recti (straight tubules), which are lined only by Sertoli cells. The tubuli recti empty into the rete testis, which is contained within a thickening of the tunica albuginea. From the rete testis extend 8 to 15 ducts called tubuli efferentes (or ductuli efferentes) that are lined by cuboidal and columnar cells, some of which are often ciliated. Motile cilia may aid in transporting the spermatozoa into the epididymis, which is the initial portion of the excretory duct system. The coiled epididymis lies in close association with the exterior surface of the testis and becomes continuous distally with the vas deferens. After passing through the inguinal canal into the peritoneal cavity, the vas deferens crosses in front of the ureter, where it becomes enlarged to form an ampulla. Distal to the ampulla, the seminal vesicle joins the vas deferens, and from this point the vas deferens continues to the urethra as the ejaculatory duct. The ejaculatory duct penetrates the prostate gland and becomes continuous with the prostatic urethra. Two small bulbourethral glands (Cowper's glands) are connected via ducts to the penile urethra, which represents the distal extension of the prostatic urethra. The corpora cavernosum and corpus spongiosum of the penis are diagrammed in a longitudinal plane of section. [Figure modified after M. Dym. From L. Weiss and R. O. Greep, *Histology* (4th ed.), McGraw-Hill Book Company, 1977].

MALE REPRODUCTIVE SYSTEM
Testis

Externally, the testis is encapsulated by the tunica albuginea, which consists of thick connective tissue. Internally, the testis is composed primarily of long, tortuous seminiferous tubules (approximately 32 in rodents; more than 200 in humans). The seminiferous tubules (**ST** in all figures) are highly coiled and are associated with loosely organized sheaths of connective tissue (**CT**) that occupy the interstices of the tubules. Contained within the connective tissue are aggregations of endocrine cells called interstitial cells of Leydig, which produce the hormone testosterone. Capillaries and lymphatic vessels traverse the connective tissue in close association with the interstitial cells and tubules, such that nutrients, hormones, and metabolites can effectively enter and exit the testicular tissue. A transverse section through the seminiferous tubules (Figs. 1 and 3) reveals their organization, which consists of an investing tunic of fibrous connective tissue, a basal lamina, and stratified epithelium (**SE**). The thickness of the fibrous tunic varies among different species, and in most cases contains cells with characteristics of smooth muscle cells. It is thought that these cells may be responsible for the contractile movements commonly observed in the seminiferous tubules. In rodents, the fibrous tunic is very thin. The stratified epithelium of the testis consists of two cell types: germ cells in different maturational stages of spermatogenesis and Sertoli cells, which are closely associated with the germ cells during their development. As the germ cells differentiate into spermatozoa, they come to lie in the tubule lumen, where their long tails (**Ta**) are often visible.

Fig. 1, ×80; Fig. 2, ×60; Fig. 3, ×100

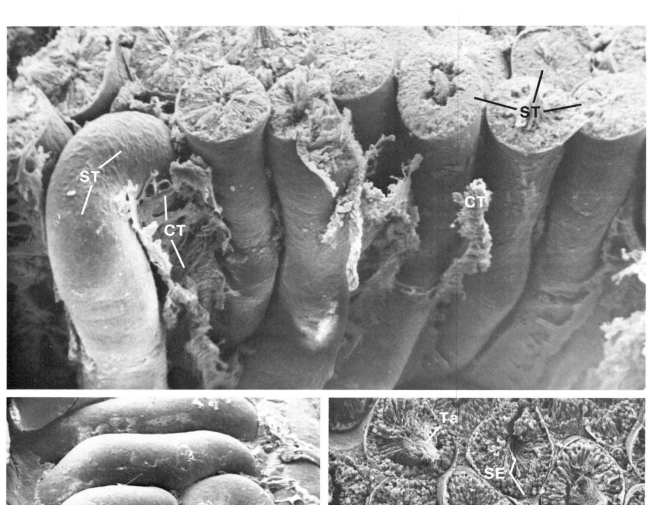

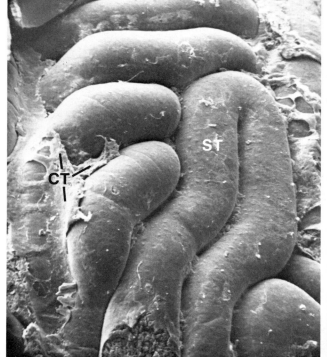

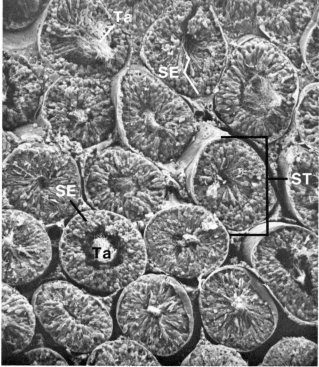

MALE REPRODUCTIVE SYSTEM
Testis

A transverse section of a seminiferous tubule is shown in Fig. 1; note that the most undifferentiated cells, the spermatogonia (**Sg**), are located at the periphery of the tubule, in close association with the basal lamina of the tubule. Some of the spermatogonia may undergo a progressive differentiation to become mature sperm; others may remain in their relatively undifferentiated state, serving to maintain (by mitosis) the population of spermatogonia for further cycles of spermatogenesis. When a spermatogonium that continues to differentiate enters prophase of the first meiotic division, it is termed a primary spermatocyte (**Sc**). After the first division, the cell is smaller and is classified as a secondary spermatocyte. The second meiotic division results in the formation of even smaller cells, the spermatids (**Sd**), containing a haploid number of chromosomes. As spermatids undergo further differentiation, a drastic change in morphology takes place, resulting in the formation of spermatozoa, the tails (**Ta**) of which are observed in the lumen of the seminiferous tubule. Somatic cells radiate from the tubule periphery toward the lumen and invest developing germ cells with cytoplasmic processes during the process of spermatogenesis. These cells are often termed sustentacular, or Sertoli, cells (**Se**). Because the process of spermatogenesis may also spread along the longitudinal axis of the seminiferous tubule as a "spermatogenic wave," transverse sections at different positions along the seminiferous tubule may vary greatly in their appearance. In the transverse section shown in Fig. 2, many of the germinal cells are in later stages of spermatogenesis. The spermatids, which are undergoing a transition into spermatozoa (a process known as spermiogenesis), are easily distinguished by their long tails (**Ta**), which extend into the tubule lumen. During differentiation, a large portion of the spermatid cytoplasm, called the residual cytoplasm, becomes detached from the head region of the spermatozoan, presumably for streamlining purposes. It is subsequently phagocytized by surrounding Sertoli cells, or moves progressively toward the distal end of the sperm tail (**arrows** in Fig. 3), where it is later shed in the epididymis and vas deferens.

Fig. 1, ×580; Fig. 2, ×330; Fig. 3, ×1650

263

MALE REPRODUCTIVE SYSTEM
Testis

A transverse section of a seminiferous tubule is illustrated in the light photomicrograph shown in Fig. 1. An area of the tubule wall, similar to that outlined by the rectangle in Fig. 1, is diagrammed in Fig. 2. Germ cells in different stages of spermatogenesis as well as Sertoli cells can be identified on the basis of nuclear size, nuclear configuration, and the cellular location within the wall of the tubule. Examples of spermatogonia (**Sg**), spermatocytes (**Sc**), and spermatids (**Sd**) are represented in both figures. Throughout the maturational process, the germ cells occupy the intercellular spaces (**IS**) among Sertoli cells (**Se**), and are intimately associated with their cytoplasmic extensions. The plasma membranes of adjacent Sertoli cells are locally fused by junctional complexes (**JC**) that divide the intercellular space into a basal compartment, in which spermatogonia predominate, and an adluminal compartment, in which the remaining spermatocytes and spermatids reside. As spermatogonia differentiate into spermatocytes, they are transferred from the basal compartment into the adluminal compartment, a process that involves the dissolution and reformation of junctional complexes. Spot-like junctional complexes (desmosomes) are also present between Sertoli cells and germ cells. The junctional complexes (particularly the zonula occludens), basal lamina (**BL**), and myoid cell layer (**ML**) functionally constitute the "blood-testis barrier," which restricts substances of high molecular weight from entering the adluminal compartment. Nutrients from surrounding capillaries (**Ca**) and hormones can gain access to the adluminal compartment only by diffusing through the junctional complexes or, indirectly, by passing through the Sertoli cell cytoplasm. The resulting compartmentalization of the tubule wall creates a specialized microenvironment for cells as they progress through various stages of spermatogenesis.

×470

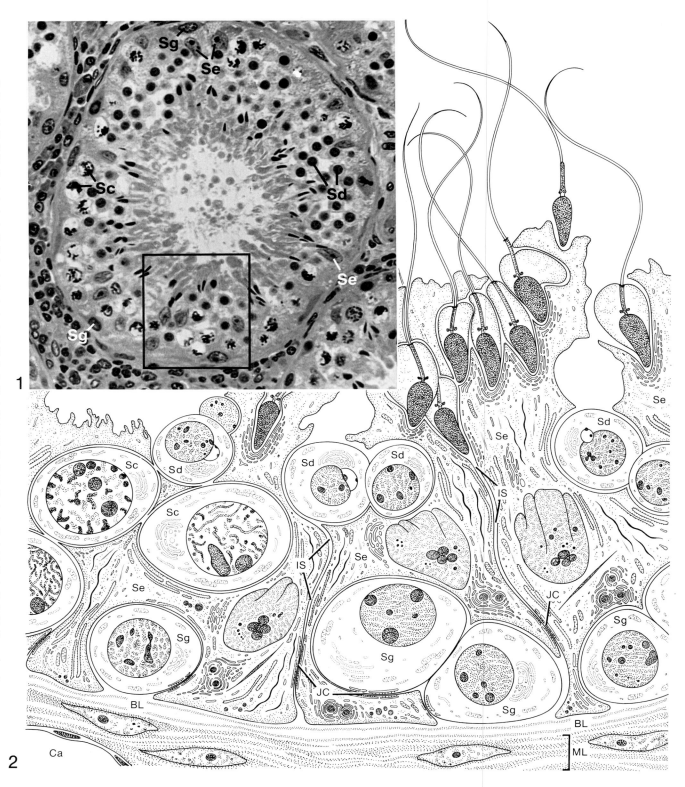

MALE REPRODUCTIVE SYSTEM
Testis

As observed in Figs. 1 to 4, the Sertoli cells (**Se**) possess a highly elaborate shape—a result of their close association with developing germ cells (**GC**). The germinal epithelium occupies deep recesses in the Sertoli cell cytoplasm which conforms to the changing shape of the germ cells during spermatogenesis. Because the developing germ cells do not undergo complete cytokinesis during mitosis and meiosis, a syncytium is formed, the cells of which are connected by thin cytoplasmic bridges (**CB** in Figs. 1 and 2). A dividing cell (**DC**) is observed in Fig. 3; note that the surrounding germ cells possess flattened, planar areas on their surface where adjacent cells were closely apposed. The small, spherical particles often distributed along the borders of the planar areas may be the sites of membrane specializations that had existed between adjacent cells. The Sertoli cell sends forth long, slender tendrils (**Te** in Figs. 1 to 3) that ramify over the surface of the germ cells. Since the syncytial germ cells are tightly packed together, leaving little intervening space, the formation of thin cytoplasmic extensions of Sertoli cells most likely represents a mechanism by which the cell can maintain intimate cytoplasmic association with each germ cell during spermatogenesis. In Fig. 5, many of the germ cells of the syncytium were removed to reveal the recesses (**Re**) in the remaining Sertoli cells. Although the Sertoli cells extend inward through the entire radius of the tubule, Fig. 5 shows that only the basal portions of the cells remain in close association with the basal lamina (**BL**) of the tubule wall. The Sertoli cells form specialized junctional complexes with one another that seal off the basal compartment from the adluminal compartment. The area enclosed within the rectangle in Fig. 5 is enlarged in Fig. 6. Germ cells in the adluminal compartment were removed in this area, leaving a recess in which the junctions (**Ju**) between adjacent Sertoli cells (**Se**) can be observed.

Fig. 1, ×1600; Fig. 2, ×1780; Fig. 3, ×1735
Fig. 4, ×1335; Fig. 5, ×845; Fig. 6, ×4670

MALE REPRODUCTIVE SYSTEM
Testis

The Sertoli cells (**Se**) characteristically radiate from the basal end of the seminiferous tubule toward the tubule lumen, as shown in Fig. 1. The syncytium of germ cells (**GC**) often develops as a column of cells that extends inward between the long trunk-like extensions of the Sertoli cells. Thin tendrils (**Te**) and flattened cytoplasmic extensions (**Ex**) of Sertoli cells are observed in association with the syncytial cells shown in Figs. 1 to 4. Since Sertoli cells are intimately, and perhaps symbiotically, associated with germ cells during the entire developmental process, they most likely play a key role in germ cell maturation. In response to follicle-stimulating hormone (FSH), the Sertoli cells in rats and rabbits have been shown to secrete an androgen-binding protein (ABP) that combines with testosterone. This creates a local concentration of androgen, which may be necessary for spermatogenesis to occur. In addition, indirect evidence suggests that Sertoli cells may produce small amounts of estrogen in response to FSH stimulation. FSH also promotes the mitotic division of spermatogonia. Those spermatogonia that continue to differentiate are transferred from the basal compartment to the adluminal compartment. Most dividing cells in the syncytium remain connected by cytoplasmic bridges (**CB** in Fig. 4). The former connections (**arrows** in Figs. 2 and 3) of the cytoplasmic bridges are observed in areas where the separation of cells has occurred during specimen preparation. The intercellular bridges may promote synchronous division of the cells in the syncytium; not all spermatogonia, however, divide at the same time. A number of them may degenerate, never reaching maturity.

Fig. 1, ×1075; Fig. 2, ×1375;
Fig. 3, ×2820; Fig. 4, ×2740

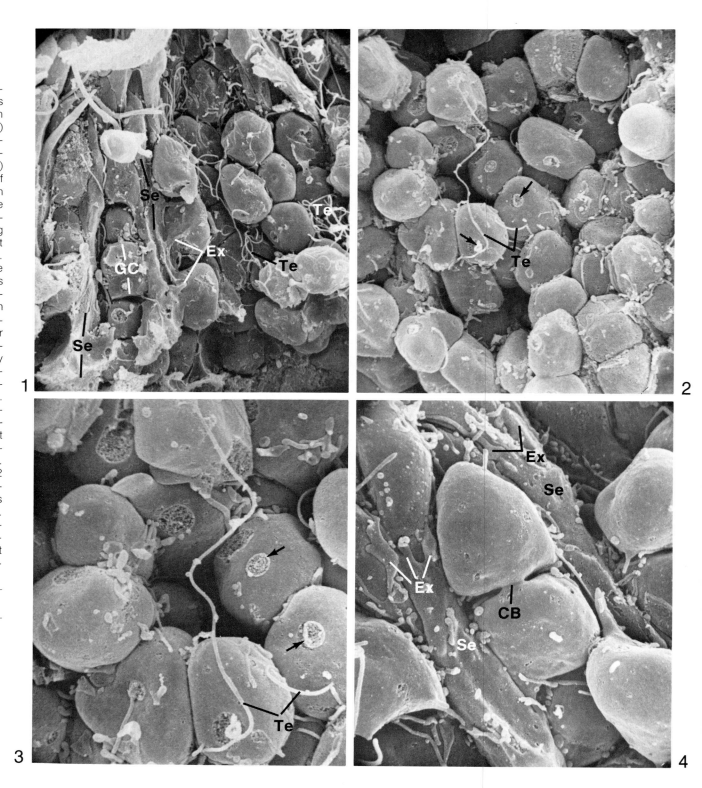

266

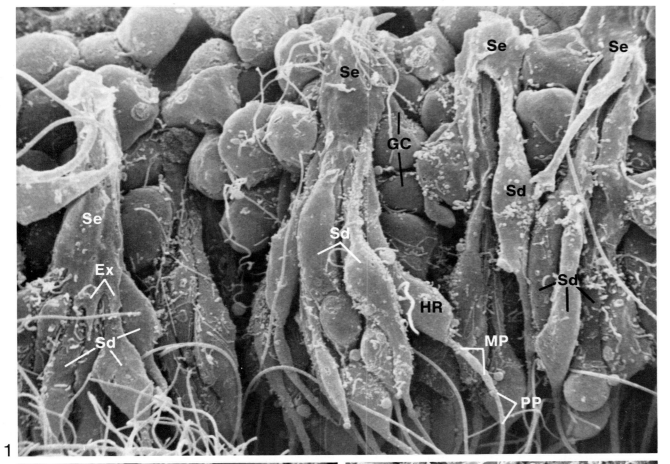

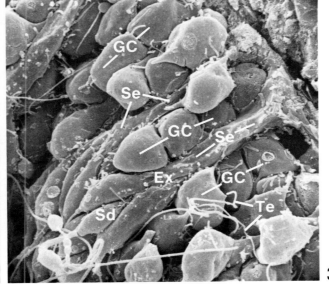

1

2

3

MALE REPRODUCTIVE SYSTEM
Testis

The second meiotic division of the secondary spermatocyte results in the formation of spermatids, each with a haploid number of chromosomes. Through the process of spermiogenesis, the spermatids undergo morphological differentiation into spermatozoa, and at fertilization the combination of chromosomes contributed by the sperm and egg restores the number of chromosomes to a diploid amount. In these figures, germ cells in the late stages of spermatogenesis are readily distinguished. During spermiogenesis the spermatids (**Sd**) elongate, forming a head region (**HR**) and a tail region; the latter consists of a middle piece (**MP**), a principal piece (**PP**), and an end piece. Groups of spermatids that were derived from the same syncytium are organized into conically shaped bundles. The developing tails extend into the tubule lumen, whereas the head regions taper toward the apex of the bundle, where they are invested by the most distal extensions (**Ex**) of Sertoli cells (**Se**). The lower left area in Fig. 1 is enlarged in Fig. 2 to reveal details of the association between spermatids and Sertoli cells. The cytoplasm of the Sertoli cell forms flattened extensions that appear to form a sleeve around the heads of the spermatids. Also visible are the long, tendril-like extensions (**Te**) of the cell. Syncytial columns of germ cells (**GC**) in earlier stages of spermatogenesis are also observed between the trunk-like extensions of Sertoli cells in Figs. 1 and 3.

Fig. 1, ×1560; Fig. 2, ×2200; Fig. 3, ×1090

MALE REPRODUCTIVE SYSTEM
Testis

The surface morphology of a mature human spermatozoon is illustrated in Figs. 1 and 2, and the internal organization is diagrammed in Fig. 3. The spermatozoon is divided into a head region (**HR**) and a tail region; the latter is further divided into a neck (**Nk**), middle piece (**MP**), principal piece (**PP**) and end piece (**EP**). Head shape varies among different species. In the human, the spermatozoon head is usually flattened, as viewed face on in Figs. 1 and 2. A lateral view of the head is illustrated in the diagram, but the neck region has been drawn face on to reveal details of its internal organization. The acrosome (**Ac**) encloses the anterior two-thirds of the nucleus (**N**). The neck of the spermatozoon is located between the head and the first gyre of a helical mitochondrial derivative in the middle piece. The major component of the neck is the connecting piece, which consists of nine striated columns (**StC**), each of which appears as a segmented band in longitudinal section. Distally, the striated columns are continuous with nine bundles of outer dense fibers (**ODF**) that extend through the middle piece. Proximally, one of the striated columns is expanded to form a capitellum (**Cap**), which is fused with the basal plate (**BP**). Lateral to the basal plate the nuclear envelope (**NE**) may be folded into the neck region during the late stages of spermiogenesis. A centriole (**Ce**) is also located in the neck. The middle piece characteristically contains a number of mitochondria (**M**), situated end to end, forming a helical wrapping around the flagellar microtubules (**Mt**). The mitochondria provide the energy for flagellum motility. A cytoplasmic enlargement, containing smooth vesicles, vacuoles, and flattened saccules, is usually located halfway along the length of the middle piece, and represents a portion of the residual cytoplasm. A dense ring, the annulus (**An**), marks the end of the middle piece. The long principal piece is characterized by a fibrous sheath (**FS**) that surrounds the outer dense fibers. The fibrous sheath also forms two opposed longitudinal columns (**LC**) that occupy the former locations of two of the outer dense fibers. The fibrous sheath terminates at the distal end of the principal piece. The characteristic microtubule arrangement of the flagellum (**F**) continues into the end piece and terminates near its distal end. [Fig. 3 after T. L. Lentz, *Cell Fine Structure*, W. B. Saunders Company, 1971.]

Fig. 1, ×15,130; Fig. 2, ×6540

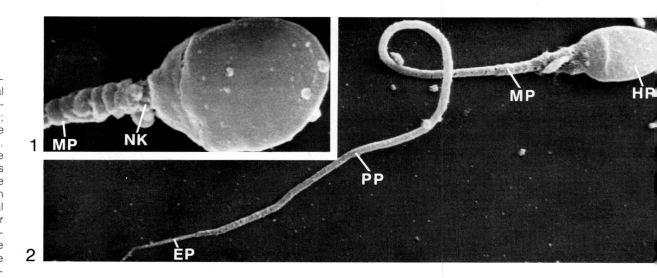

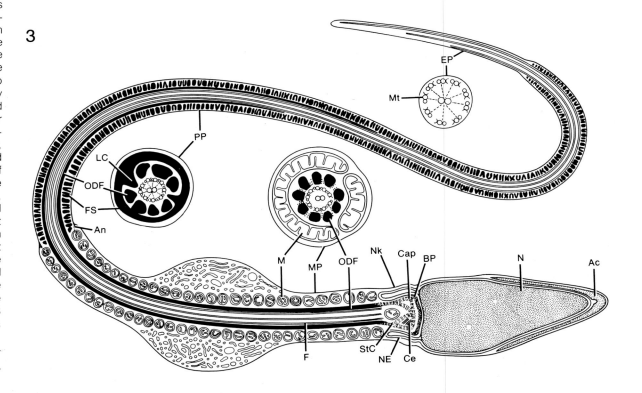

MALE REPRODUCTIVE SYSTEM
Ductus Epididymidis

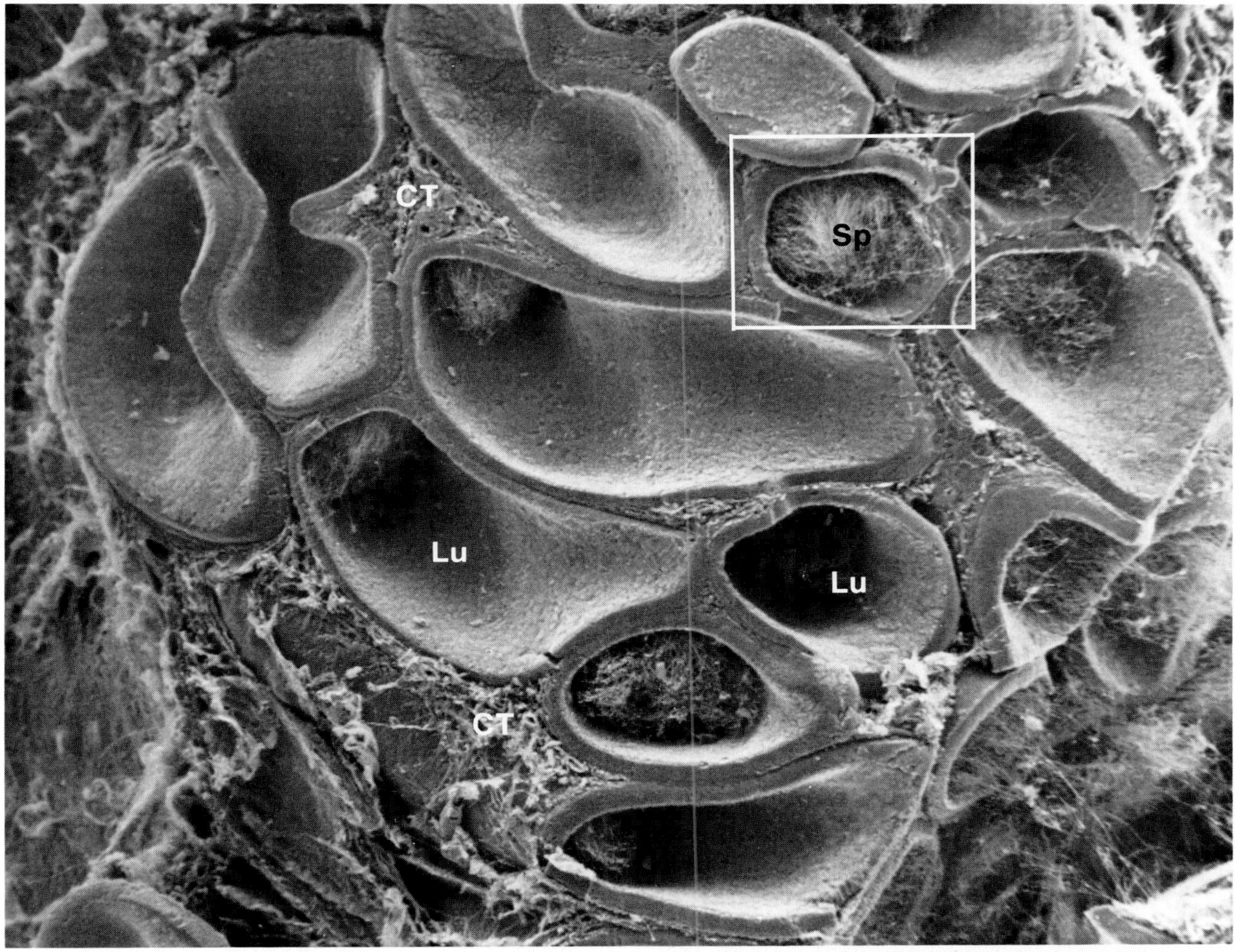

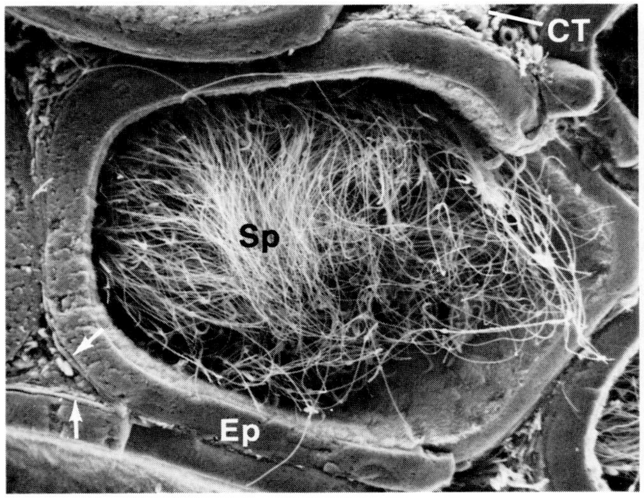

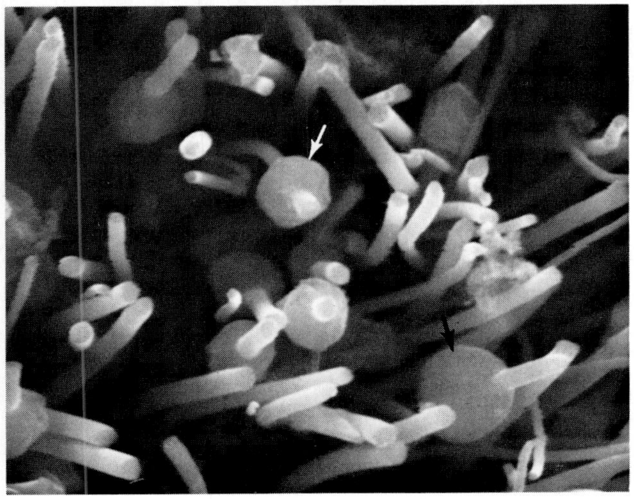

After passing through a system of intra-testicular ducts, the spermatozoa enter the epididymis, which is the initial portion of the excretory duct system that makes up the male urogenital tract. The epididymis, as shown in Fig. 1, consists of a single, highly coiled tube. When cryofractured, the lumen (**Lu**) is revealed in many regions along its length. The region enclosed by the rectangle in Fig. 1 is further enlarged in Fig. 2 to reveal details of the wall of the epididymis. Dense aggregations of spermatozoa (**Sp**) are frequently observed within the lumen. The epididymis is lined by pseudostratified columnar epithelium (**Ep**), which is supported by a basal lamina. The basal lamina is surrounded by smooth muscle fibers (**arrows**) that produce peristaltic waves of contraction along the longitudinal axis of the epididymis so as to propel the spermatozoa along its length. The interstices of the coiled portions of the epididymis are occupied by connective tissue (**CT**) that is rich in blood capillaries. When spermatozoa within the epididymis lumen are viewed in detail (Fig. 3), small spheres (**arrows**) located at the distal end of many of the tails are observed. These are thought to represent residual cytoplasm that has migrated along the length of the tail. The residual cytoplasm is shed sometime during passage through the epididymis and vas deferens.

Fig. 1, ×120; Fig. 2, ×305; Fig. 3, ×3845

MALE REPRODUCTIVE SYSTEM
Ductus Epididymidis

A lateral view of the wall of the epididymis is shown in Fig. 1 (rodent), and a corresponding region is illustrated in the light micrograph in Fig. 2 (monkey). The pseudostratified epithelium (**Ep**) lining the epididymis characteristically possesses long microvilli (**Mv**) that were initially termed stereocilia because of their close resemblance to cilia. For size comparison, a tail (**Ta**) of one of the many spermatozoa found in the epididymis is observed next to the microvillous border. Residual cytoplasm (**RC**) is also present at the distal end of the tail. The cell borders (**CB**) of the epithelium appear as ridges in Fig. 1. The vacuolated interior of the fractured cells may represent the location of intracellular organelles, such as the Golgi apparatus or lysosomes, which are known to be prominent in these cells. Also visible is the layer of smooth muscle (**SM** in Figs. 1 and 2) underlying the epithelium. In Fig. 2, the irregular distribution of cell nuclei (**Nu**) is characteristic of pseudostratified epithelium. As illustrated in Fig. 3, spermotozoa (**Sp**) are often embedded in the microvillous (**Mv**) border of the epithelial cells, and, when removed, depressions (**arrows** in Fig. 4) often remain in their place. The head region of the rat spermatozoon is characteristically hook-shaped. Although the precise function of the epididymis is not known, it is here that the spermatozoa become motile and fertile. The epithelial cells apparently perform specific functions at different regions along the epididymis. Many of the cells liberate a secretory product (**SP** in Figs. 5 and 6) into the lumen of the epididymis, but evidence of resorptive activity has also been observed. It is believed that many of the cytoplasmic fragments eliminated during spermatogenesis are resorbed and digested in the epididymis. Studies have indicated that approximately 99% of the fluid exiting the seminiferous tubules is resorbed in the epididymis.

Fig. 1, ×2705; Fig. 2, ×445; Fig. 3, ×2500
Fig. 4, ×715; Fig. 5, ×380; Fig. 6, ×2355

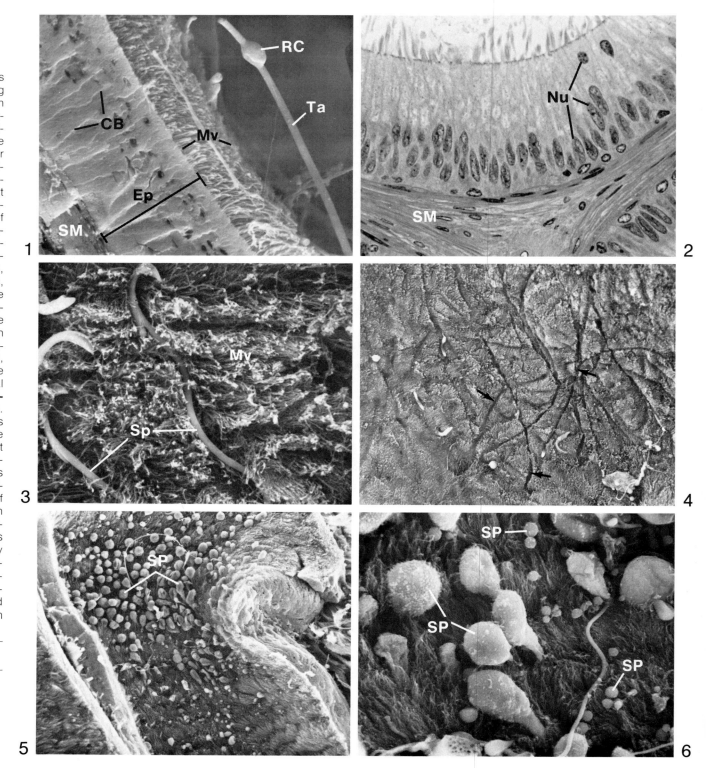

MALE REPRODUCTIVE SYSTEM
Ductus Deferens

After pursuing a tortuous course, the epididymis becomes continous with the ductus deferens. A transverse section of this long, thick-walled extratesticular duct is shown in Fig. 1. The majority of the tubular wall consists of three layers of smooth muscle (**SM**), which are spirally arranged around the longitudinal axis of the vas deferens. The presence of smooth muscle is responsible for powerful peristaltic contractions that serve to propel spermatozoa along the length of the vas deferens during the process of ejaculation. The smooth muscle is surrounded peripherally by an adventitia (**Ad**) layer consisting of connective tissue through which blood vessels (**BV**), lymphatic vessels, and nerves travel in a course parallel to the tube. The lumen (**Lu**) of the vas deferens is narrow. Along most of its extent, the mucosal lining (Figs. 1 to 3) consists of pseudostratified columnar epithelium (**Ep**) that possesses sterocilia (**St**). The epithelium is anchored by a basement membrane to an underlying lamina propria (**LP**) containing connective tissue rich in elastic fibers. The mucosa characteristically forms longitudinal folds along the length of the duct. As observed from the luminal surface (Fig. 3), a type of secretory material (**SM**) is often associated with the prominent microvillous (**Mv**) border of the epithelial cells. Although the precise nature of the secretion is not known, it may play a role in the nourishment and final maturation of spermatozoa (**Sp**) during their passage through the lumen. Before the vas deferens penetrates the prostate, it becomes dilated to form the ampulla. It is at the final portion of the ampulla where the seminal vesicles become continuous with the vas deferens. Distal to the ampulla, the vas deferens enters the prostate, where it is termed the ejaculatory duct, and subsequently enters the prostatic urethra.

Fig. 1, ×70; Fig. 2, ×665; Fig. 3, ×860

271

MALE REPRODUCTIVE SYSTEM
Seminal Vesicle

The seminal vesicles are accessory glands of the male reproductive tract and arise as evaginations of the ductus deferens during embryonic development. The elongated saccular organs secrete a yellow, viscid liquid that is rich in fructose content. This carbohydrate is thought to be utilized as an energy source by the motile spermatozoa of the ejaculate. The seminal vesicle shown here in transverse section was fractured open and the majority of the secretory product that normally occupies the lumen was removed by washing to reveal details of the mucosal surface. The wall of the seminal vesicle consists of an outer connective tissue layer, smooth muscle, and a mucosa. The mucosa consists of pseudostratified epithelium in association with an underlying layer of connective tissue. The mucosa forms an extensive system of branched folds that give rise to primary, secondary, and tertiary divisions, which increase the potential secretory surface area. The folding is complex and often difficult to interpret in histological sections. In scanning electron micrographs, it appears that the folding produces long, tube-like structures (**Tu**) that often branch and project into the lumen. These tube-like structures contain secretory material (**SM**) identical to that found in the lumen proper. Much of the extensive folding of the mucosa appears to originate as invaginations of the epithelium, such that the lumen of the tube-like structure is continuous with the lumen proper of the seminal vesicle. Consequently, after material is secreted into the tubes by the epithelial lining, it then passes into the lumen of the seminal vesicle at places (**arrows**) where the epithelium had originally invaginated during the organ's morphogenesis. The secretory product can be seen in many of these locations. The cells lining the unfolded epithelial surface (**ES**) secrete directly into the lumen of the seminal vesicle.

×110

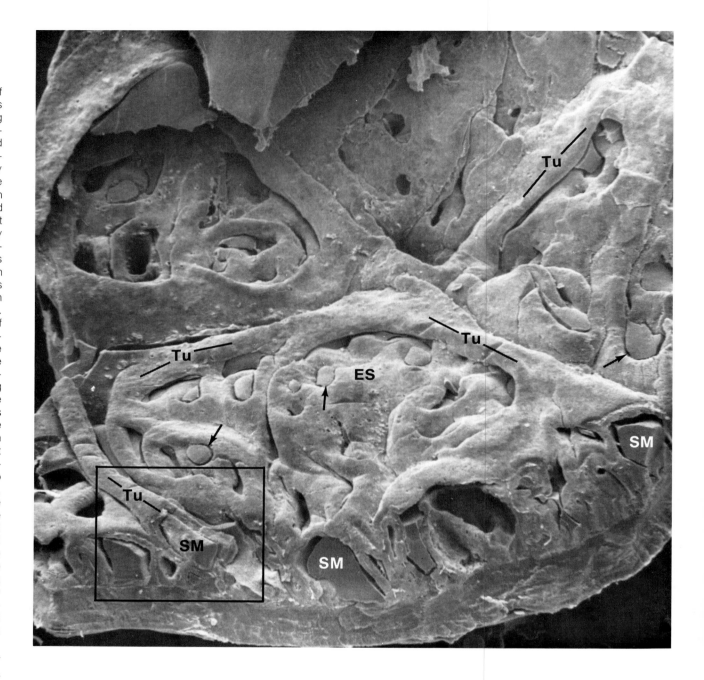

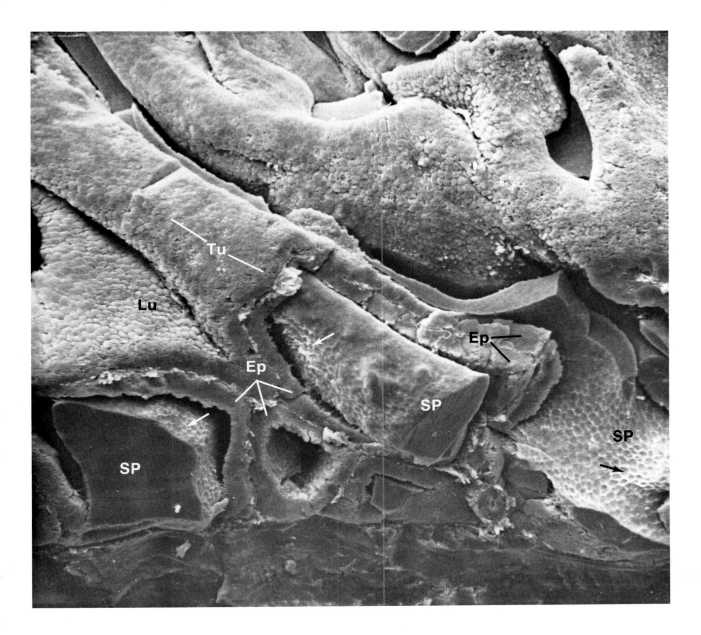

MALE REPRODUCTIVE SYSTEM
Seminal Vesicle

This figure is an enlargement of the area of the seminal vesicle enclosed by the rectangle in the previous figure. The complex folding of the mucosal surface enchances the secretory capacity of the seminal vesicle. It is apparent that the tube-like structures (**Tu**) are made up of extensively folded epithelium (**Ep**) that forms two layers of cells, one facing the lumen proper (**Lu**) of the seminal vesicle, and the other facing the interior of the tube. The latter produces a secretory product (**SP**) that fills the interior of the tube. At locations where the secretory product has been separated from the epithelium, imprints (**arrows**) formed by each of the secretory cell's apical surface are present on the secretory product.

×535

MALE REPRODUCTIVE SYSTEM
Seminal Vesicle

This view into the interior of the distal end of the seminal vesicle reveals the large out-pocketings (**Ou**) formed by invaginations of the mucous membrane. The topography of the interior surface in this region appears highly folded (**Fo**), and the secretion that normally fills the lumen of the seminal vesicle was removed during tissue preparation.

×60

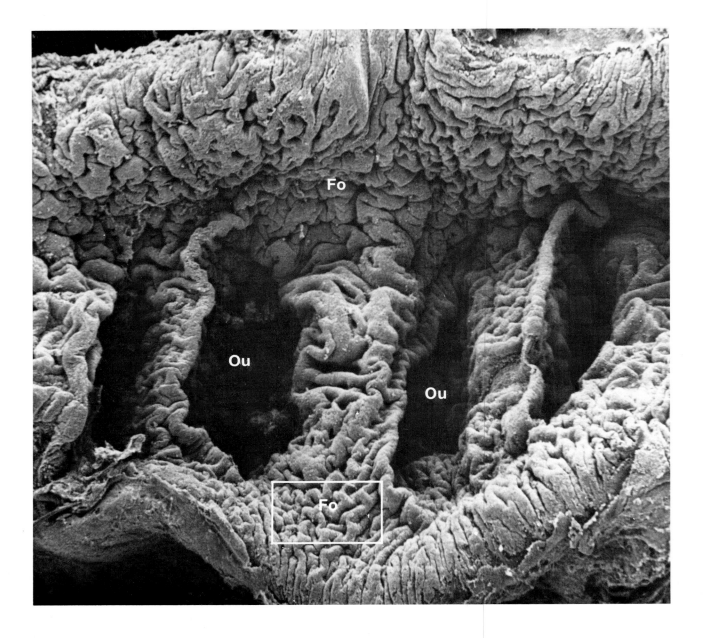

MALE REPRODUCTIVE SYSTEM
Seminal Vesicle

Fig. 1 is an enlargement of the area enclosed by the rectangle in the previous figure. Many of the folds (**Fo**) on the mucosal surface anastomose and branch. A number of epithelial cells are in the process of liberating a secretory product (**arrows**) on the surface. Details of the epithelial surface are shown in Figs. 2 to 5. As secretory product accumulates within them, the epithelial cells appear to become rounded at their apical border (Fig. 3), and there is a decrease in the number of surface microvilli (**Mv**). Cells that are in the process of releasing their secretory product (**SP**) can be seen in Figs. 2 to 4. Other cells have already released their secretory product and appear collapsed (**Co**). Some of the secretory product appears in the form of filamentous material; however, the majority of the secretion has been removed by washing. When the epithelial surface is viewed at higher magnification, areas can be found (**arrows** in Figs. 4 and 5) where the apical plasma membrane has apparently fused with the membrane of underlying secretory granules. The fusion of the membranes allows the secretory material (**SM** in Fig. 5), which was enclosed by the granule membrane, to exit from the cell by the process of exocytosis.

Fig. 1, ×340; Fig. 2, ×1710
Fig. 3, ×3085; Fig. 4, ×4895; Fig. 5, ×10,680

FEMALE REPRODUCTIVE SYSTEM

Introduction

The menstrual cycle in humans is variable in length, but has an average duration of approximately 28 days. Although transitions are gradual, the cycle is divided into four stages on the basis of events that occur concurrently in the ovary and uterus in response to changes in the concentration of various hormones. The *menstrual phase* begins with the first signs of menstrual discharge and lasts for a period of 1–4 days, during which time the entire functionalis layer of the uterine endometrium is sloughed. This is followed by what has been variously referred to as a *proliferative, reparative, estrogenic,* or *follicular phase* that begins at the end of menstruation and terminates with ovulation, which occurs about mid-cycle (approximately day 14 of the menstrual cycle). During this period, the functionalis layer is regenerated from the remaining basal layer of the uterine endometrium, a process that occurs under the influence of estrogen. Estrogen is produced in the ovary by cells of the ovarian follicles, which are undergoing maturation at this time. Early follicle maturation is, in turn, stimulated by the rising levels of follicle-stimulating hormone (FSH), which is released from basophil cells in the pars distalis of the anterior pituitary gland. The final stages of follicle maturation require the synergistic action of FSH and another hormone, the leuteinizing hormone (LH), which is released from basophil cells

in the pars distalis. The release of FSH and LH from the pituitary is controlled by separate polypeptides (FSH-releasing factor and LH-releasing factor), which are produced by neurons of the hypothalamic nuclei. These releasing factors, liberated from free nerve endings, gain access to the basophil and acidophil cells via the portal circulation of the pituitary. The secretion of these releasing factors from nerve endings is cyclical, and hence is thought to be responsible for the regularity of events in the ovary and uterus. At the end of the proliferative (follicular) phase (around day 14), a surge in the level of LH appears to be one of the key events in triggering ovulation of the oocyte. Those follicular cells, which surrounded the oocyte before its ovulation, remain in the ovary and, under the influence of LH, form the corpus luteum. The formation of the corpus luteum, about day 14, marks the beginning of the *secretory phase* (also called *progravid* or *luteal phase*), which lasts until approximately day 26 or 27. In addition to the formation of a functioning corpus luteum, this period is characterized by secretion from uterine glands and early development of the embryo in preparation for implantation in the uterus (if fertilization has occurred). During this time the corpus luteum is maintained by the continued secretion of LH (the hormone prolactin serves this function in rodents). Because the

lutein cells synthesize and secrete both progesterone and estrogen into the blood stream, the corpus luteum has been considered a miniature endocrine gland. The hormones that it produces are responsible for changes that take place during this period in the oviduct and uterus. Toward the end of this period (if pregnancy does not occur), the corpus luteum ceases to function, and the resulting decreased levels of estrogen and progesterone produce the onset of the *premenstrual phase* (lasting only 1 or 2 days). During this stage, the deprivation of progesterone and estrogen initiates changes in the uterine vasculature, which bring about menstruation. Nonprimate mammals, however, do not menstruate; their sexual cycle is called an estrous cycle.

A more detailed account of the series of events involved in maturation of the egg (oocyte), as well as the differentiation and multiplication of the cells that make up the follicle, will bring to light a number of important points concerning the production of estrogen and progesterone and their subsequent effects on target tissues, such as the uterus and oviduct.

In humans, the germ cells, or oogonia, are formed early in embryonic development from yolk sac endoderm cells that migrate into the genital ridges of the forming gonad. The entire complement of several hundred thousand germ cells is present in the ovarian cortex at the time of birth. Each oogonium is surrounded by a layer of simple squamous epithelium and constitutes a primordial follicle. The primordial follicles remain relatively quiescent until puberty, at which time the circulating blood levels of follicle-stimulating hormone begin to rise dramatically. In response to the increased levels of FSH, the primordial follicle becomes activated to initiate the maturation process. The first sign of maturation is a change in the size and shape of the simple squamous follicle cells that surround the oogonium; this change marks the beginning of their transformation into cuboidal cells. At this point, the primordial follicle is considered a primary (unilaminar) follicle. The follicle (granulosa) cells subsequently enlarge and proliferate to form a secondary (multilaminar) follicle that consists of stratified cuboidal epithelium, which is invested by a layer of ovarian connective tissue called the theca folliculi. Such secondary follicles are transformed into mature Graafian follicles by the progressive formation of a large fluid-filled antrum, an increase in the number of granulosa cells, meiotic division of the oocyte, and an increased vascularity of the theca. The secondary oocyte in the Graafian follicle is blocked in metaphase of the second maturation division. If fertilization occurs after ovulation, the second maturation division is completed, a sec-

ondary polar body is formed, and a mature ovum results. Fertilization, if it occurs, usually does so in the upper part of the oviduct. Under these conditions, the embryo is usually in a blastocyst stage with a trophoblast layer by the time it reaches the uterine cavity to become implanted in the uterine endometrium by about day 21 of the cycle. The embryonic trophoblast cells early differentiate into chorionic villi of the fetal placenta and secrete human chorionic gonadotropin (HCG). This hormone is a glycoprotein that stimulates the corpus luteum to continue its production of progesterone and estrogen. If, however, pregnancy does not occur, the corpus luteum eventually ceases to produce its hormones after about 12 days and gradually becomes converted into scar tissue (a corpus albicans).

Estrogen is most concentrated in plasma just before ovulation, and rises again at midluteal phase. Much of the circulating estrogen is protein bound and may be oxidized into the inactive glucuronide or sulfate forms in the liver. Estrogen facilitates development of ovarian follicles, repair of the functionalis layer of the endometrium, and causes increased motility of the oviducts. It also causes increased uterine blood flow and enhances excitability of uterine muscles. As estrogen levels in the blood rise, a negative feedback effect tends to decrease FSH secretion or release

from the hypothalamic–pituitary system. In sufficient quantity, estrogens also appear to trigger a burst of LH secretion by a positive feedback mechanism, resulting in ovulation.

After ovulation, apparently in response to the luteinizing hormone (LH), the granulosa and theca cells of the Graafian follicle are transformed into a corpus luteum that becomes highly vascularized. LH activates adenyl cyclase in the forming corpus luteum, and, in turn, increased cyclic AMP initiates a reaction that includes protein synthesis and steroid secretion. Progesterone, produced by cells of the corpus luteum, is responsible for changes associated with the secretory or progestational uterus—in particular, secretion by uterine glands. Large amounts of progesterone inhibit LH secretion. Secretion of FSH and LH is inhibited by high circulating estrogen and progesterone levels during the luteal phase of the menstrual cycle. Release of LH is inhibited by the secretion of a LH-inhibiting factor produced by neurons in the hypothalamus.

Atresia, or degeneration of follicles, can occur during any period of their development. Characteristic initial signs of this event are the appearance of pyknotic follicle cell nuclei; the cells become detached and float in the antrum of the follicles. The factors responsible for follicular atresia are not well understood.

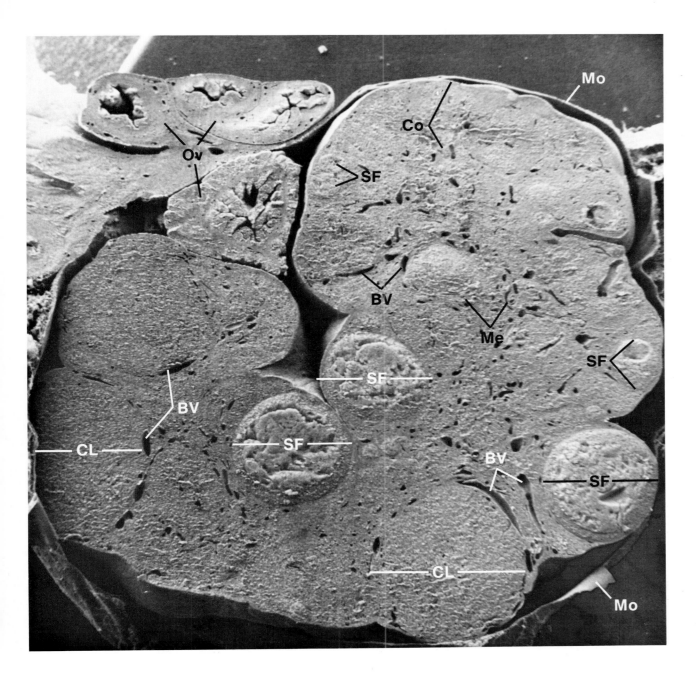

FEMALE REPRODUCTIVE SYSTEM
Ovary

The rat ovary, shown here in section, is rather indistinctly divided into a central medulla (**Me**) and an outer cortex (**Co**). The ovary is composed of an extensive connective tissue component of cells and fibers (primarily reticular and collagenic) called the stroma, which forms a packing between the developing germ cells. The medulla is principally composed of a loosely packed connective tissue stroma that contains many blood vessels (**BV**), lymphatic vessels, and nerves. The outer surface of the ovary consists of a single layer of cells, squamous or cuboidal in shape, which is continuous with the peritoneum at the mesovarium (**Mo**). The cortex of the ovary contains many germ cells, each surrounded by a single layer of flattened follicle cells. When the primordial follicle is stimulated by hormones, the follicle cells change their shape and proliferate. This event results in the formation of multilaminar secondary follicles (**SF**) of different sizes. Also visible in this section are several corpora lutea (**CL**), which have formed from the follicular cells remaining after the egg is ovulated. The distal end of the oviduct (**Ov**) is located close to the ovary and is sectioned at several different levels in this micrograph.

×55

FEMALE REPRODUCTIVE SYSTEM
Ovary

The free surfaces of the simple squamous cells that form the mesovarium are illustrated in Fig. 1. Each cell is characterized by a bulge that denotes the internal position of its nucleus. In addition, the cell surfaces are populated with a large number of finger-like projections called microvilli (**Mv**), which are variable in length (**arrows**). The rat ovarian cortex is covered by a layer of epithelium (**Ep** in Fig. 3). Connective tissue in the outer portion of the cortex is so compactly arranged that the region is called the tunica albuginea (**TA**). The connective tissue stroma (**St** in Figs. 2 and 3) in the ovarian cortex internal to the tunica albuginea contains nests of primordial and primary follicles (**PF** in Figs. 2 and 3) as well as larger secondary or multilaminar follicles (**SF** in Figs. 2 and 3). Small cavities (**arrows** in Fig. 3), sometimes present between the follicle cells (**FC**) of secondary follicles, are called Call-Exner vacuoles and contain fluid in the living condition. As the follicle matures, cavities containing the fluid eventually coalesce to produce the antrum of an antral, or Graafian, follicle. A section of a young secondary follicle is illustrated in Fig. 4. The central depression represents the position of the primary oocyte, which was removed during sectioning. The oocyte surface is invested by a distinct layer called the zona pellucida (**ZP**), which appears fibrillar in this preparation. The zona pellucida is surrounded by a multilayer of cells called follicular, or granulosa, cells (**GC**). The most peripheral layer of the granulosa cells is separated from the surrounding connective tissue stroma (**St**) by a distinct basement membrane called the membrana limitans externa (**arrows** in Fig. 4). The developing follicle illustrated in Fig. 5 is sectioned so as to illustrate the primary oocyte (**PO**), a developing antrum (**An**) containing follicular liquor, granulosa cells (**GC**), and the surrounding theca folliculi (**TF**).

Fig. 1, ×8595; Fig. 2, ×720
Fig. 3, ×555; Fig. 4, ×560; Fig. 5, ×420

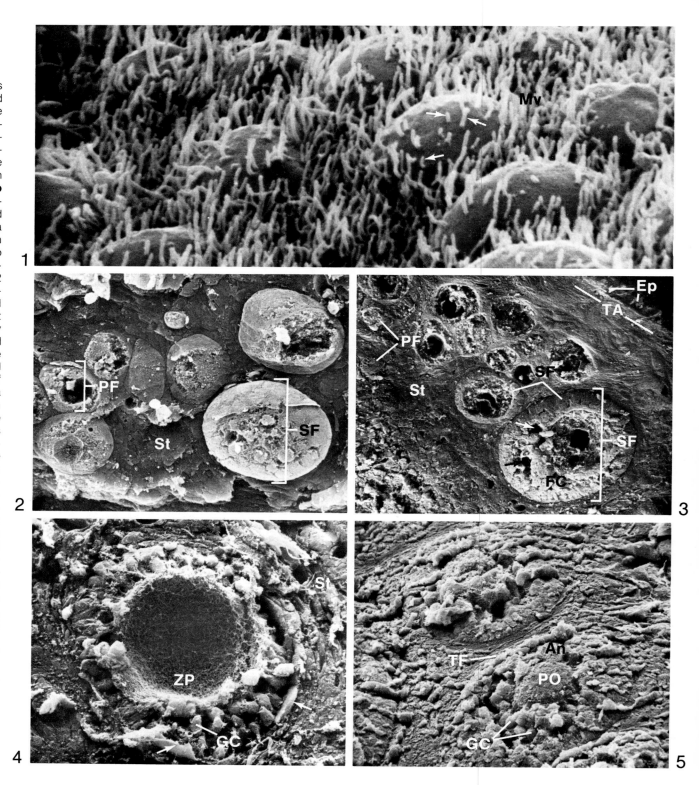

280

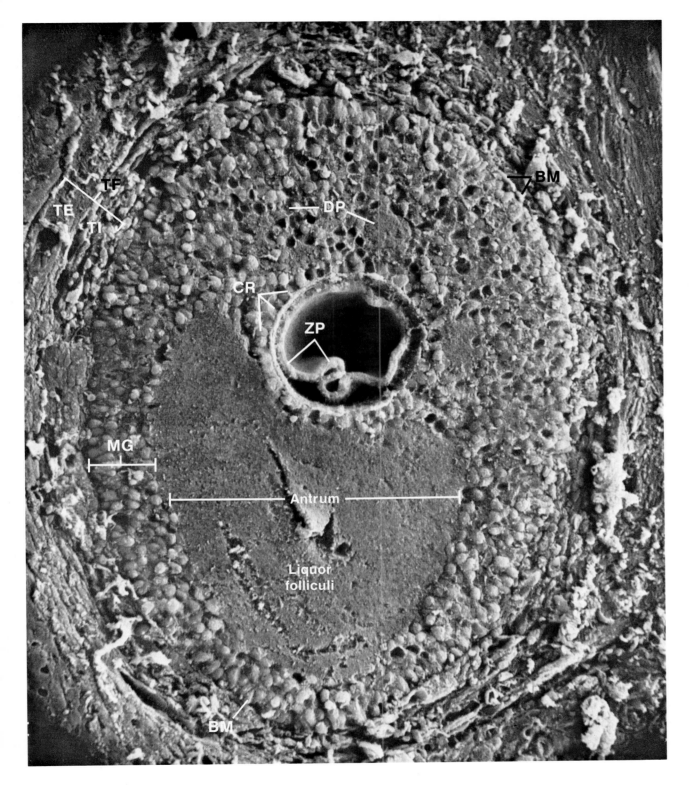

FEMALE REPRODUCTIVE SYSTEM
Ovary

Continued growth of the secondary follicle results in the eventual formation of the mature Graafian follicle, shown here in section. During this developmental period, the oocyte enlarges, and the follicle cells, or granulosa cells, continue to proliferate. In this preparation the germ cell, or oocyte, was removed, so that only a cavity remains to denote its former position. The zona pellucida (**ZP**) surrounding the oocyte is present, but locally folded. A rather large cavity filled with a secretory product occupies a large part of the Graafian follicle. This cavity, the follicular antrum, contains a viscid secretory product called the liquor folliculi, which contains proteins, hyaluronic acid, and sulfated mucopolysaccharides secreted primarily by the granulosa cells. Estrogen, produced by cells in the theca interna, is also found in the follicular liquor. The antrum is surrounded by follicle cells, which are collectively called the membrana granulosa (**MG**). Those follicle cells that invest the germ cell and suspend it in the antrum collectively constitute the discus proligerus (**DP**), and the follicle cells immediately surrounding the zona pellucida constitute the corona radiata (**CR**). During early stages of development, the ovarian stroma adjacent to the follicle becomes condensed into a layer called the theca folliculi (**TF**). By the time an antrum appears in the follicle, the theca folliculi is divided into a theca interna (**TI**) and a theca externa (**TE**). The theca interna is quite vascular, contains steroid-secreting cells, and is separated from the membrana granulosa by a thick basement membrane (**BM**). The theca externa contains connective tissue cells and fibers. Overall, a period of 10–14 days is required for a follicle to reach full maturity in the human, at which time it may be 10 mm or more in diameter.

×420

FEMALE REPRODUCTIVE SYSTEM
Ovary

Components of the Graafian follicle are illustrated in the figures on this page. In Fig. 1 and the corresponding light photomicrograph of a similar region (Fig. 2), the oocyte (**Oo**) is sectioned in a plane that reveals the zona pellucida (**ZP**), its centrally located nucleus, and surrounding follicle cells, which form the corona radiata (**CR**) and discus proligerus (**DP**). The oocyte nucleolus (**NI**) is apparent in Fig. 2. A portion of the antrum (**An**) is located in the lower parts of both figures. The cells of the membrana granulosa (**MG** in Fig. 3) form a stratified layer around the antrum (**An** in Fig. 3), and are separated from the surrounding theca (**Th**) layer (Fig. 4) by a basement membrane (**BM** in Fig. 5). Granulosa cells possess numerous thin cytoplasmic extensions (**arrows** in Fig. 5) that are closely associated with adjacent follicle cells to form desmosomes, gap junctions, and occluding (tight) junctions. These processes may even extend inward, through the zona pellucida, to make contact with the oocyte. These features suggest that the follicle cells play an important role in supplying the egg with nutrients required for its metabolism. Furthermore, the occluding junctions between processes probably play a role in creating a follicular fluid chemically different from blood. Follicle cell junctions appear to be selective in their permeability, allowing low-molecular-weight proteins such as albumin to accumulate in the follicle antrum, but apparently restricting the passage of high-molecular-weight proteins into the antrum. This condition suggests the presence of a blood-follicle barrier similar to the blood-testis barrier in the male reproductive system. The theca interna, shown in Figs. 3 and 4, contains capillaries (**Ca**) and elongate, or spindle-shaped, cells that characteristically appear vacuolated after chemical fixation (**arrow** in Fig. 4). These cells are thought to produce estrogen. Toward the ovarian surface the theca interna may become locally thickened in the form of a cone that is thought to play the role of "pathmaker" for the movement of the growing follicle to the ovarian surface.

Fig. 1, ×675; Fig. 2, ×490
Fig. 3, ×990; Fig. 4, ×1960; Fig. 5, ×3010

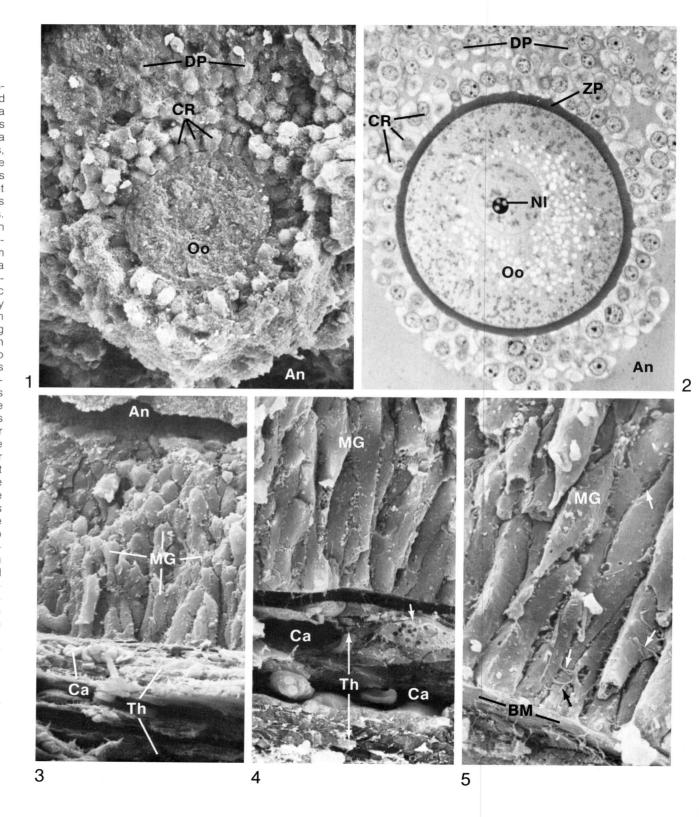

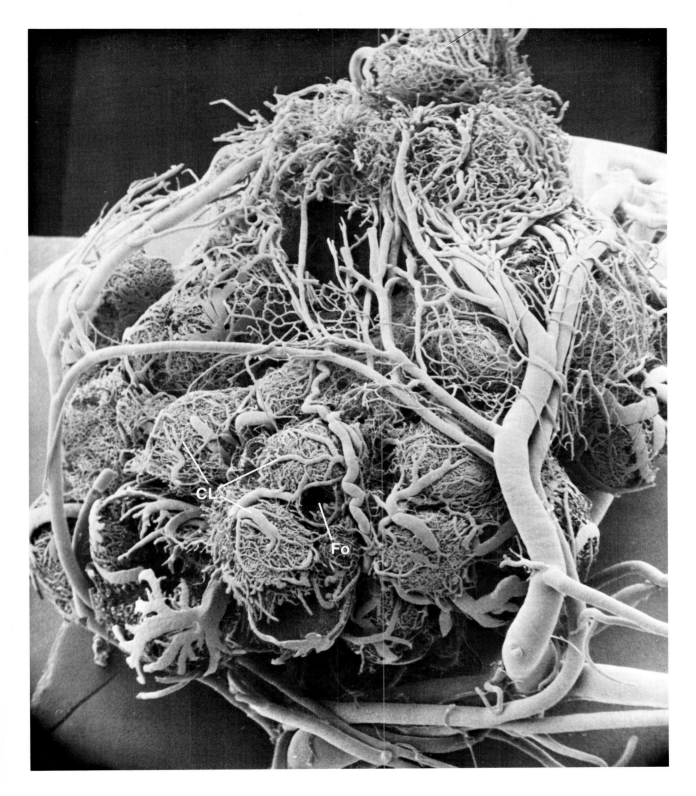

FEMALE REPRODUCTIVE SYSTEM
Ovary

Shown here is a cast of the circulatory system of an entire rat ovary. Branches of the ovarian and uterine artery enter the medulla through the hilus of the ovary and subdivide into a number of spiral vessels. The spiral vessels extend to the cortical-medullary boundary where they branch and anastomose into plexuses. Branches from the plexuses enter the cortex to form extensive capillary networks associated with the outer layer of the growing and mature follicles (**Fo**) and with the highly vascular corpora lutea (**CL**). Many of the small vessels organized into capillary plexuses are arranged into vascular spheres that represent the multiple corpora lutea present in this animal.

×45

283

FEMALE REPRODUCTIVE SYSTEM
Ovary

The extent of the vascular system in the ovary is not readily appreciated in sections. Casts of the ovarian vasculature, however, do reveal the remarkable distribution of vessels, as shown by a cut through the interior of an ovarian vascular cast in Fig. 1. The ovarian vasculature must be capable of adapting to marked changes in ovarian size associated with transient follicle development, ovulation, and subsequent formation of the corpora lutea. Besides being responsible for the distribution of ovarian stimulating hormones throughout the cortex, the vasculature regulates blood pressure throughout the ovary. During follicular growth, there are many tightly coiled arteries; after ovulation, however, they undergo progressive extension, eventually uncoiling and becoming sinuous. The ability of ovarian arteries to coil and uncoil at different times during ovarian function allows the vasculature to accommodate changes in ovarian size and blood pressure. The ovarian hormones are thought to exert a trophic action on the ovarian arteries comparable to that observed in certain parts of the endometrial and myometrial vasculature. During follicle development, a capillary network is established early in the theca, and it becomes more dilated as the follicle continues to develop. A cast of the capillary plexus within the theca interna of a developing follicle is illustrated in Fig. 2. The capillaries in the theca interna form an interconnecting network and serve to nourish the follicle cells and egg, since vessels are not present in the granulosa. The permeability of the theca capillaries also increases before ovulation. After ovulation, these capillaries and other ovarian vessels grow into the differentiating corpus luteum, providing an extensive blood supply for this endocrine gland as shown in the cast of the corpus luteum vasculature in Fig. 3. These examples provide a striking demonstration of the close relationship between the blood vasculature and organ function.

Fig. 1, ×35; Figs. 2 and 3, ×110

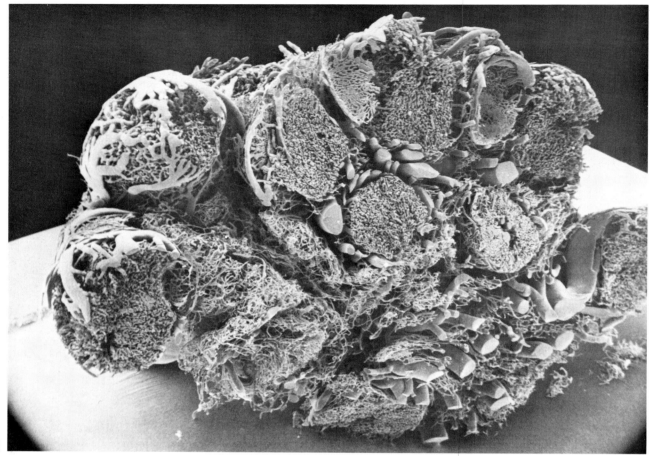

1

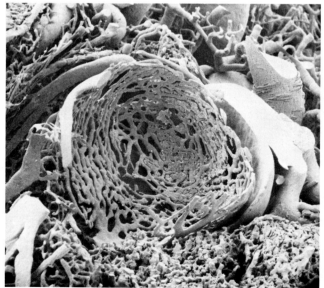

2

3

284

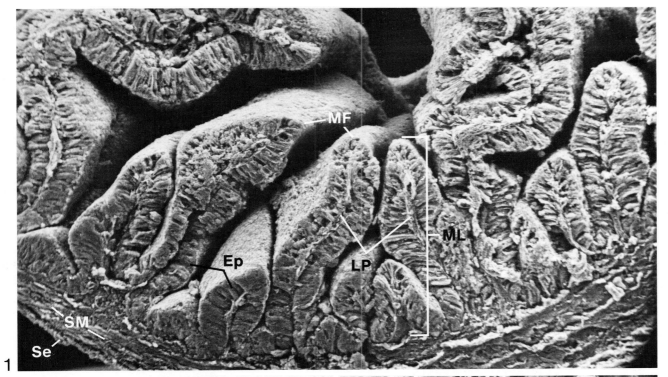

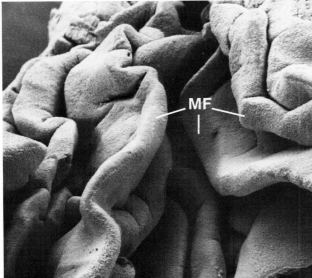

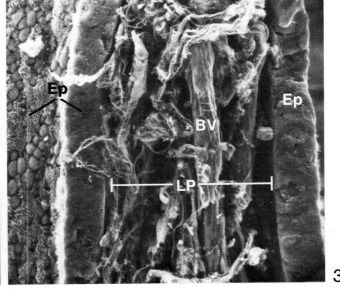

FEMALE REPRODUCTIVE SYSTEM
Oviduct

The paired fallopian tubes, or oviducts, are muscular tubes that convey the ovulated oocyte to the uterine cavity. Each oviduct opens at one end, close to the ovary. This end of the oviduct, the infundibulum, has finger-like extensions (fimbriae) that collectively form an expanded funnel around a large surface area of the ovary. One of the fimbriae, called the ovarian fimbria, is much larger than the others and is often connected directly to the ovary. During ovulation, the vessels in the fimbriae become engorged with blood, bringing them closer to the ovarian surface. This appears to increase the likelihood that an egg ovulated into the peritoneal space will pass into the infundibulum of the oviduct. The infundibulum is continuous with the ampulla, the longest segment of the oviduct. The ampulla, in turn, narrows into an isthmus region that extends toward the uterus. That portion of the oviduct which penetrates the uterine wall is called the intramural portion, or pars interstitialis. Fertilization usually occurs in the upper portion of the oviduct, and early divisions of the embryo take place during the several days required for its transport to the uterine cavity. The ovum is propelled through the oviduct by peristaltic muscular contractions and also by the rhythmic beat of cilia that project into the lumen. The wall of the fallopian tube has three layers, as shown in a section of the ampulla in Fig. 1. These include the inner mucosa layer (**ML**), an intermediate muscularis layer consisting of circularly and longitudinally arranged smooth muscle (**SM**), and an outer serosal layer (**Se**). The highly folded inner mucosa layer consists of an epithelial layer (**Ep**) that covers a thin connective tissue lamina propria (**LP**). The surface of the mucosal folds (**MF**) are shown in Fig. 2 as viewed from the lumen of the ampulla. A section through one of the mucosal folds is shown in Fig. 3, in which it is possible to observe the epithelial (**Ep**) layer in both surface and lateral views. The lamina propria (**LP**) forms a packing for the interior of the mucosal fold and contains blood vessels (**BV**), lymphatic vessels, and nerves. All illustrations of the oviduct are from the rabbit.

Fig. 1, ×185; Fig. 2, ×60; Fig. 3, ×480

FEMALE REPRODUCTIVE SYSTEM
Oviduct

The folded mucosal surface, as viewed from within the oviduct, is illustrated in Fig. 1. The epithelium consists of simple columnar cells, many of which have cilia (**Ci**) associated with their free surfaces. Ciliated cells are especially numerous on the fimbriae of the infundibulum and in the ampulla. Nonciliated, or secretory, cells (**SC**) produce a secretion that serves to maintain a moist environment in the oviduct lumen and which may also be a nutritive substance required by the ovum. In the rabbit, the secretory product of the nonciliated cells forms an outer envelope for the ovum. Details of the shape and packing of the columnar epithelial cells are provided by preparations in which a portion of the epithelial sheet has been removed; both ciliated (**Ci**) and secretory cells (**SC**) are visible in Figs. 2 and 3. The epithelium is attached to the underlying lamina propria by a basement membrane (**BM**), which is apparent where the epithelial layer has been removed. Folds (**Fo** in Fig. 3) are also associated with the lateral borders of the epithelial cells.

Fig. 1, ×375; Fig. 2, ×1095; Fig. 3, ×2200

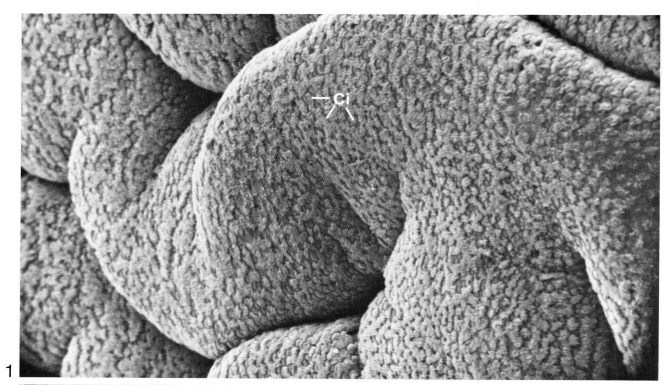

1

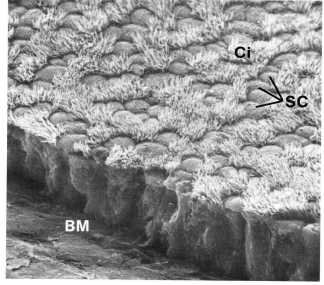

2

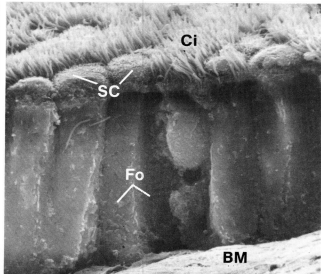

3

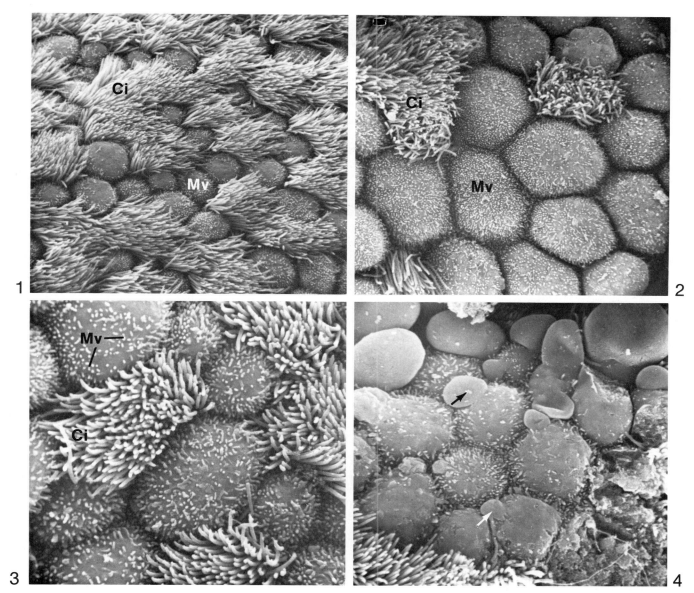

FEMALE REPRODUCTIVE SYSTEM
Oviduct

The surface features of the epithelium in the rabbit ampulla are illustrated in the figures on this page. Both ciliated (**Ci**) and secretory cells are abundant in the infundibulum and ampulla, but the relative proportion of secretory cells appears to be greater in the ampulla. The surfaces of the secretory cells are irregularly populated with short microvilli (**Mv**), and some have a single centrally located cilium. The rounded or irregularly shaped excrescences observed on the surface (**arrows** in Fig. 4) of some secretory cells probably indicate a stage in the release of secretory product into the lumen. The presence and length of cilia as well as the activity of the secretory cells are all under hormonal control. The beat of cilia is toward the uterus, facilitating the transport of the ovum in this direction. The number of ciliated cells and the length of cilia appear to be greatest in the follicular phase of the menstrual cycle. In the luteal phase of the menstrual cycle, ciliated cells markedly decrease in number, and the secretory cells may become inactive or degenerate. Castration of female rabbits, for example, eventually results in the loss of cilia and the inactivity of secretory cells. But when estrogen is injected into castrated female animals, there is noticeable recovery of ciliated and secretory cells. The development of ciliated and secretory cells in the oviduct, especially in the infundibulum and ampulla, is thus dependent upon the presence and level of estrogen. Progesterone has been reported to increase the frequency of ciliary beat at a time when an ovum would likely be present in the oviduct. Later in the cycle, progesterone appears to be responsible for loss of cilia and inactivity of the secretory cells.

Fig. 1, ×2050; Fig. 2, ×2890;
Fig. 3, ×4910; Fig. 4, ×3560

FEMALE REPRODUCTIVE SYSTEM
Uterus

The uterus is a tubular structure containing a central cavity or lumen (**Lu**) which is surrounded by layers of epithelium, connective tissue, and smooth muscle. The inner layer of the uterus, the endometrium (**En**), includes a surface epithelium of simple columnar cells and an underlying connective tissue stroma. As can be seen here, the endometrium may be extensively folded. The epithelial layer invaginates the surrounding stroma to form many simple or branched tubular glands. The openings into these glands are denoted by the **arrows**. The endometrium is surrounded by a thick layer called the myometrium (**My**), which consists of smooth muscle bundles as well as connective tissue. Sections of large blood vessels (**BV**) are present in the middle layer of the myometrium. The external uterine surface may be covered in some areas by a thin connective tissue layer, the adventitia, but over a considerable extent the covering consists of a thin serosal layer (**Se**) that forms the visceral peritoneum. The uterus serves to receive the embryo after its transit through the oviduct, participates in the implantation of the embryo, and develops those vascular relationships required for maintaining the embryo throughout intrauterine development.

×50

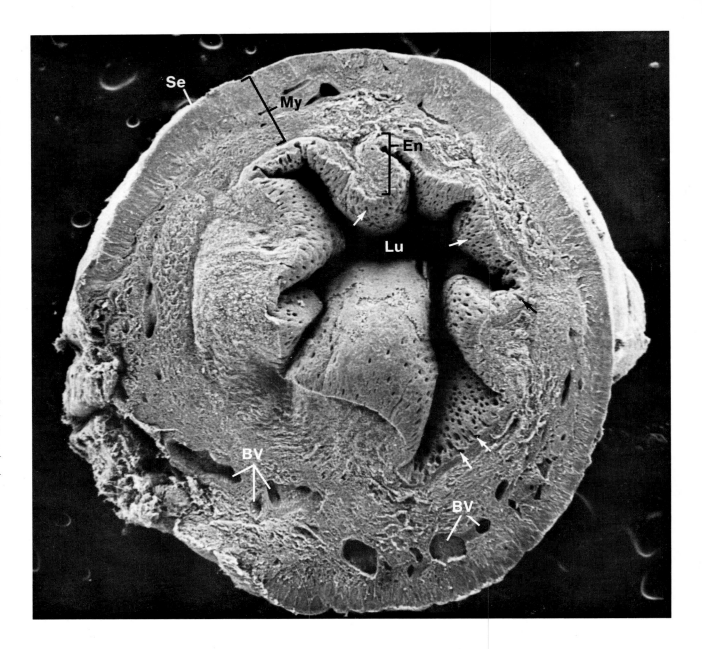

288

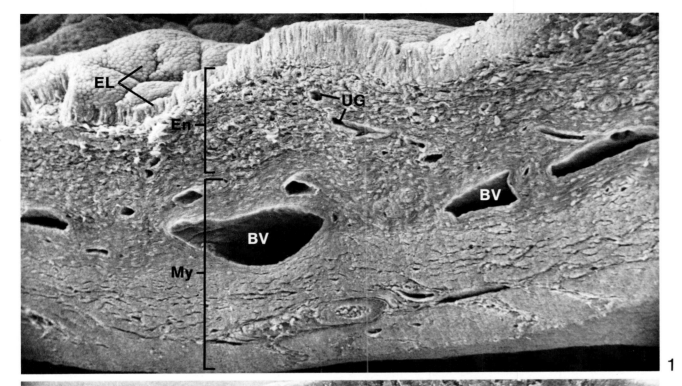

FEMALE REPRODUCTIVE SYSTEM
Uterus

A section through the wall of the rat uterus is illustrated in Fig. 1. The approximate thickness of the endometrium (**En**) is denoted; the tissue consists of an epithelial cell layer (**EL**) and an underlying connective tissue stroma. Sections of a uterine gland (**UG**) are identified in the stroma of the endometrium. The middle portion of the myometrium (**My**), which contains large blood vessels (**BV**), is termed the stratum vasculare. During pregnancy smooth muscle cells in the myometrium increase in number and undergo marked hypertrophy. There is also an increase in the connective tissue elements (collagen). Smooth muscle cells may undergo intermittent contraction and relaxation in the nonpregnant female, but during pregnancy uterine muscle is relatively inactive. Although the hormone relaxin appears to quiet uterine musculature, increased myometrial contractions occur in response to oxytocin, a hormone liberated from the neurohypophysis. Prostaglandins generally act to stimulate smooth muscle, and they also stimulate endometrial contractions. For this reason, they are sometimes used as abortifacients. A portion of the endometrium is illustrated at higher magnification in Fig. 2. The surface epithelium (**Ep**) and uterine glands (**UG**) are identified. The stroma (**St**), or connective tissue elements of the endometrium, consists of fibers (**Fi**) and many fibroblasts. Numerous capillaries and small blood vessels (**BV**) are present in the endometrial stroma.

Fig. 1, ×260; Fig. 2, ×895

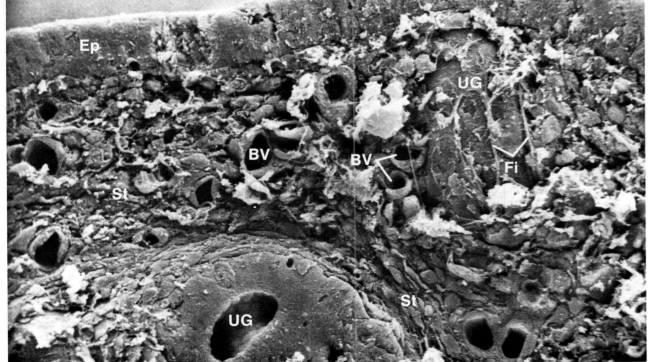

FEMALE REPRODUCTIVE SYSTEM
Uterus

The numerous uterine glands open (**arrows**) onto the surface of the highly folded endometrium (**En**) of the rabbit uterus, as shown in Fig. 1. In humans, uterine glands regenerate under the influence of estrogen from the narrow basalis layer during a period of approximately 10 days following menstruation. The columnar cells that line the tubular glands are continuous with the epithelial cells that line the lumen of the uterus. After ovulation and under the influence of progesterone, the glands begin actively secreting glycogen, mucus, lipid and other substances that serve to provide a moist and nourishing environment at a time when a fertilized egg is likely to be undergoing implantation in the uterine cavity. At this time, the glands often become coiled and their lumina become dilated. An ostium (**Os**), the opening into one of the tubular uterine glands, is shown in Fig. 2. The epithelial cells lining the uterine cavity possess a concentration of short microvilli (**Mv**) that uniformly populate the cell surface. A number of the epithelial cells bulge into the lumen, and the microvilli associated with the free surfaces of these cells are longer (**Mv** in Fig. 3). Cilia are also present on the surface of some cells lining the uterine cavity.

Fig. 1, ×55; Fig. 2, ×2990; Fig. 3, ×14,420

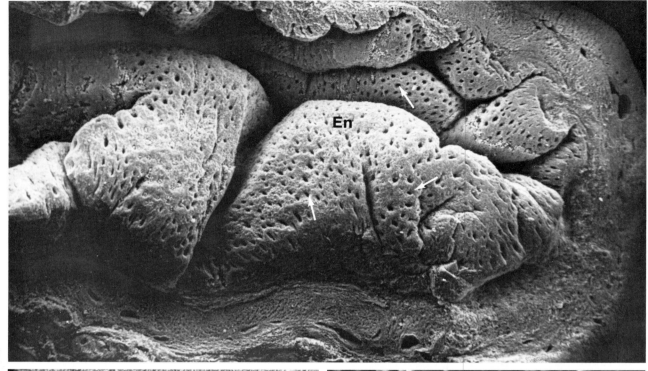

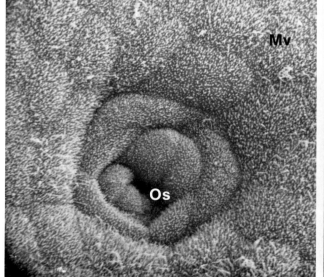

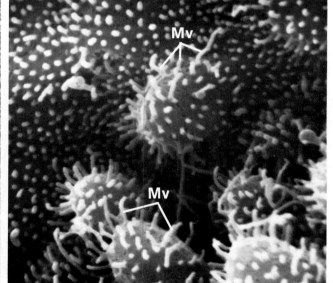

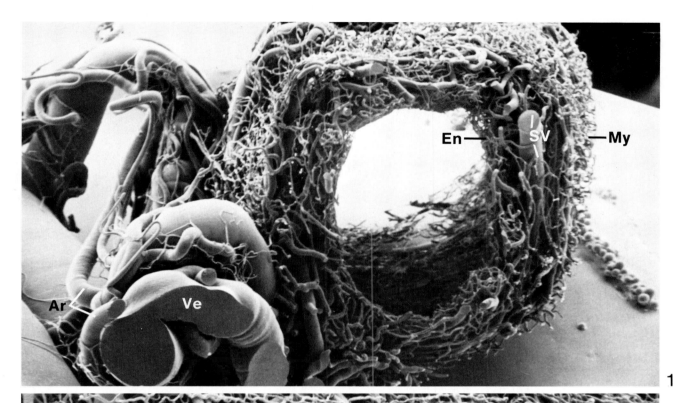

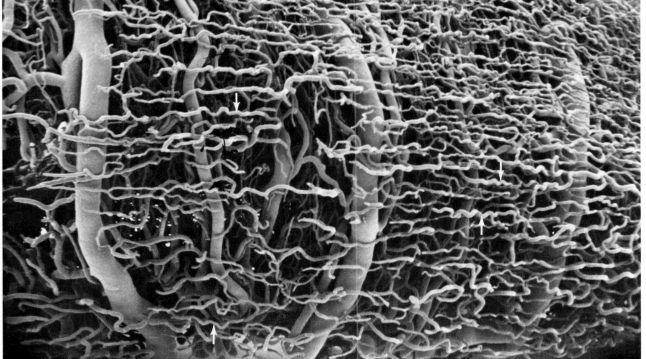

FEMALE REPRODUCTIVE SYSTEM
Uterus

As demonstrated by a cast of the rat uterine vasculature (Fig. 1), the uterine arteries (**Ar**) and veins (**Ve**) course longitudinally along the outer surface of the uterus. At various intervals, branches of the arteries penetrate the uterine perimetrium and distribute radially in the myometrium between the circular and longitudinally arranged bundles of smooth muscle. The region of the myometrium in which the large branches are located is termed the stratum vasculare (**SV**). Further subdivision of these vessels gives rise to separate vascular plexuses that supply the endometrial (**En**) and myometrial (**My**) areas. The vascular replica of the outer surface of the uterus (Fig. 2) is viewed such that the long axis of the uterine tube is oriented horizontally. The larger vessels in the myometrium are observed to course radially. Divisions of these vessels form long spiralling capillaries (**arrows**) passing parallel to the long axis of the uterine tube. These capillaries most likely supply the smooth muscle in the outer myometrium as well as the outer serosal surface. The spiralling of these capillaries is probably a result of the constricted state of the smooth muscle at the time the casting medium was injected, and may reflect the ability of muscle vasculature to accommodate shortening and lengthening of the muscle. This vascular conformation is very similar to that observed previously in vascular casts of skeletal muscle (see Fig. 2 on p. 145). Uterine vessels in general undergo dilation in response to injected estrogen. Thus estrogen is influential in producing early vascular changes, including antimesometrial hyperemia, edema, and mitosis. Progesterone intensifies the estrogen effect particularly with respect to the development of capillaries around the uterine glands. Without a preliminary estrogen priming, however, progesterone alone does not markedly influence the architecture of the uterine vasculature.

Figs. 1 and 2, ×45

FEMALE REPRODUCTIVE SYSTEM
Uterus

Additional details of the uterine vasculature of the rat are illustrated by the casts shown in these figures. The large vessels that form the stratum vasculare (**SV**) are observed in a longitudinal section of the uterine wall (Fig. 1). Their divisions supply the myometrial (**My**) layer in which they reside as well as the endometrial (**En**) lining of the uterus. In the rat, a plexus of vessels forms at the base of the endometrium, which consists primarily of veins. A second endometrial plexus, called the subepithelial plexus, extends to more superficial regions of the endometrium. This plexus is derived from radial arteries that originate as branches from both myometrial and basal endometrial vessels. The subepithelial plexus, as viewed from the luminal surface of the uterus in Figs. 2 and 3, is composed of anastomosing capillary loops, some of which form a basket-like network around the location of the uterine glands. The holes (✳) in the luminal surface of these casts correspond to openings in the tubular glands which extend toward the epithelial surface (compare with the scanning micrograph of the opening of a uterine gland in Fig. 2 on p. 290). Although the uterine vessels in the rat are affected by progesterone and estrogen, the process of menstruation does not occur in rodents, in contrast to humans. This is attributed primarily to differences between the endometrial vasculature of these two species. In the uterus of the human female, a separate system of arterial vessels divides to supply the basal portion of the endometrium, whereas another system of coiled arterioles forms a rich capillary bed in the superficial functionalis layer of the endometrium. In response to decreases in progesterone and estrogen levels, the coiled arterioles undergo intermittent contractions, causing a deprivation of nutrients to the functionalis layer over variable time periods. This results in an eventual sloughing of the entire functionalis layer, with intermittent bleeding, and constitutes the basis for menstruation. The remaining vessels in the basal portion of the endometrium serve to revascularize the area as a new functionalis is formed during the proliferative phase of the menstrual cycle.

Fig. 1, ×140; Fig. 2, ×145; Fig. 3, ×530

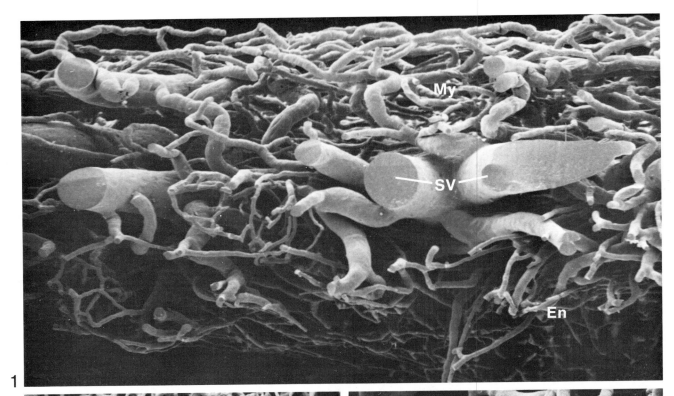

1

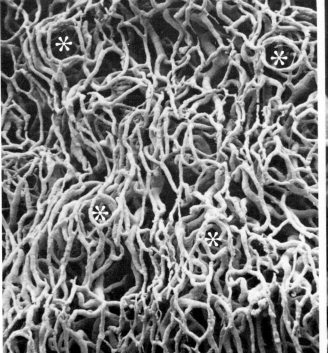

2

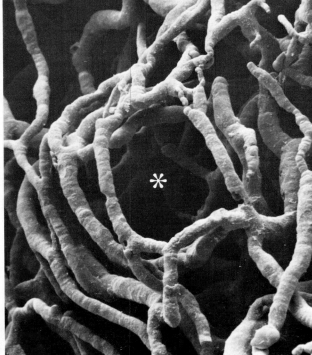

3

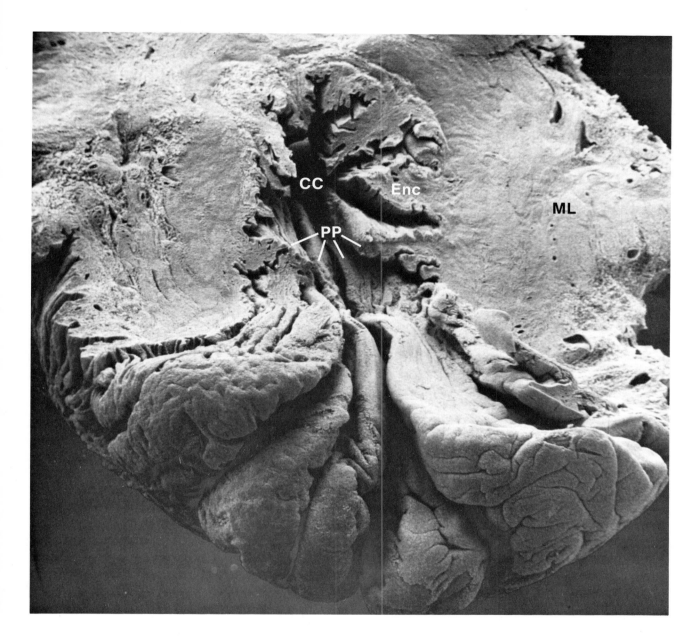

FEMALE REPRODUCTIVE SYSTEM
Cervix

The cervix is the lower extension of the uterus, through which the cervical canal (**CC**) communicates with the vagina. The inner mucous membrane, which consists of both epithelium and lamina propria, is called the endocervix (**Enc**), and is continuous with the endometrium of the uterus above and the mucous membrane of the vagina below. The endocervix is extensively folded along the cervical canal to produce the plica palmatae (**PP**). Long tubular glands, often branched, extend from the epithelial layer into the adjacent lamina propria. Mucus is secreted by cervical glands in response to estrogen, and is maximal at the time of ovulation. The consistency and chemical properties of the secreted mucus are also affected by hormones. During most of the menstrual cycle, the mucus is highly viscous and detrimental to spermatozoan viability. At mid-cycle, however, under the influence of estrogen, the mucus becomes more hydrated and less viscous, a condition more favorable to sperm migration. Cervical glands sometimes become obstructed and are then known as nabothian cysts. The muscular layer (**ML**), or myometrium, of the cervix is continuous with the myometrium of the uterus, and consists of rather ill-defined longitudinal and circular smooth muscle bundles with large amounts of intervening collagenic fibers. Dilation of the cervix at parturition is accomplished primarily by a softening of its intercellular substance. In response to the hormones of pregnancy, there are also increases in vascularity and amount of tissue fluid during this time. Sufficient amounts of smooth muscle are present in the wall of the cervix to permit contraction to its former size after birth. A perimetrium (serosa, peritoneum) consisting of connective tissue covered by a single layer of mesothelial cells surrounds much of the uterus and cervix. The lower end of the cervix projects into the vagina as the portio vaginalis. At the lower end of the cervical canal, there is a transition from simple columnar epithelium into stratified squamous epithelium (the exocervix), which covers the portio vaginalis.

×45

FEMALE REPRODUCTIVE SYSTEM
Cervix

Several of the furrows and folds (plicae palmatae) in the endocervix are illustrated in Fig. 1; the apical borders of the epithelial cells are barely apparent at this magnification. A portion of the epithelial cell layer was mechanically removed in Fig. 2, and as a result, it is possible to visualize the columnar form of the epithelial cells (**EC**) as well as the underlying basal lamina (**BL**). All of the columnar cells in the epithelial sheet attach basally to this thin basal lamina. The free surfaces of the columnar epithelial cells are illustrated in Figs. 3 and 4. Two types of cells can be identified on the basis of differences in their apical surface specializations. A number of the epithelial cells are ciliated (**CC**), and the ciliary beat of these cells is directionally biased toward the vagina. Other epithelial cells are secretory (**Se** in Fig. 3) in function and are covered with microvilli (**Mv** in Fig. 4) over much of their apical surface. These cells secrete mucus, and the apertures (**arrows**) seen in many of the cell surfaces in Fig. 4 probably correspond to places where secretory product has been released. The cervical endometrium does not undergo marked cyclic change during the menstrual cycle.

Fig. 1, ×130; Fig. 2, ×280;
Fig. 3, ×650; Fig. 4, ×2605

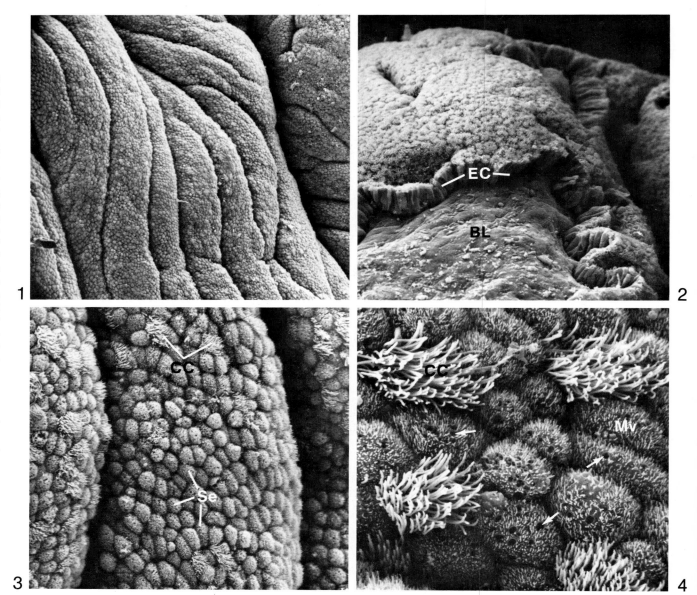

FEMALE REPRODUCTIVE SYSTEM
Vagina

The vagina (Fig. 1) is a fibromuscular tube consisting of an inner mucosal layer (**ML**) that is adjacent to the lumen (**Lu**). A middle muscularis layer, and an outer adventitial layer constitute the remaining portions of the vaginal wall. The muscularis layer consists of smooth muscle bundles that are both circularly and longitudinally arranged. The adventitial layer consists of dense connective tissue that may be confluent with the connective tissue of surrounding organs. The mucosal layer consists of stratified squamous epithelium and an underlying lamina propria containing blood vessels (**BV**), nerves, and lymphatics. Both the epithelium (**Ep**) and the lamina propria (**LP**), as shown in Fig. 2, may exhibit transverse folds. The surface of the epithelium is kept moist by a lubricating mucus secreted from uterine glands, cervical glands, and the glands of Bartholin in the vestibule. The wall of the vagina, however, lacks glands. As in other organs, the lamina propria of the vagina is often infiltrated with neutrophils and lymphocytes (often organized into lymphatic nodules), which appear in response to the presence of foreign substances that may have penetrated the epithelial layer. The stratified squamous epithelium is not normally keratinized in humans, but may become so during vitamin A deficiency. The highly flattened outer cells in the protective stratified squamous epithelium (**SC** in Fig. 3) are continuously shed by the process of desquamation (**arrows**) and are replaced by division of generative cells in the basal layer of the epithelium. In a number of mammals (e.g., rat and guinea pig), the vaginal epithelium undergoes cyclic changes, but the epithelium varies little during the menstrual cycle of the human female. The width of the epithelium is dependent upon levels of female sex hormones; in their absence the epithelium becomes atrophic. Glycogen is abundant in the middle and superficial cells of the epithelium and is greatest in amount at the time of ovulation. Glycogen is thought to provide nourishment for sperm and to serve as a substrate for bacteria that reside in the vagina.

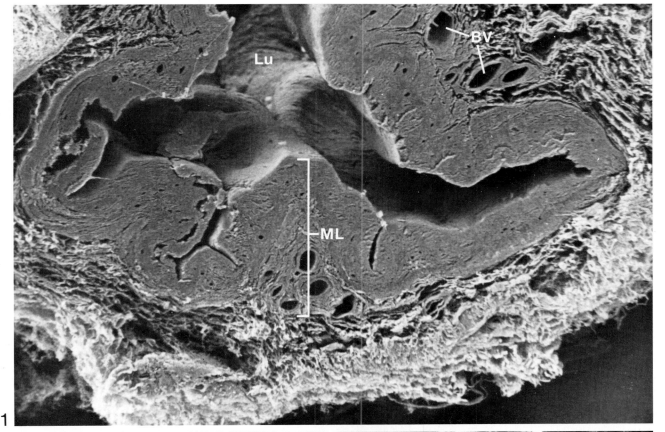

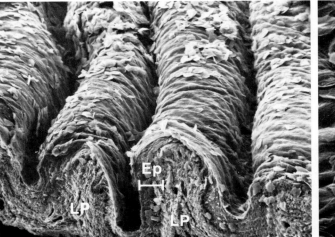

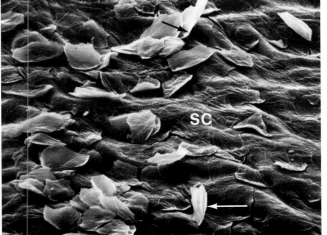

Fig. 1, ×75; Fig. 2, ×90; Fig. 3, ×405

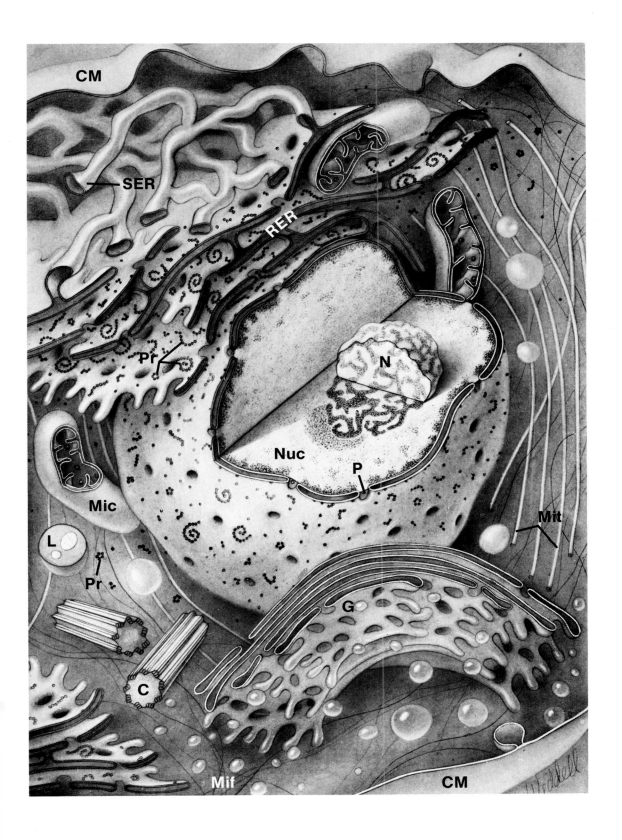

APPENDIX A
Cell Organelles

The accompanying diagram serves to illustrate many of the cell organelles in three-dimensional view. The cell membrane (**CM**), or plasmalemma, is depicted at the top and bottom of the diagram. The centrally located nucleus (**Nuc**) is invested by a porous nuclear envelope (**P**) that consists of inner and outer nuclear membranes, which are continuous at the pore margin. Heterochromatin is attached to the inner layer of the nuclear envelope, but much of the more dispersed and finely granular chromatin (euchromatin) is distributed throughout the nucleus. The nucleolus (**N**) consists of a dense, twisted band (the nucleolonema) surrounded by a less dense material that forms the pars amorpha. The nucleolus is the site of synthesis of ribosomal RNA. Most of the proteins for export from the cell are synthesized in association with the rough-surfaced (granular) endoplasmic reticulum (**RER**), which consists of many membranous stacks that enclose spaces called cisternae. Polyribosomes (**Pr**) are present on the outer surface of these membranes as well as on the cytoplasmic side of the outer layer of the nuclear envelope. Polyribosomes (**Pr**) may also be present in the hyaloplasm and be unassociated with membranes. These structures appear to be involved, for the most part, in the synthesis of proteins, many of which are for utilization within the cell. The smooth-surfaced endoplasmic reticulum (**SER**), or agranular endoplasmic reticulum, sometimes takes the form of branched and anastomosing tubules that may in some cases be continuous with the granular endoplasmic reticulum. The Golgi apparatus (**G**) consists of a variable number of closely apposed, flattened membranous saccules enclosing cisternae as well as membrane-bound vesicles and vacuoles. Mitochondria (**Mic**) are enclosed by two membranes, the inner one of which is folded, forming cristae that partially compartmentalize the mitochondrion. Mitochondria contain DNA, ribosomes, and other enzymes necessary for synthesizing a variety of proteins, many of which are involved in aerobic respiration. Lysosomes (**L**) exist in a variety of forms. They are membrane-bound and contain potent hydrolytic enzymes. One of the lysosomes (secondary lysosome) present in the drawing contains two enclosed structures undergoing intracellular digestion. Paired centrioles (**C**) are oriented at right angles to each other, and their wall is made up of nine triplet tubules. Other cellular constituents illustrated in the drawing include microtubules (**Mit**) and microfilaments (**Mif**). The remainder of the cytoplasm, or hyaloplasm, contains a vast array of enzymes and other chemicals in a sol-gel consistency. [Reproduced by permission. From M. Hogan, J. A. Alvarado, and J. E. Weddell, *Histology of the Human Eye*, W. B. Saunders Company, 1971.]

APPENDIX A
Cell Organelles

As improvements in the scanning electron microscope have increased resolution, it has become possible to examine intracellular structures, such as cell organelles. The morphological information that can be gained depends upon the method of exposing the cell interior. Fig. 1 shows a cat pancreatic cell that was cracked open to reveal the vesicular endoplasmic reticulum (**VER**). The inner surface of the reticulum (✽) is also apparent. The nucleus (**Nu**) is exposed in a manner that illustrates the interior chromatin, the inner nuclear membrane (**IM**), and the outer nuclear membrane (**OM**). This area is enlarged in Fig. 2 to illustrate the many nuclear pores (**NP**) present on the inner nuclear membrane. The outer nuclear membrane is covered with rough granules, approximately 450 Å in diameter, which probably correspond to ribosomes. In some places, small openings are observed (**arrows**) on the outer membrane that are surrounded by several granules gathered like rosettes. Ion-etching of cracked open cells reveals additional information; the cytoplasm is easily etched away, whereas the internal membranes are more resistant to etching. The nucleus in Fig. 3 has been ion-etched to reveal nuclear pores on the inner nuclear membrane that appear somewhat larger than those in Fig. 2, which may be due to the etching process. [Figs. 1 to 3 courtesy of K. Tanaka, A. Iino, and T. Naguro. From Scanning electron microscopic observation on intracellular structures of ion-etched materials, *Arch. Histol. Jap.* 39(3):165–175, 1976. These micrographs will also appear in *SEM—Atlas of Histology* by K. Tanaka et al., Igaku-shoin, Tokyo (in preparation).]

Fig. 1, 14,460; Fig. 2, 28,480; Fig. 3, 17,800

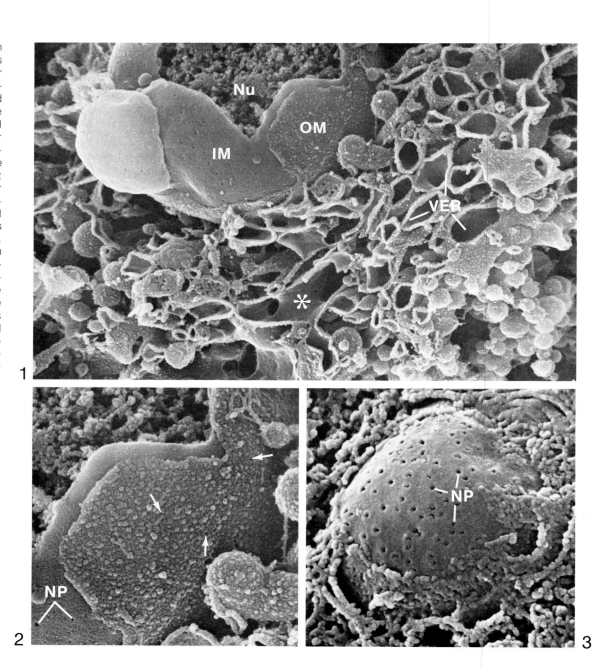

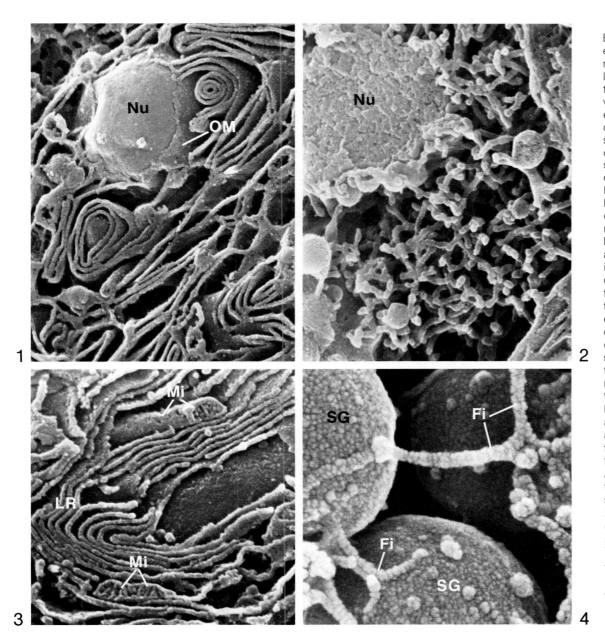

1

2

3

4

Endoplasmic reticulum is one of the organelles that is most effectively revealed by means of ion-etching. Lamellar (Fig. 1), tubular (Fig. 2), and vesicular (see previous page) types of endoplasmic reticulum can be revealed in the rabbit pancreatic cells. The endoplasmic surface is densely studded with clustered granules, which most likely represent ribosomes and polysomes. In Fig. 1, the outer membrane (**OM**) of the nucleus (**Nu**) seems to be continuous with the lamellar endoplasmic reticulum. In Fig. 2, extensive branching of the tubular endoplasmic reticulum is visible adjacent to the nucleus. Mitochondria (**Mi**) and lamellar endoplasmic reticulum (**LR**) can be seen in Fig. 3. Mitochondria cristae are visible, but cannot always be clearly defined by ion-etching. The inner and outer membranes of the cristae cannot be differentiated, and thus they have the appearance of a thick, solid plate. Fine filaments are commonly observed to connect endoplasmic reticulum to secretory granules or to connect secretory granules to one another. Shown in Fig. 4 are terminations of such filamentous (**Fi**) structures on the surface of secretory granules (**SG**). These filaments have a characteristic cross-banding with intervals of approximately 100 Å. It is likely that these structures are microtubules and may assist in controlling the movement and formation of secretory granules within the cytoplasm, but this has not yet been confirmed. [Figs. 1 to 4 courtesy of K. Tanaka, A. Iino, and T. Naguro. From, Scanning electron microscopic observation on intracellular structures of ion-etched materials, *Arch. Histol. Jap.* 39(3):165–175, 1976. These micrographs will also appear in *SEM—Atlas of Histology* by K. Tanaka et al., Igaku-shoin, Tokyo (in preparation).]

Fig. 1, 9,790; Fig. 2, 11,570;
Fig. 3, 17,800; Fig. 4, 111,250

A variety of plasma membrane specializations exist between neighboring cells in a sheet, as characterized by epithelium. One type of intercellular junction is termed a zonula occludens (**ZO**) (occluding or tight junction). It is a belt-like specialization that surrounds cells near their apical, or free, border (see inset B), and is characterized by the often punctate fusion of the outer leaflets of the plasma membrane. As a result, the extracellular space is obliterated, so that a barrier is provided to the passage of fluids, electrolytes, and macromolecules through the extracellular compartments. Adhering junctions provide for cellular attachments involving an intercellular substance present in an intercellular space of rather uniform thickness. Variations in the size and shape of the intercellular junctions result in their classification as maculae (focal junctions), bands or fasciae, and belts or zonulae. The zonula adherens (**ZA**) forms a complete band around the juxtaluminal border of cells (see inset B). The intercellular space is about 200 Å in diameter, and the cytoplasm adjacent to the junction is electron dense and finely granular. Cytoplasmic filaments about 60 Å in diameter (actin) are associated with the zonula adherens. The macula adherens (termed desmosome, **MA-DESM**) is characterized by a dense intercellular substance that often forms a line in the center of the 200 Å intercellular space. In addition, bundles of cytoplasmic tonofilaments loop at right angles through an electron-dense cytoplasmic band that appears attached to the inner leaflet of the plasma membrane. A macula adherens (**MA**) with a button-like shape, but lacking tonofilaments, is present in some epithelia. A hemidesmosome (**HEMI-DESM**), or half-desmosome, is a special type of desmosome that is frequently found at the base of epithelial cells, where they are anchored to the underlying connective tissue via the basement membrane (**BM**). The basement membrane, or basal lamina, consists of a homogeneous, periodic acid-Schiff positive (amorphous) layer immediately adjacent to the plasma membrane and, in addition, a layer of fine collagenic (reticular) fibrils is associated with the external surface of the amorphous layer of the basement membrane. The gap junction (**GJ**), or nexus, provides for ion communication between cells in many epithelia. In this type, the intercellular space is narrowed to about 20 Å. Hexagonal subunits are built into and between the plasma membranes of adjacent cells in this region. The center of the hexagonal subunit appears to contain a "pore" about 10–15 Å wide through which ions and small molecules (molecular weight less than about 1100 daltons) may pass between adjacent cells. [Reproduced by permission. From M. Hogan, J. A. Alvarado, and J. E. Weddell, *Histology of the Human Eye,* W. B. Saunders Company, 1971.]

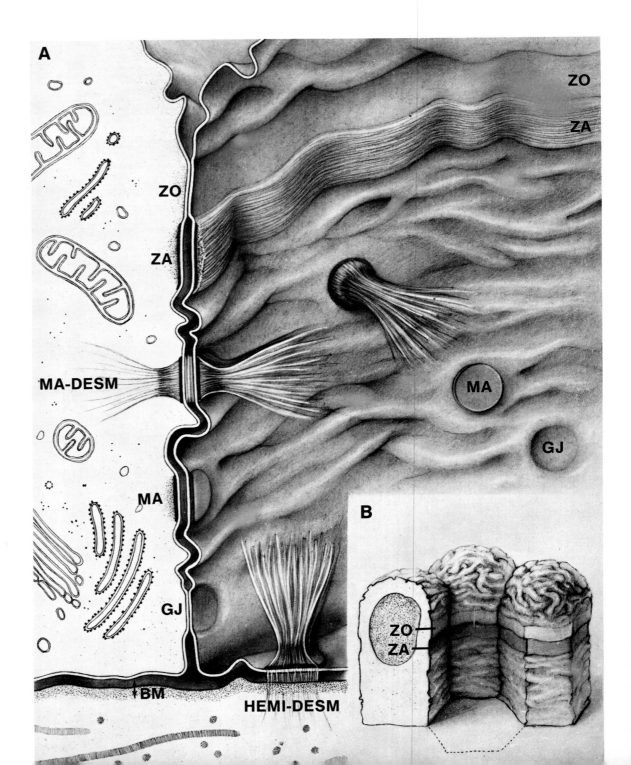

The animals used as sources of tissues and organs illustrated in this book included albino rats, guinea pigs, and rabbits. Anesthetized animals were perfused via the heart with 500–1000 ml of cold 2.5% glutaraldehyde in 0.1 M cacodylate buffer (pH 7.4). Optimum fixation was evidenced by complete blanching and stiffening of the organs. Lungs were inflated either with air or with fixative during vascular perfusion. Specific tissues or organs were removed from the animal after perfusion, and appropriate tissue slices were prepared and stored overnight in cold buffer. After several buffer rinses, the samples were then postfixed in 1% buffered osmium tetroxide solution for 2 hours. Following postfixation, the double thiocarbohydrazide (TCH) osmium tetroxide ligand-binding procedure (Malick and Wilson, 1975) was employed. This technique causes the binding of additional layers of TCH-osmium tetroxide to the tissue or organ surfaces, so that tissue surfaces that are normally prone to acquire an electrical charge when viewed in the electron beam of the scanning electron microscope are rendered electron conductive. Specimens were then dehydrated in a graded series (30, 50, 75, 95, 100%) of ethanol. Tissue and organ slices were exposed to absolute ethanol for 1 hour, during which time the ethanol was changed every 20 minutes. Following dehydration, the absolute ethanol was removed by the critical-point drying method described by Anderson (1951), using liquid carbon dioxide. After drying, the specimens were mounted on aluminum stubs with copper conducting tape. They were then coated with a thin ($\sim$100 Å) layer of gold, a sputter-coating method being used on most and an evaporative-coating method for a few. Specimens were viewed in a Cambridge S4 scanning electron microscope.

In addition to ensuring optimum tissue fixation and proper application of the double TCH-osmium tetroxide ligand-binding method, perhaps the most important aspect of the preparative procedures is the manner chosen to expose the interior of the specimen. For each specimen, various methods were employed to ensure that specific internal constituents of the tissue or organ were revealed. The methods used included mechanical tearing or cutting with a sharp razor blade after fixation. This was done while the tissue was in buffer or, after dehydration, in absolute ethanol. Internal exposure of tissues and organs was found to produce a different surface topography when performed in an aqueous medium (buffer) compared to performing this operation in the dehyrated and dried state. Some specimens were also cryofractured during the final stages of dehydration. Cryofracture consists of submerging the tissue in liquid nitrogen and fracturing the specimen with a chilled razor blade or scissors. Tissues possessing a dense cellular packing (e.g., nervous tissue, retina, internal ear) were mechanically torn with forceps after critical-point drying. Tissues requiring internal exposure in a dehydrated state were not subjected to postfixation and the double TCH technique until after rehydration. In such tissues as the thyroid, seminal vesicle, and gallbladder, the extracellular product was removed by cutting and subsequent washing of the tissue in 0.9% sodium chloride before fixation. The specimens were then preserved by immersion fixation.

MICROVASCULATURE REPLICAS

It is frequently useful to display the microvasculature of an organ in three dimensions. To produce a vascular replica, the animal is perfused with cold Tyrode's Ringer solution containing 0.5% heparin to remove all blood. The vasculature is then infused with Batson's #17 anatomical corrosion compound (available from Polysciences, Warrington, PA). The thoroughly mixed solution is introduced into the circulatory system through a cannula at a rate of approximately 2 ml/min. (*Caution:* the compound should be mixed and perfused in a hood.) After polymerization of the compound is complete (approximately 1–2 hours), the organs and tissues to be studied are dissected from the animal. If interior vascularization of a tissue or organ is to be studied, the specimens are cut open with a sharp razor blade. Either entire organs or slices of organs are placed in separate solutions of potassium hydroxide (340 g/liter) and macerated in an oven at 60°C. Maceration of osseous tissues (bones) is done in dilute (25%) clorox solution when it is necessary to obtain casts of both the inorganic matrix and the microvasculature. The casts are periodically washed with distilled water every 2–4 hours with minimal disruption. After each washing, the casts are placed in a fresh maceration solution. A total maceration time of 4–7 days is usually required, depending upon organ size and tissue constituents. When maceration is complete, no trace of tissue remains, only a hardened polymerized cast of the microvasculature. The casts are washed carefully in distilled water 8–12 times to ensure that all tissue remnants and hydroxide are removed. The washed replicas are then carefully dehydrated in ethanol, dried by critical-point method, attached to specimen stubs, coated with metal, and studied in the scanning electron microscope.

BIBLIOGRAPHY

CHAPTER 1 EPITHELIUM

Anderson, J. H., and A. B. Taylor. 1973. Scanning and transmission electron microscopic studies of jejunal microvilli of the rat, hamster and dog. *J. Morphol.* 141(3):281–291.

Andrews, P. M., and K. R. Porter. 1973. The ultrastructural morphology and possible functional significance of mesothelial microvilli. *Anat. Rec* 177:409–426.

Andrews, P. M., and K. R. Porter. 1974. A scanning electron microscopic study of the nephron. *Am. J. Anat.* 140(1):81–115.

Baradi, A. F., and S. N. Rao. 1976. A scanning electron microscope study of mouse peritoneal mesothelium. *Tissue & Cell* 8(1):159–162.

Buss, H., H. H. Dahm, and R. Lindenfelser. 1973. Surface of the endocardium of the rat heart: A scanning electron microscopic study. *Beitr. Pathol.* 148(4):340–359.

Dirksen, E. R. 1974. Ciliogenesis in the mouse oviduct. A scanning electron microscope study. *J. Cell Biol.* 63(3):899–904.

Edanaga, M. 1974. A scanning electron microscope study of the endothelium of the vessels. I. Fine structure of the endothelial surface of aorta and some other arteries in normal rabbits. *Arch. Histol. Jap.* 37(1):1–14.

Ketelbant-Balasse, P., F. Rodesch, P. Neve, and J. M. Pasterls. 1973. Scanning electron microscope observations of apical surfaces of dog thyroid cells. *Exp. Cell Res.* 79(1):111–119.

Morks, R., and B. Bhogal. 1974. Scanning electron microscopy of the normal epidermis and the epidermis in vitro. *Br. J. Dermatol.* 90:387–396.

Parakkal, P. F. 1974. Cyclical changes in the vaginal epithelium of the rat seen by scanning electron microscopy. *Anat. Rec.* 178(3):529–537.

Peine, C. J., and F. N. Low. 1975. Scanning electron microscopy of cardiac endothelium of the dog. *Am. J. Anat.* 142(2):137–158.

Wheeler, E. E., J. B. Gavin, and P. B. Herdson. 1973. A study of endocardial endothelium using freeze-drying and scanning electron microscopy. *Anat. Rec.* 175(3):579–584.

CHAPTER 2 CONNECTIVE TISSUE

Collard, J. G., and J. H. M. Temmink. 1976. Surface morphology and agglutinability with concanavalin A in normal and transformed murine fibroblasts. *J. Cell Biol.* 68:101–112.

Goodall, R. J., and J. E. Thompson. 1971. A scanning electron microscope study of phagocytosis. *Exp. Cell Res.* 64(1):1.

Kessler, S., and C. Kuhn. 1975. Scanning electron microscopy of mast cell degranulation. *Lab. Invest.* 32(1):71–77.

Krüger, P. G., and G. D. Bloom. 1973. Granule alteration and vacuole formation, two primary structural features of the histamine release process in mast cells. *Experientia* 29:329–330.

Leake, E. S., M. J. Wright, and Q. N. Myrvik. 1975. Differences in surface morphology of alveolar macrophages attached to glass and to millipore filters: A scanning electron microscopy study. *J. Reticuloendothel. Soc.* 17(6):370–379.

Motta, P. 1975. Scanning electron microscopic observation of mammalian adipose cells. *J. de Microscopië* 22(1):15–20.

Orenstein, J. M., and E. Shelton. 1976. Surface topography of leukocytes in situ: Cells of mouse peritoneal milky spots. *Exp. Mol. Pathol.* 24(3):415–423.

Orenstein, J. M., and E. Shelton. 1977. Membrane phenomena accompanying erythrophagocytosis: A scanning electron microscope study. *Lab. Invest.* 36(4):363–374.

Overton, J. 1976. Scanning microscopy of collagen in the basement lamella of normal and regenerating frog tadpoles. *J. Morphol.* 150(4):805–824.

Polliack, A., N. Lampen, B. D. Clarkson, E. De Harven, Z. Bentwich, F. P. Siegal, and H. G. Kunkel. 1973. Identification of human B and T lymphocytes by scanning electron microscopy. *J. Exp. Med.* 138(3):607–624.

Tizard, I. R., and W. L. Holmes. 1974. Degranulation of sensitised rat peritoneal mast cells in response to antigen, compound 48/80 and polymyxin B. *Int. Arch. Allergy* 46:867–879.

Wortis, H. H. 1975. Scanning electron micrographs of T and B cells. *Birth Defects* 11(1):558–562.

CHAPTER 3 SKELETAL TISSUE

Cartilage

Clarke, I. C. 1972. The microevaluation of articular surface contours. *Ann. Biomed. Eng.* 1(1):31–43.

Clarke, I. C. 1974. Articular cartilage: A review and scanning electron microscope study. II. The territorial fibrillar architecture. *J. Anat.* 118(2):261–280.

Enna, C. D., and M. Zimny. 1974. A scanning electron microscopy study of articular cartilage obtained from contracted joints of denervated hands. *Hand* 6(1):65–69.

Ghadially, F. N., R. L. Ailsby, and A. F. Oryschak. 1974. Scanning electron microscopy of superficial defects in articular cartilage. *Ann. Rheum. Dis.* 33(4):327–332.

Ghadially, F. N., J. A. Ghadially, A. F. Oryschak, and N. K. Yong. 1977. The surface of dog articular cartilage: A scanning electron microscope study. *J. Anat.* 123(2):527–536.

Hork, Y. D., Z. Bozdech, and V. Horn. 1974. Ultrastructure of the synovial membrane and articular cartilage in transmission and scanning electron microscopy. *Cas. Lek. Cesk.* 113(26):777–781.

Hork, Y. D., Z. Bozadech, and V. Horn. 1974. Proceedings: Ultrastructure of the synovial membrane and articular cartilage in haemophilia in the transmission and scanning electron microscope. *Folia Morphol.* (Praha) 22(4):330–332.

Hough, A. J., W. G. Banfield, F. C. Mottram, and L. Sokoloff. 1974. The osteochondral junction of mammalian joints: An ultrastructural and microanalytic study. *Lab. Invest.* 31(6):685–695.

Minns, R. J., and F. S. Stevens. 1977. The collagen fibril organization in human articular cartilage. *J. Anat.* 123(2):437–457.

Mow, V. C., and W. M. Lai. 1974. Some surface characteristics of articular cartilage. I. A scanning electron microscopy study and a theoretical model for the dynamic interaction of synovial fluid and articular cartilage. *J. Biomech.* 7(5):449–456.

Mow, V. C., W. M. Lai, J. Eisenfeld, and I. Redler. 1974. Some surface characteristics of articular cartilage. II. On the stability of articular surface and a possible biomechanical factor in etiology of chondrodegeneration. *J. Biomech.* 7(5):457–468.

Puhl, W. 1974. Micromorphology of the surface of normal articular cartilage. *Z. Orthop.* 112(2):262–272.

Redler, I. 1974. A scanning electron microscopic study of human normal and osteoarthritic articular cartilage. *Clin. Orthop.* 103:262–268.

Sosnierz, M., B. Biakas, and A. Wejsflog. 1973. Scanning electron microscopic studies (SEM) of the cartilage of the femoral head in primary coxarthrosis. *Chir. Narzadow. Ruchu. Ortop. Pol.* 38(2):133–138.

Zimny, M. L., and I. Redler. 1972. Scanning electron microscopy of chondrocytes. *Acta Anat.* (Basel), 83(3):398–402.

Zimny, M. L., and I. Redler. 1974. Chondrocytes in health and disease. In O. Johari and I. Corvin (eds.), *Scanning Electron Microscopy/1974.* Chicago: IIT Research Institute, pp. 805–812.

Zimny, M. L., and I. Redler. 1974. Morphological variations within a given area of articular surface of cartilage. *Z. Zellforsch. Mikrosk. Anat.* 147(2):163–167.

Bone

Belanger, L. F. 1969. Osteocytic osteolysis. *Calc. Tiss. Res.* 4:1.

Boyde, A. 1972. *The Biochemistry and Physiology of Bone. Volume I. Structure and function* (G. H. Bourne, ed.), New York: Academic Press, pp. 259–310.

Boyde, A., P. G. Howell, and S. J. Jones. 1974. Measurement of lacunar volume in bone using a stereological grid counting method evolved for the scanning electron microscope. *J. Microsc.* (Oxford) 101(3):261–266.

Horn, V., and M. Y. Dvo-R-Ak. 1974. Ultrastructure of functional bone components in scanning and transmission electron microscopy. *Z. Mikrosk. Anat. Forsch.* 88(5):836–848.

Hough, A. J., W. G. Banfield, F. C. Mottram, and L. Sokoloff. 1974. The osteochondral junction of mammalian joints: An ultrastructural and microanalytic study. *Lab. Invest.* 31(6):685–695.

Jones, S. J., and A. Boyde. 1974. The organization and gross mineralization patterns of the collagen fibres in Sharpey fibre bone. *Cell Tissue Res.* 148(1):83–96.

Jones, S. J., A. Boyde, and J. B. Pawley. 1975. Osteoblasts and collagen orientation. *Cell Tissue Res.* 159(1):73–80.

Jones, S. J., and A. Boyde. 1977. Some morphological observations on osteoclasts. *Cell Tiss. Res.* 185:387–397.

Laczkó, J., G. Lévai, S. Varga, and J. Gyarmati. 1975. Preparation of the tibial growth organ of young rats for scanning electron microscopy. *Mikroskopie* 31(3/4):57–65.

Muto, H. 1974. Scanning electron microscopic studies of the cut surface of integument bone. *Acta Anat. Nippon* 49(4):257–269.

Swedlow, D. B., P. Frasca, R. A. Harper, and J. L. Katz. 1975. Scanning and transmission electron microscopy of calcified tissues. *Biomater. Med. Devices Artif. Organs* 3(1):121–153.

Whitehouse, W. J. 1974. The quantitative morphology of anisotropic trabecular bone. *J. Microsc.* (Oxford) 2:153–168.

Whitehouse, W. J. 1975. Scanning electron micrographs of cancellous bone from the human sternum. *J. Pathol.* 116:213–224.

Whitehouse, W. J., and E. D. Dyson. 1974. Scanning electron microscope studies of trabecular bone in the proximal end of the human femur. *J. Anat.* 118:417–444.

Zeidman, H., and F. W. Cooke. 1971. Scanning electron microscope analysis of fractures in bovine bone. *Surg. Forum* 22:444–445.

CHAPTER 4 CIRCULATING BLOOD, BLOOD VESSELS, AND BONE MARROW

Becker, R. P., and P. P. DeBruyn. 1976. The transmural passage of blood cells into myeloid sinusoids and the entry of platelets into the sinusoidal circulation: A scanning electron microscope investigation. *Am. J. Anat.* 145:183–206.

Chisholm, G. M., J. L. Gainer, and G. E. Stoner, 1973. Scanning electron microscope studies of aortic structure. *Angiologica* 10:10–14.

Clarke, J. A., A. N. Broers, and A. J. Salsburg. 1971. High-resolution scanning electron microscopy of the surface of red blood cells. *J. Microsc.* (Oxford) 93:233–236.

Collatz, C. B. 1974. Repair in arterial tissue: Scanning electron microscope and light microscope study on the endothelium of rabbit thoracic aorta following noradrenaline in toxic doses. *Virchows Arch.* 363:33–46.

Collatz, C. B., and C. Barbarsch. 1973. Repair in arterial tissue: Scanning electron microscope and light microscope study on the endothelium of rabbit thoracic aorta following a single dilation injury. *Virchows Arch.* 360:93–106.

Cope, D. A., and M. R. Roach. 1975. Scanning electron microscope study of human cerebral arteries. *Can. J. Physiol. Pharmacol.* 53:651–659.

Edanaga, M. 1974. Scanning electron microscope study on the endothelium of the vessels. I. Fine structure of the endothelial surface of aorta and some other arteries in normal rabbits. *Arch. Histol. Jap.* 37:1–14.

Edanaga, M. 1974. Scanning electron microscope study on the endothelium of vessels. II. Fine surface structure of the endocardium in normal rabbits and rats. *Arch. Histol. Jap.* 37:301–312.

Fishman, J. A., G. B. Ryan, and M. J. Karnovsky. 1975. Endothelial regeneration in the rat carotid artery and the significance of endothelial denudation in the pathogenesis of myointima thickening. *Lab. Invest.* 32:339–351.

Fujimoto, S., K. Yamamoto, and Y. Takeshige. 1975. Electron microscopy of endothelial microvilli of large arteries. *Anat. Rec.* 183:259–266.

Irino, S., T. Ono, K. Watanabe, K. Toyota, J. Uno, N. Takasugi, and T. Murakami. 1975. Scanning electron microscopic studies on microvascular architecture, sinus wall, and transmural passage of blood cells in the bone marrow by a new method of injection replica and non-coated specimens. O. Johari and I. Corvin (eds.), *Scanning Electron Microscopy/1975.* Chicago: IIT Research Institute, pp. 267–274.

Kawamura, J., S. D. Gertz, and T. Sunaga. 1974. Scanning electron microscope observations on the luminal surface of the rabbit common carotid artery subjected to ischemia by arterial occlusion. *Stroke* 5:765–774.

Peine, C. J., and F. N. Low. 1975. Scanning electron microscopy of cardiac endothelium of the dog. *Am. J. Anat.* 142:137–157.

Polliack, A., N. Lampen, B. D. Clarkson, E. DeHarven, Z. Bentwich, F. P. Siegal, and H. G. Kunkel. 1973. Identification of human B and T lymphocytes by scanning electron microscopy. *J. Exp. Med.* 138:607–624.

Rahman, Y. E., B. J. Wright, and E. A. Cerny. 1973. Studies on the mechanism of erythrocyte aging and destruction. II. Membrane fragmentation in rat erythrocytes after in vitro treatment with lysophosphatides: Scanning electron microscope studies. *Mech. Ageing Dev.* 2:151–162.

Smith, U., J. W. Ryan, D. D. Michie, and D. W. Smith. 1971. Endothelial projections as revealed by scanning electron microscopy. *Science* 173:925–927.

Tokunaga, J., M. Osaka, and T. Fujita. 1973. Endothelial surface of rabbit aorta as observed by scanning electron microscopy. *Arch. Histol. Jap.* 36:129–141.

Weiss, L. 1976. The hematopoietic microenvironment of the bone marrow: An ultrastructural study of the stroma in rats. *Anat. Rec.* 186:161–184.

Webster, W. S., S. P. Bishop, and J. C. Geer. 1974. Experimental aortic intimal thickening. II. Endothelialization and permeability. *Am. J. Pathol.* 76:265–284.

CHAPTER 5 LYMPHOID ORGANS

Bentwich, Z., and H. G. Kunkel. 1973. Specific properties of human B and T lymphocytes and alterations in disease. *Transplant Rev.* 16:29–50.

Criswell, B. S., R. R. Rich, J. Dardano, and S. L. Kimzey. 1975. Scanning electron microscopy of normal and mitogen-stimulated mouse lymphoid cells. *Cell Immunol.* 19:336–348.

Fujita, T. 1974. A scanning electron microscope study of the human spleen. *Arch. Histol. Jap.* 37(3):187–216.

Fujita, T., M. Miyoshi, and T. Murakami. 1972. Scanning electron microscope observation of the dog mesenteric lymph node. *Z. Zellforsch.* 133:147–162.

Hwang, W. S., T. Y. Ho, S. C. Luk, and G. T. Simon. 1974. Ultrastructure of the rat thymus. A transmission, scanning electron microscope, and morphometric study. *Lab. Invest.* 31(5):473–487.

Irino, S., T. Ono, K. Watanabe, K. Toyota, J. Uno, N. Takasugi, and T. Murakami. 1975. Scanning electron microscope studies on microvascular architecture, sinus wall, and transmural passage of blood cells in the bone marrow by a new method of injection replica and non-coated specimens. In O. Johari and I. Corvin (eds.), *Scanning Electron Microscopy/1975.* Chicago: IIT Research Institute, pp. 267–274.

Kay, M. M., B. Belohradsky, K. Yee, J. Vogel, D. Butcher, J. Wybran, and H. H. Fundenberg. 1974. Cellular interactions: Scanning electron microscopy of human thymus-derived rosette-forming lymphocytes. *Clin. Immunol. Immunopathol.* 2(3):301–309.

Komitowski, D., N. Paweletz, and K. Goerttler. 1974. Scanning electron microscopic studies of dissociated thymic cells in early stages of Rauscher leukemia. *Z. Krebsforsch.* 82(4):269–275.

Lenz, H. 1974. Surface alterations of the epithelium of the hyperplastic pharyngeal tonsil. A scanning electron microscopic study. *Arch. Otorhinolaryngol.* (NY), 206(2):113–126.

Lin, P. S., A. G. Cooper, and H. H. Wortis. 1973. Scanning electron microscopy of human T-cell and B-cell rosettes. *N. Engl. J. Med.* 289(11):548–551.

Lin, P. S., and D. F. Wallach. 1974. Surface modification of T-lymphocytes observed during rosetting. *Science* 184(143):1300–1301.

Linthicum, D. S., S. Sell, R. M. Wagner, and P. Trefts. 1974. Scanning immunoelectron microscopy of mouse B and T lymphocytes. *Nature* 252(5479):173–175.

Lipscomb, M. F., K. V. Holmes, and S. Vitetta. 1975. Cell surface immunoglobulin. XII. Localization of immunoglobulin on murine lymphocytes by scanning immunoelectron microscopy. *Eur. J. Immunol.* 5:255–259.

Luk, S. C., C. Nopajaroonsri, and G. T. Simon. 1973. The architecture of the normal lymph node and hemolymph node. A scanning and transmission electron microscopic study. *Lab. Invest.* 29(2):258–265.

Murakami, T., T. Fujita, and M. Miyoshi. 1973. Closed circulation in the rat spleen as evidenced by scanning electron microscopy of vascular casts. *Experientia* 29(11):1374–1375.

Polliack, A., N. Lampen, B. D. Clarkson, E. De Harven, Z. Bentwich, F. P. Siegal, and H. G. Kunkel. 1973. Identification of human B and T lymphocytes by scanning electron microscopy. *J. Exp. Med.* 138(3):607–624.

Song, S. H., and A. C. Groom. 1974. Scanning electron microscope study of the splenic red pulp in relation to the sequestration of immature and abnormal red cells. *J. Morphol.* 144(4):439–451.

Tokunaga, J., M. Edanaga, T. Fujita, and K. Adachi. 1974. Freeze cracking of scanning electron microscope specimens: A study of the kidney and spleen. *Arch. Histol. Jap.* 37:165–182.

Weiss, L. 1974. A scanning electron microscopic study of the spleen. *Blood* 43(5):665–691.

CHAPTER 6 NERVOUS TISSUE

General

Bruni, J. E., R. E. Clattenburg, and D. B. Montemurro. 1974. Ependymal tanycytes of the rabbit third ventricle: Scanning electron microscope study. *Brain Res.* 73:145–150.

Clementi, F., and D. Marini. 1972. The surface fine structure of the walls of the cerebral ventricles and of the choroid plexus in cat. *Z. Zellforsch.* 123:82–95.

Cloyd, M. W., and F. N. Low. 1974. Scanning electron microscopy of the subarachnoid space in the dog. I. Spinal cord levels. *J. Comp. Neurol.* 153:325–367.

Coates, P. W. 1977. The third ventricle of monkeys: Scanning electron microscopy of surface features in mature males and females. *Cell Tissue Res.* 177(3):307–316.

Collins, P., and G. M. Morriss. 1975. Changes in the surface features of choroid plexus of the rat following the administration of acetazolamide and other drugs which affect CSF secretion. *J. Anat.* 120:571–579.

Dierick, K., and G. De Waele. 1975. Scanning electron microscopy of the wall of the third ventricle of the brain of *Rana temporaria.* II. Electron microscopy of the ventricular surface of the pars ventralis of the tuber cinereum. *Cell Tissue Res.* 159:81–90.

Duce, I. R., J. F. Reeves, and P. Keen. 1976. A scanning electron microscope study of the development of free axonal sprouts at the cut ends of dorsal spinal nerve roots in the rat. *Cell Tissue Res.* 170:507–514.

Friedmann, I. 1973. *The Ultrastructure of Sensory Organs.* New York: American Elsevier Publishing Company, Inc.

Hamberger, A., H. A. Hansson, and A. Sellstrom. 1975. Scanning and transmission electron microscopy on bulk prepared neuronal and glial cells. *Exp. Cell Res.* 92:1–10.

Krstic, R. 1974. Observations of the nodes of Ranvier of rat sciatic nerve fibers under the scanning and transmission electron microscope. *Period. Biol.* 76:105–108.

Lewis, E. R., and Y. Y. Zeevi. 1969. Scanning electron mi-

croscopy of neural networks. *Proc. Electron Microsc. Soc. Am.* 27:8–9.

Meller, K., and W. Tetzlaff. 1975. Neuronal migration during the early development of the cerebral cortex: Scanning electron microscope study. *Cell Tissue Res.* 163:313–325.

Mestres, P., and W. Breipohl. 1976. Morphology and distribution of supraependymal cells in the third ventricle of the albino rat. *Cell Tissue Res.* 168:303–314.

Phillips, M. I., L. Balhorn, M. Leavitt, and W. Hoffman. 1974. Scanning electron microscope study of the rat subfornical organ. *Brain Res.* 80:95–110.

Scott, D. E., G. Krobisch-Dudley, W. K. Paull, and G. P. Kozlowski. 1977. The ventricular system in neuroendocrine mechanisms. III. Supraependymal neuronal networks in the primate brain. *Cell Tissue Res.* 179:235–254.

Scott, D. E., D. H. Van Dyke, W. K. Paull, and G. P. Kozlowski. 1974. Ultrastructural analysis of the human cerebral ventricular system. III. The choroid plexus. *Cell Tissue Res.* 150:389–398.

Eye

Anderson, D. R. 1971. Scanning electron microscopy of primate trabecular network. *Am. J. Ophthalmol.* 71:91–101.

Ben-Shaul, Y., and A. A. Moscona. 1975. Scanning electron microscopy of embryonic neural retina cell surfaces. *Dev. Biol.* 44(2):386–393.

Bill, A. 1970. Scanning electron microscope studies of the canal of Schlemm. *Exp. Eye Res.* 10:214–218.

Bill, A., and B. Svedbergh. 1972. Scanning electron microscopic studies of the trabecular meshwork and the canal of Schlemm: An attempt to localize the main resistance to outflow of aqueous humor in man. *Acta Ophthalmol.* (Kbh) 50(3):295–320.

Breipohl, W., N. Bornfeld, G. J. Bijvank, H. Laugwitz, and M. Pfautsch. 1973. Scanning electron microscopy of the retinal pigment epithelium in chick embryos and chicks. *Z. Zellforsch. Mikrosk. Anat.* 146(4):543–552.

Davanger, M. 1975. The suspensory apparatus of the lens. The surface of the ciliary body: A scanning electron microscopic study. *Acta Ophthalmol.* (Kbh) 53(1):19–33.

Dickson, D. H. 1975. Fine structure and mechanics of the anterior border of the primate iris: A scanning and transmission electron microscope study. *Can. J. Ophthalmol.* 10(2):227–238.

Farnsworth, P. N., S. C. Fu, P. A. Burke, and I. Bahia. 1974. Ultrastructure of rat eye lens fibers. *Invest. Ophthalmol.* 13(4):274–279.

Hansson, H. A. 1970. Scanning electron microscopy of the lens of adult rat. *Z. Zellforsch.* 107:187–198.

Hansson, H. A. 1970. Scanning electron microscopy of the zonular fibers in the rat eye. *Z. Zellforsch.* 107:191–209.

Harding, C. V., M. Bagchi, A. Weinsteder, and V. Peters. 1974. A comparative study of corneal epithelial cell surfaces utilizing the scanning electron microscope. *Invest. Ophthalmol.* 13(12):906–912.

Hollenberg, M. J., J. P. H. Wyse, and B. J. Lewis. 1976. Surface morphology of lens fibers from eyes of normal and microphthalmic (Browman) rats. *Cell Tissue Res.* 167:425–438.

Lim, W. C., and W. A. Webber. 1975. A scanning electron-microscopic study of the posterior and anterior surfaces of the rat iris in pupillary dilation and constriction. *Exp. Eye Res.* 20(5):445–462.

Martin, C. L. 1974. Development of pectinate ligament structure of the dog: Study by scanning electron microscopy. *Am. J. Vet. Res.* 35(11):1433–1439.

Meller, K., and W. Tetzlaff. 1976. Scanning electron microscopic studies on the development of the chick retina. *Cell Tissue Res.* 170:145.

Nelson, G. A., and J. P. Revel. 1975. Scanning electron microscopic study of cell movements in the corneal endothelium of the avian embryo. *Dev. Biol.* 42(2):315–333.

Pfister, R. R. 1975. The normal surface of conjunctiva epithelium. A scanning electron microscopic study. *Invest. Ophthalmol.* 14(4):267–279.

Polack, F. M., and P. S. Binder. 1975. Detachment of Descemet's membrane from grafts following wound separation: Light and scanning electron microscopic study. *Am. Ophthalmol.* 7(1):47–54.

Svedbergh, B., and A. Bill. 1972. Scanning electron microscopic studies of the corneal endothelium in man and monkeys. *Acta Ophthalmol.* (Kbh) 50(3):321–336.

Worthen, D. M., and M. G. Wickham. 1973. Fine structure and function of ocular tissues: The anterior chamber angle and limbus. *Int. Ophthalmol. Clin.* 13(3):109–129.

Organ of Corti

Bredberg, G., H. W. Ades, and H. Engström. 1972. Scanning electron microscopy of the normal and pathologically altered organ of Corti. *Acta Otolaryngol.* (Suppl.) 301:3–48.

Dallos, P. 1973. *The Auditory Periphery.* New York: Academic Press, pp. 196–211.

Engström, B. 1974. Scanning electron microscopy of the inner ear structure of the organ of Corti and its neural pathways. *Acta Otolaryngol.* (Suppl.) 319:57–66.

Hoshino, T., and A. Kodama. 1977. The contact between the cochlear sensory cell hairs and the tectorial membrane. In O. Johari and I. Corvin (eds.), *Scanning Electron Microscopy/1977.* Chicago: IIT Research Institute, pp. 409–414.

Hunter-Duvar, I. M. 1977. Morphology of the normal and the acoustically damaged cochlea. In O. Johari and I. Corvin (eds.), *Scanning Electron Microscopy/1977.* Chicago: IIT Research Institute, pp. 421–428.

Lim, D. J. 1969. Three dimensional observation of the inner ear with the scanning electron microscope. *Acta Otolaryngol.* (Suppl.) 255:5–38.

Lim, D. J. 1976. Morphological and physiological correlates in cochlear and vestibular sensory epithelia. In O. Johari and R. Becker (eds.), *Scanning Electron Microscopy/1976.* Chicago: IIT Research Institute, pp. 269–276.

Lim, D. J. 1977. Current review of SEM techniques for inner ear sensory organs. In O. Johari and I. Corvin (eds.), *Scanning Electron Microscopy/1977.* Chicago: IIT Research Institute, pp. 401–408.

Soudijn, E. R. 1976. Scanning electron microscopy of the organ of Corti. *Ann. Otol. Rhinol. Laryngol.* 86(Suppl.):16–58.

Takiguchi, T. 1973. Morphological study of the organ of corti by scanning electron microscopy. *J. Otolaryngol. Jap.* 76(5):578–585.

Tanaka, T., N. Kosaka, T. Takiguchi, T. Aoki, and S. Takahara. 1973. Observations on the cochlea with SEM. In O. Johari and I. Corvin (eds.), *Scanning Electron Microscopy/1973.* Chicago: IIT Research Institute, pp. 427–434.

Vestibular Sensory Receptors

Dohlman, G. F. 1969. The shape and function of the cupula. *J. Laryngol.* 83:43–53.

Dohlman, G. F. 1971. The attachment of the cupulae, otolith and tectorial membranes to the sensory cell areas. *Acta Otolaryngol.* 71:89–105.

Engström, H., B. Bergström, and H. W. Ades. 1972. Macula utricule and macula sacculi in the squirrel monkey. *Acta Otolaryngol.* (Suppl.) 301:75–126.

Harada, Y. 1973. The scanning electron microscopic observation of the vestibular organ and electrical activity of isolated individual semicircular ampullae. *Adv. Oto-Rhino-Laryngol.* 19:50–65.

Hillman, D. E. 1972. Observations on morphological features and mechanical properties of the peripheral vestibular receptor system in the frog. In A. Brodal and O. Pompeiano (eds.), *Progress in Brain Research* (Vol. 37). Amsterdam: Elsevier Publ. Co., pp. 329–339.

Hillman, D. E. 1976. Morphology of peripheral and central vestibular systems. In R. Llinas and W. Precht (eds.), *Frog Neurobiology.* Berlin: Springer-Verlag, pp. 452–480.

Hillman, D. E., and E. R. Lewis. 1971. Morphological basis for a mechanical linkage in otolithic receptor transduction in the frog. *Science* 174:416–419.

Hunter-Duvar, I. M. 1975. Hearing and hair cells. *Can. J. Otolaryngol.* 4(1):152–160.

Lim, D. J. 1973. Ultrastructure of the otolithic membrane and cupola. *Adv. Oto-Rhino-Laryngol.* 19:35–49.

Lim, D. J. 1973. Formation and fate of the otoconia: Scanning

and transmission electron microscopy. *Ann. Otol. Rhinol. Laryngol.* 82(1):23–35.

Lindeman, H. H. 1969. Studies on the morphology of the sensory regions of the vestibular apparatus. *Ergebn. Anat. EntwGesh.* 42(1):1–113.

Lindeman, H. H. 1973. Anatomy of the otolith organs. *Adv. Oto-Rhino-Laryngol.* 20:405–433.

Rosenhall, U., and B. Engström. 1974. Surface structures of the human vestibular sensory regions. *Acta Otolaryngol.* (Suppl.) 319:3–18.

Ross, M. D., and D. R. Peacor. 1975. The nature and crystal growth of otoconia in the rat. *Ann. Otol. Rhinol. Laryngol.* 84(1 pt. 1):22–36.

Stahle, J., B. Engström, and L. Högberg. 1973. Inner ear microsurgery using lasers. *Adv. Oto-Rhino-Laryngol.* 19:88–100.

Wersäll, J. 1956. Studies on the structure and innervation of the sensory epithelium of the cristae ampullares in the guinea pig. A light and electron microscope investigation. *Acta Otolaryngol.* (Suppl.) 126:1–85.

Wersäll, J., B. Björkroth, A. Flock, and P.-G. Lundquist. 1973. Experiments on ototoxic effects of antibiotics. *Adv. Oto-Rhino-Laryngol.* 20:14–41.

Olfactory and Gustatory

Beidler, L. M. 1970. Physiological properties of mammalian taste receptors. In G. E. Wolstenholme and J. Knight (eds.), *Taste and Smell in Vertebrates* (Ciba Foundation Symposium). London: J. and A. Churchill, pp. 51–70.

Bronshtein, A. A. 1962. Experimental histochemical study of the olfactory epithelium of vertebrates during exposure to aromatic substances. *Doklady Akademii Nauk SSSR* 145:–661.

Dethier, V. G. 1975. Transductive coupling in chemosensory systems. In F. O. Schmitt, D. M. Schneider, and D. M. Crothers (eds.), *Functional Linkage in Biomolecular Systems.* New York: Raven Press, pp. 253–271.

Farbmann, A. I. 1965. Fine structure of the taste bud. *J. Ultrastruct. Res.* 12:328–350.

Graziadei, P. 1969. The ultrastructure of vertebrate taste buds. In C. Pfaffmann (ed.), *Olfaction and Taste.* New York: Rockefeller Univ. Press, pp. 315–330.

Henkin, R., P. Graziadei, and D. Bradley. 1969. The molecular basis of taste and its disorders. *Ann. Intern. Med.* 71:791–821.

Jaeger, C. B., and D. E. Hillman. 1976. Morphology of gustatory organs. In R. Llinas and W. Precht (eds.), *Frog Neurobiology.* Berlin: Springer-Verlag, pp. 588–606.

Mozell, M. W., and M. Jagodowicz. 1973. Chromatographic separation of odorants by the nose: Retention times measured across in vivo olfactory mucosa. *Science* 181:1247–1249.

Murray, R. G. 1973. The ultrastructure of taste buds. In J. Friedmann (ed.), *The Ultrastructure of Sensory Organs*. Amsterdam: North-Holland Publ. Co., pp. 1–39.

Reese, T. S., and M. W. Brightman. 1970. Olfactory surface and central olfactory connexions in some vertebrates. In G. Wolstenholme and J. Knight, (eds.), *Taste and Smell in Vertebrates* (Ciba Foundation Symposium). London: J. and A. Churchill, pp. 115–149.

Vinnikov, J. A., and L. K. Titova. 1957. *Morphology of Olfactory Organs*. Ed. Medgiz. Moscow.

CHAPTER 7 MUSCLE TISSUE

Ashraf, M., and C. M. Bloor. 1976. The nature of myocardial Z line ridges seen with the scanning electron microscope. *J. Mol. Cell. Cardiol.* 8:489–495.

Ashraf, M., L. H. Livingston, and C. M. Bloor. 1976. Scanning electron microscopy of T tubules of myocardial cells. In O. Johari and R. P. Becker (eds.), *Scanning Electron Microscopy/1976*. Chicago: IIT Research Institute, pp. 631–636.

Ashraf, M., and H. D. Sybers. 1975. Scanning electron microscopy of the heart after coronary occlusion. *Lab. Invest.* 32:157–162.

Ashraf, M., and H. D. Sybers. 1975. Preparation of myocardial tissue: normal and infarcted. In M. A. Hayat (ed.), *Principles and Techniques of Scanning Electron Microscopy*. New York: Van Nostrand Reinhold, pp. 1–20.

Bacaner, M., J. Broadhurst, T. Hutchinson, and J. Lilley. 1973. Scanning transmission electron microscope studies of deep-frozen unfixed muscle correlated with spatial localization of intracellular elements by fluorescent x-ray analysis. *Proc. Natl. Acad. Sci.* (USA) 70:3423–3427.

Cohen, C. 1975. The protein switch of muscle contraction. *Sci. Am.* Dec. 1975, pp. 36–45.

Geissinger, H. D., and I. Grinyer. 1976. Correlated scanning electron microscopy in transmission (TREM) and reflection (REM) on sections of skeletal muscle. *Mikroskopie* 32:329–333.

Murray, J., and A. Weber. 1974. The cooperative action of muscle proteins. *Sci. Am.* Feb. 1974, pp. 59–71.

Sakuragawa, N., T. Sato, and T. Tsubaki, 1973. Scanning electron microscopic study of skeletal muscle: Normal, dystrophic, and neurogenic atrophic muscle in mice and humans. *Arch. Neurol.* 28:247–251.

Simpson, F. O., D. G., Rayns, and J. M. Ledingham. 1974. Fine structure of mammalian myocardial cells. *Adv. Cardiol.* 12:15–33.

Somlyo, A. P. 1975. Structural characteristics, mechanisms of contraction, innervation and proliferation of smooth muscle cells: Ultrastructure and function of vascular smooth muscle. *Adv. Exp. Med. Biol.* 57:1–80.

Sommer, J. R., and R. A. Waugh. 1976. The ultrastructure of the mammalian cardiac muscle cell—with special emphasis on the tubular membrane systems. A review. *Am. J. Pathol.* 82:192–232.

Squire, J. M. 1973. General model of myosin filament structure. III. Molecular packing arrangements in myosin filaments. *J. Mol. Biol.* 77:291–323.

Sybers, H. D., and M. Ashraf. 1973. Preparation of cardiac muscle for scanning electron microscopy. In O. Johari and I. Corvin (eds.), *Scanning Electron Microscopy/1973*. Chicago: IIT Research Institute, pp. 342–348.

Sybers, H. D., and M. Ashraf. 1974. Scanning electron microscopy of cardiac muscle. *Lab. Invest.* 30:441–450.

CHAPTER 8 INTEGUMENTARY SYSTEM

Barriere, H., P. Litoux, and C. Geraut. 1974. Study of psoriatic horny cells with the scanning electron microscope. *Dermatologica* 149:257–265.

Brown, J. A. 1972. Scanning electron microscopy of human dermal fibrous tissue. *J. Anat.* 113(2):159–168.

Holbrook, K. A., and G. F. Odland. 1975. The fine structure of developing human epidermis: Light, scanning, and transmission electron microscopy of the periderm. *J. Invest. Dermatol.* 65(1):16–38.

Hunter, J. A., and B. Finlay. 1973. Identification of elastic tissue in human skin viewed in the scanning electron microscope. *J. Microsc.* 98(1):41–47.

Marks, R., and B. Bhogal. 1974. Scanning electron microscopy of the normal epidermis and the epidermis in vitro. *Br. J. Dermatol.* 90(4):387–396.

Millington, P. F., and I. A. Brown. 1970. Scanning electron microscope studies of some internal surfaces in human skin. *Z. Zellforsch.* 106:209–219.

Orfanos, C. E. 1972. Scanning electron microscopy of the surface of the skin and skin appendages: Disease-induced changes and influence of exogenous lesions. *Arch. Dermatol. Forsch.* 244:80–85.

Stofft, E., G. Muller, and I. E. Richter. 1974. Scanning electron microscopy studies on the lower side of the human corium. *Anat. Anz.* 136(4):359–371.

Wolf, J. 1974. Relationship between desmosomes and serrations of primary relief of cornified epidermal cells. *Folia Morphol.* (Phaha) 22(2):145–150.

Wong, C. K., and H. R. Vickers. 1972. A study of vertical sections of normal human skin with the scanning electron microscope. *Dermatologica* 145(6):371–376.

CHAPTER 9 THE DIGESTIVE SYSTEM

Anderson, J. H., and A. B. Taylor. 1973. Scanning and transmission electron microscopic studies of jejunal microvilli of the rat, hamster and dog. *J. Morphol.* 141(3):281–291.

Andrews, P. M. 1976. Microplicae: Characteristic ridge-like folds of the plasmalemma. *J. Cell Biol.* 68:420–429.

Asquith, P., A. G. Johnson, and W. T. Cooke. 1970. Scanning electron microscopy of normal and celiac jejunal mucosa. *Am. J. Digest. Dis.* 15:511–521.

Baratz, R. S., and A. I. Farbman. 1975. Morphogenesis of rat lingual filiform papillae. *Am. J. Anat.* 143:283–302.

Booth, W. V., M. Zimmy, H. J. Kaufman, and I. Cohn, Jr. 1973. Scanning electron microscopy of small bowel strangulation obstruction. *Am. J. Surg.*, 125(1):129–133.

Capoferro, R. 1974. A scanning electron microscopic study of normal and X-irradiated guinea-pig gastric mucosa. *Scand. J. Gastroenterol.* 9(6):533–538.

Hattori, T., and S. Fujita. 1974. Fractographic study on the growth and multiplication of the gastric gland of the hamster. The gland division cycle. *Cell Tissue Res.* 153(2):145–149.

Ohashi, Y., Kita, S., and T. Murakami. 1976. Microcirculation of the rat small intestine as studied by the injection replica scanning electron microscope method. *Arch. Histol. Jap.* 39(4):271–282.

Piasecki, C. 1975. Observations on the submucosa plexus and mucosal arteries of the dog's stomach and the first part of the duodenum. *J. Anat.* 119:133–148.

Takagi, T., S. Takebayashi, K. Tokuyasu, and K. Tsuji. 1974. Scanning electron microscopy on the human gastric mucosa; fetal, normal and various pathological conditions. *Acta Pathol. Jap.* 24(2):233–247.

Taylor, A. B., and J. H. Anderson. 1972. Scanning electron microscope observations of mammalian intestinal villi, intervillus floor and crypt tubules. *Micron* 3:430–453.

Toner, P. G., and K. E. Carr. 1969. The use of scanning electron microscopy in the study of the intestinal villi. *J. Pathol.* 97:611–617.

Urakami, Y. 1975. Endoscopical and histological studies on gastric metaplasia of the surface epithelium of the duodenal mucosa. *Jap. J. Gastroenterol.* 72(3):221–231.

CHAPTER 10 GLANDS OF DIGESTION

Liver

Brooks, S. E., and G. H. Haggis. 1973. Scanning electron microscopy of rat's liver. Application of freeze-fracture and freeze-drying techniques. *Lab. Invest.* 29(1):60–64.

Compagno, J., and J. W. Grisham. 1974. Scanning electron microscopy of extrahepatic biliary obstruction. *Arch. Pathol.* 97:348–351.

Elias, H., and J. C. Sherrick. 1969. *Morphology of the Liver.* New York: Academic Press.

Grisham, J. W., W. Nopanitaya, J. Compagno, and A. E. H. Nägel. 1975. Scanning electron microscopy of normal rat liver: Surface structure of its cell and tissue components. *Am. J. Anat.* 144:295–322.

Itoshima, T., T. Kobayashi, Y. Shimada, and T. Murakami. 1974. Fenestrated endothelium of the liver sinusoids of the guinea pig as revealed by scanning electron microscopy. *Arch. Histol.* Jap. 37(1):15–24.

Itoshima, T., Y. Shimada, N. Hayashi, and T. Murakami. 1976. A scanning electron microscopy of subcellular structures of the human hepatic cell. *Arch. Histol. Jap.* 39(1):15.

Katsumi, M., A. Richardson, W. Mayr, and N. Javitt. 1977. Subcellular pathology of rat liver in cholestasis and choleresis induced by bile salts. *Lab. Invest.* 36(3):249–258.

Layden, T. J., J. Schwarz, and J. L. Boyer. 1975. Scanning electron microscopy of the rat liver. *Gastroenterology* 69(3):724–738.

Mall, F. P. 1906. A study of the structural unit of the liver. *Am. J. Anat.* 5:227–308.

Miyai, K., J. Abraham, S. Linthicum, and R. Wagner. 1976. Scanning electron microscopy of hepatic ultrastructure. *Lab. Invest.* 35(4):369–376.

Motta, P., and K. R. Porter. 1974. Structure of rat liver sinusoids and associated tissue spaces as revealed by scanning electron microscopy. *Cell Tissue Res.* 148(1):111–125.

Motta, P., and K. R. Porter. 1974. Study of the spaces of Disse and of the rat liver sinusoids by scanning electron microscopy. *Bull. Assoc. Anat.* 58:631–638.

Motta, P. 1975. Scanning electron microscope study of the rat liver sinusoid. *Cell Tissue Res.* 164:371–385.

Motta, P., and G. Fumagalli. 1975. Structure of rat bile canaliculi as revealed by scanning electron microscopy. *Anat. Rec.* 182(4):499–513.

Nemchausky, B. A., T. J. Layden, and J. C. Boyer. 1977. Effects of chronic choleretic infusions of bile acids on the membrane of the bile canaliculus. *Lab. Invest.* 36(3):259–267.

Rappaport, A. M. 1958. The structural and functional unit in the human liver (liver acinus). *Anat. Rec.* 130:673–690.

Rappaport, A. M. 1963. Acinar units and the pathophysiology of the liver. In C. Rouiller (ed.), *The Liver: Morphology, Biochemistry, Physiology* (Vol. I). New York: Academic Press, pp. 265–328.

Rappaport, A. M., Z. J. Borowy, W. M. Lougheed, and W. N. Lotto. 1954. Subdivision of hexagonal liver lobules into a structural and functional unit. *Anat. Rec.* 119:11–34.

Sasse, D., and A. Schenk. 1975. A three-dimensional presentation of the functional liver unit. *Acta Anatomica* 93(1):78–87.

Skaaring, P., and F. Bierring. 1976. On the intrinsic innervation of normal rat liver. Histochemical and scanning electron microscopical studies. *Cell Tissue Res.* 171:141–155.

Gallbladder

Bills, P. M., and D. Lewis. 1975. A structural study of gallstones. *Gut* 16:630–637.

Mueller, J. C., A. L. Jones, and J. A. Long. 1972. Topographic and subcellular anatomy of the guinea pig gallbladder. *Gastroenterology* 63(5):856–868.

Pancreas

Fujita, T., and T. Murakami. 1973. Microcirculation of monkey pancreas with special reference to the insuloacinar portal system. A scanning electron microscope study of vascular casts. *Arch. Histol. Jap.* 35(4):255–263.

Zimny, M. L., and W. G. Blackard. 1975. Surface structure of isolated pancreatic islet cells. *Cell Tissue Res.* 164:467–471.

CHAPTER 11   RESPIRATORY SYSTEM

Andrews, P. M. 1974. A scanning electron microscopic study of the extrapulmonary respiratory tract. *Am. J. Anat.* 139:299–324.

Breipohl, W., C. Herberhold, and R. Kerschek. 1977. Microridge cells in the larynx of the male white rat. Investigations by reflection scanning electron microscopy. *Arch. Oto-Rhino-Laryngol.* 215:1–9.

Castleman, W. L., D. L. Dungworth, and W. S. Tyler. 1974. Intrapulmonary airway morphology in three species of monkey: A correlated scanning and transmission electron microscopic study. *Am. J. Anat.* 142:107–122.

Greenwood, M. F., and P. Holland. 1972. The mammalian respiratory tract surface. A scanning electron microscopic study. *Lab. Invest.* 27(3):296–304.

Groniowski, J., M. Walski, and W. Biczysko. 1972. Application of scanning electron microscopy for studies of the lung parenchyma. *J. Ultrastruct. Res.* 38:473–481.

Kilburn, K. H. 1974. Functional morphology of the distal lung. In G. H. Bourne and J. F. Danielli (eds.), *International Review of Cytology* (Vol. 37). New York: Academic Press.

Kuhn, C., III, and E. H. Finke. 1972. The topography of the pulmonary alveolus: Scanning electron microscopy using different fixatives. *J. Ultrastruct. Res.* 38:161–173.

Nowell, J. A., and W. S. Tyler. 1971. Scanning electron microscopy of the surface morphology of mammalian lungs. *Am. Rev. Respir. Dis.* 103:313–328.

Smolich, J. J., B. F. Stratford, J. E. Maloney, and B. C. Ritchie. 1977. Postnatal development of the epithelium of larynx and trachea in the rat: Scanning electron microscopy. *J. Anat.* 124(3):657–673.

Sperry, D. G., and R. J. Wassersug. 1976. A proposed function for microridges on epithelial cells. *Anat. Rec.* 185:247–253.

Wang, N. S., H. W. Taeusch, W. M. Thurlbeck, and M. E. Avery. 1973. A combined scanning and transmission electron microscopic study of alveolar epithelial development of the fetal rabbit lung. *Am. J. Pathol.* 73:365–376.

Wang, N., and W. M. Thurlbeck. 1970. Scanning electron microscopy of the lung. *Human Path.* 1(2):227–231.

CHAPTER 12   URINARY SYSTEM

Kidney

Andrews, P. M. 1975. Scanning electron microscopy of human and rhesus monkey kidneys. *Lab. Invest.* 32(5):510–518.

Andrews, P. M. 1977. A scanning and transmission electron microscope comparison of puromycin aminonucleoside-induced nephrosis to hyperalbuminemia-induced proteinuria with emphasis on kidney podocyte pedicel loss. *Lab. Invest.* 36(2):183–197.

Andrews, P. M., and K. R. Porter. 1974. A scanning electron microscopic study of the nephron. *Am. J. Anat.* 140(1):81–115.

Andrews, P. M., and K. R. Porter. 1975. Scanning electron microscopy of the nephrotic kidney. *Virchows Arch.* B 17:195–211.

Arakawa, M., and J. Tokunaga. 1974. Further scanning electron microscope studies of the human glomerulus. *Lab. Invest.* 31(5):436–440.

Arakawa, M., J. Tokunaga, T. Shimotori, and Y. Kinoshita. 1974. A scanning electron microscope study of the glomerulus of normal and nephritic rabbits. *Virchows Arch. Cell Pathol.* 17(2):97–112.

Beeuwkes, R., and J. V. Bonventre. 1975. Tubular organization and vascular-tubular relations in the dog kidney. *Am. J. Physiol.* 229(3):695–713.

Bulger, R. E., F. L. Siegel, and R. Pendergrass. 1974. Scanning and transmission electron microscopy of the rat kidney. *Am. J. Anat.* 139(4):483–501.

Bulger, R. E., F. L. Siegel, and R. Pendergrass. 1976. Proximal tubule tendrils: Fact or artefact. *Am. J. Anat.* 146:323–330.

Burke, J. A. 1974. Scanning electron microscopy of the mammalian urinary tract. *Ped. Res.* 8:454.

Camazine, S. M., G. B. Ryan, E. R. Unanue, and M. J. Karnovsky. 1976. Isolation of phagocytic cells from the rat renal glomerulus. *Lab. Invest.* 35(4):315–326.

Carroll, N., G. W. Crock, C. C. Funder, C. R. Green, K. N. Ham, and J. D. Tange. 1974. Scanning electron microscopy of the rat renal papilla. *J. Anat.* 117(3):447–452.

Evan, A. P., and K. D. Gardner. 1976. Comparison of human polycystic and medullary cystic disease with diphenyl-amine-induced cystic disease. *Lab. Invest.* 35(1):93–101.

Fourman, J., and D. B. Moffat. 1972. *The Blood Vessels of the Kidney.* Oxford: Blackwell.

Fujita, T., J. Tokunaga, and M. Edanaga. 1976. Scanning electron microscopy of the glomerular filtration membrane in the rat kidney. *Cell Tissue Res.* 166:299–314.

Hagege, J., M. Gabe, and G. Richet. 1974. Scanning of the apical pole of distal tubular cells under differing acid-base conditions. *Kidney Int.* 5(2):137–146.

Huecker, H., and H. Frenzel. 1975. Scanning electron microscopy of the distal nephron and calyx of the human kidney. *Virchows Arch. Cell Pathol.* 18(2):157–164.

Lee, M. M. 1974. Morphologic effects of procaine amide on mouse kidney as observed by scanning and by transmission electron microscopy. *Lab. Invest.* 31(4):324–331.

Lehtonen, E., I. Virtanen, and J. Wartiovaara. 1973. Visualization of human glomerular changes by scanning electron microscopy. *Virchows Arch. Zellpathol.* 13(3):259–265.

Limas, C., C. Umas, and M. Gessel. 1976. Effects of indomethacin on renomedullary interstitial cells. *Lab. Invest.* 34(5):522–528.

Osvaldo, L., and H. Latta. 1966. Interstitial cells of the renal medulla. *J. Ultrastruct. Res.* 15:589.

Pfaller, W., and J. Klima. 1976. A critical reevaluation of the structure of the rat uriniferous tubule as revealed by scanning electron microscopy. *Cell Tissue Res.* 166:91–100.

Shaaring, P., and J. Kjaergaard. 1974. Scanning electron microscopy of podocytes of the rat kidney. *Acta Anat.* (Basel) 87(3):394–403.

Spinelli, F. 1974. Structure and development of the renal glomerulus as revealed by scanning electron microscopy. *Int. Rev. Cytol.* 39:345–378.

Tokunaga, J., M. Edanaga, T. Fujita, and K. Adachi. 1974. Freeze cracking of scanning electron microscope specimens: A study of the kidney and spleen. *Arch. Histol. Jap.* 37(2):165–182.

Webber, W. A., and J. Lee. 1974. The ciliary pattern of the parietal layer of Bowman's capsule. *Anat. Rec.* 180(3):449–455.

Wheeler, E. E., and P. B. Herdson. 1973. Freeze fracturing and freeze drying of renal tissue for scanning electron microscopy. *Am. J. Clin. Pathol.* 60(2):229–233.

Ureter and Bladder

Kinoshita, H. 1972. Scanning electron microscopic observations of the normal transitional cell epithelium and transitional cell carcinoma of the urinary bladder. *Jap. J. Urol.* 63(8):649–657.

Mooney, J. K. Jr., and F. Hinman. 1974. Aging and replacement of the luminal cells in the mammalian bladder studied by scanning electron microscopy. *Invest. Urol.* 11(5):396–401.

Mooney, J. K. Jr., and F. Hinman. 1974. Surface differences in cells of proximal and distal canine urethra. *J. Urol.* 3(4):495–501.

Skoluda, D., I. E. Richter, and K. Busse. 1974. Experiments in coli cystitis: Scanning electron microscopy of the bladder urothelium. *Urol. Int.* 29(4):299–311.

Countercurrent Systems

Berliner, R. W., N. G. Levinsky, D. G. Davidson, and M. Eden. 1958. Dilution and concentration of urine and the action of antidiuretic hormone. *Am. J. Med.* 24:730.

Gottschalk. C. W., and M. Mylle. 1959. Micropuncture study of the mammalian urinary concentrating mechanism: Evidence for the countercurrent hypothesis. *Am. J. Physiol.* 196:927.

Jamison, R. L., C. M. Bennet, and R. W. Berliner. 1967. Countercurrent multiplication by the thin loops of Henle. *Am. J. Physiol.* 212:357.

Kriz, W., and A. F. Lever. 1969. Renal countercurrent mechanisms: Structure and function. *Am. Heart J.* 78(1):101–118.

Lever, A. F. 1965. The vasa recta and countercurrent multiplication. *Acta Med. Scandinav.* (Suppl.) 178:434.

Lever, A. F., and W. Kriz. 1966. Countercurrent exchange between the vasa recta and the loop of Henle. *Lancet* 1:1057.

Marsh, D. J., and L. A. Segel. 1971. Analysis of countercurrent diffusion exchange in blood vessels of the renal medulla. *Am. J. Physiol.* 221:817.

Stevenson, J. C. 1965. Ability of counterflow systems to concentrate. *Nature* 206:1215.

Ullrich, K. J., K. Kramer, and J. W. Boylan. 1961. Present knowledge of the countercurrent system in the mammalian kidney. *Prog. Cardiovas. Dis.* 3:395.

Renal Vascular Casts

Anderson, B. G., and W. D. Anderson. 1976. Renal vasculature of the trout demonstrated by scanning electron microscopy, compared with canine glomeruli vessels. *Am. J. Anat.* 145:443–458.

Evan, A. P., and W. G. Dail, Jr. 1976. Efferent arterioles in the cortex of the rat kidney. *Anat. Rec.* 187:135–146.

Moffat, D. B., and J. Fourman. 1963. The vascular pattern of the rat kidney. *J. Anat.* 97:543–553.

Murakami, T., M. Masayuki, and T. Fujita. 1971. Gomerular vessels of the rat kidney with special reference to double efferent arterioles: A scanning electron microscope study of corrosion casts. *Arch. Histol. Jap.* 33(3):179–198.

Murakami, T. 1972. Vascular arrangement of the rat renal glomerulus: A scanning electron microscope study of corrosion casts. *Arch. Histol. Jap.* 34:87–107.

Nowell, J. A., and C. L. Lohse. 1974. Injection replication of the microvasculature for SEM. In O. Johari and I. Corvin (eds.), *Scanning Electron Microscopy/1974.* Chicago: IIT Research Institute, pp. 267–274.

Spinelli, F. R., H. Wirz, C. Brucher, and G. Pehling. 1972. Nonexistence of shunts between afferent and efferent arterioles of juxtamedullary glomeruli in dog and rat kidneys. *Nephron* 9:123–128.

Trueta, J., A. E. Barclay, P. M. Daniel, K. J. Franklin, and M. M. Prichard. 1947. *Studies of the Renal Circulation.* Oxford: Blackwell Scientific Publications.

CHAPTER 13 THYROID GLAND

Fujita, H., and T. Murakami. 1974. Scanning electron microscopy on the distribution of the minute blood vessels in the thyroid gland of the dog, rat and rhesus monkey. *Arch. Histol. Jap.* 36(3):181–188.

Halmi, N. S., 1975. Thyroid gland. In *A Syllabus of Endrocrinology for Medical and Graduate Students.* Iowa City: The University of Iowa, pp. 48–64.

Hanse, J., and P. Skaaring. 1973. Scanning electron microscopy of normal rat thyroid. *Anat. Anz.* 134(3):177–185.

Ketelbant-Balasse, P., and P. Neve. 1974. Morphological modifications of apical surfaces of thyroid cells in different functional conditions. In O. Johari and I. Corvin (eds.), *Scanning Electron Microscopy/1974.* Chicago: ITT Research Institute, pp. 761–768.

Ketelbant-Balasse, P., F. Rodesch, P. Neve, and J. Pasteels. 1972. Aspects of the 1st stage of thyroid secretion using scanning electron microscope. *Ann. Endocrinol.* (Paris) 33(6):629–631.

Ketelbant-Balasse, P., F. Rodesch, P. Neve, and J. M. Pasteels. 1973. Scanning electron microscope observations of apical surfaces of dog thyroid cells. *Exp. Cell Res.* 79(1):111–119.

Kobayashi, S. 1973. Scanning electron microscope observations of the thyroid. *Arch. Histol. Jap.* 36(2):107–117.

Lavedan, J., and C. G. Theret. 1973. Functional aspects of thyroid cells studied by means of scanning electron microscopy: Intracellular detection of iodine by simultaneous X-ray microdiffraction. *C. R. Acad. Sci.* D. (Paris) 276(6):1003–1004.

CHAPTER 14 ADRENAL GLAND

Bennett, H. S., 1940. The life history and secretion of the adrenal cortex of the cat. *Am. J. Anat.* 67:151–228.

Bloodworth, J. M., and Powers, K. L., 1968. The ultrastructure of the normal dog adrenal. *J. Anat.* 102:457–476.

Brown, W. J., Barajas, L., and Latta, H., 1971. The ultrastructure of the human adrenal medulla: With comparative studies of white rat. *Anat. Rec.* 169:173–184.

Elfvin, L. G., 1967. The development of the secretory granules in the rat adrenal medulla. *J. Ultrastruct. Res.* 17:45–62.

Halmi, N. S., 1975. Adrenal glands. In *A Syllabus of Endocri-*

nology for Medical and Graduate Students. Iowa City: The University of Iowa, pp. 84–107.

Idelman, S., 1970. Ultrastructure of the mammalian adrenal cortex. Int. Rev. Cytol. 27:181–281.

Long, J. A., and Jones, A. L., 1967. Observations on the fine structure of the adrenal cortex of man. Lab. Investig. 17:355–370.

Rhodin, J. A., 1971. The ultrastructure of the adrenal cortex of the rat under normal and experimental conditions. J. Ultrastruct. Res. 34:23–71.

CHAPTER 15   MALE REPRODUCTIVE SYSTEM

Baccetti, B., and A. G. Burrini. 1973. An improved method for the scanning electron microscopy of spermatozoa. J. Microscopy 99:101–107.

Berns, D. M., R. A. Rodzen, and E. E. Brueschke, 1974. Vasa deferentia of the human and dog: Study with the SEM. In O. Johari and I. Corvin (eds.), Scanning Electron Microscopy/1974. Chicago: IIT Research Institute, pp. 647–654.

Brueschke, E. E., L. J. Znaeveld, R. Rodzen, and D. Berns. 1974. Development of a reversible vas deferens occlusive device. 3. Morphology of the human and dog vas deferens: A study with the scanning electron microscope. Fertil. Steril. 25(8):687–702.

Bustos-Obregon, E., and J. E. Flechon, 1975. Comparative SEM study of boar, bull, and ram spermatozoa. Cell Tissue Res. 161:329–341.

Connell, C. J. 1976. A scanning electron microscope study of the interstitial tissue of the canine testis. Anat. Rec. 185:389–402.

Dym, Martin. 1976. The mammalian rete testes: A morphological examination. Anat. Record 186(4):493.

Gould, K. G., and D. E. Martin. 1975. SEM/EDX: A tool of potential diagnostic value in cases of infertility. J. Reprod. Med. 14:197–200.

Gravis, C. J. 1978. A scanning electron microscope study of the Sertoli cell and spermiation in the Syrian hamster. Am. J. Anat. 151:21–38.

Hafez, E. S., and H. Kanagawa. 1973. Scanning electron microscopy of human, monkey and rabbit spermatozoa. Fertil. Steril. 24:776–787.

Hutson, J. C. 1978. The effects of various hormones on the surface morphology of testicular cells in culture. Am. J. Anat. 151:55–70.

Miyake, K., and H. Mitsuya. 1975. Observation on the shape of the infertile seminiferous tubules by scanning electron microscopy (author's transl.) Jap. J. Urol. 66(1):6–14.

Nowell, J. A., and L. J. Faulkin. 1974. Internal topography of the male reproductive system. In O. Johari and I. Corvin (eds.), Scanning Electron Microscopy/1974. Chicago: IIT Research Institute, pp. 639–646.

Russel, L. 1977. Movement of spermatocytes from the basal to the adluminal compartment of the rat testis. Am. J. Anat. 148(3):313–328.

Tung, P. S., E. Y. C. Lin, and I. B. Fritz. 1976. A scanning electron microscopic study of cultured cells prepared from rat seminiferous tubules. O. Johari and R. Becker (eds.), Scanning Electron Microscopy/1976. Chicago: IIT Research Institute 6:417–424.

Zaneveld, L. J. D., P. F. Tauber, D. Port, D. Propping, and G. F. B. Schumacher. 1974. SEM of the human, guinea pig, and rhesus monkey seminal coagulum. J. Reprod. Fertil. 40:223–225.

CHAPTER 16   FEMALE REPRODUCTIVE SYSTEM

Ovary

Bjersing, L., and S. Cajander. 1974. Ovulation and the mechanism of follicle rupture. II. Scanning electron microscopy of rabbit germinal epithelium prior to induced ovulation. Cell Tissue Res. 149(3):301–312.

Bjersing, L., and S. Cajander. 1975. Ovulation and the role of the ovarian surface epithelium. Experientia 31(5):605–608.

Carter, H. W. 1974. Cumulus cells of the hamster ovum and their interaction with spermatozoa: A correlative light and scanning electron microscope study. In O. Johari and I. Corvin (eds.), Scanning Electron Microscopy/1974. Chicago: IIT Research Institute, pp. 623–630.

Cherney, D., P. Motta, and L. Didio. 1973. Ovarian villi in rabbits studied with light, scanning and transmission electron microscopy. J. Microscopie 17:37–40.

Diercks, K. 1927. Der normale mensuelle Zyklus der menschlichen Vaginalschleimhaut. Arch. Gynäk. 130:40–69.

Gould, K. G., and B. Vet. 1973. Preparation of mammalian gametes and reproductive tract tissues for scanning electron microscopy. Fertil. Steril. 24(6):448–456.

Hafez, E. S. E. (ed.). 1975. Scanning Electron Microscopic Atlas of Mammalian Reproduction. Tokyo: Igaku Shoin.

Ludwig, H., and M. Metzger. 1976. The Human Female Reproductive Tract. New York: Springer-Verlag.

Motta, P. 1974. The fine structure of ovarian cortical crypts and cords in mature rabbits: A transmission and scanning electron microscopic study. Acta Anat. (Basel) 90(1):36–64.

Motta, P., and J. Van Blerkom. 1974. A scanning electron microscope study of the luteo-follicular complex. I. Follicle and oocyte. J. Submicrosc. Cytol. 6:297–310.

Motta, P., and J. Van Blerkom. 1975. A scanning electron microscopic study of the luteo-follicular complex. II. Events leading to ovulation. Am. J. Anat. 143(2):241–263.

Oviduct

Dirksen, E. R. 1974. Ciliogenesis in the mouse oviduct: A scanning electron microscope study. J. Cell Biol. 62(3):899–904.

Ferenczy, A. 1974. The surface ultrastructure of the human fallopian tube. A comparative morphophysiologic study. In O. Johari and I. Corvin (eds.), Scanning Electron Microscopy/1974. Chicago: IIT Research Institute, pp. 613–622.

Ferenczy, A., R. M. Richart, F. J. Agate, M. L. Purkerson, and E. W. Dempsey. 1972. Scanning electron microscopy of the human fallopian tube. Science 175(23):783–784.

Patek, E., and L. Nilsson. 1973. Scanning electron microscopic observations on the ciliogenesis of the infundibulum of the human fetal and adult fallopian tube epithelium. Fertil. Steril. 24(11):819–831.

Patek, E., L. Nilsson, E. Johannisson, M. Hellema, and J. Bout. 1973. Scanning electron microscopic study of the human fallopian tube. III. The effect of midpregnancy and of various steroids. Fertil. Steril. 24(1):31–43.

Rumery, R. E., and E. M. Eddy. 1974. Scanning electron microscopy of the fimbriae and ampullae of rabbit oviducts. Anat. Rec. 178(1):83–101.

Stalheim, O. H. V., J. E. Gallagher, and B. L. Deyoe. 1974. Scanning electron microscopy of the bovine, equine, porcine, and caprine uterine tube (oviduct). Am. J. Vet. Res. 36:1069–1075.

Uterus

Anderson, W., Y. Kang, and E. DeSombre. 1975. Estrogen and antagonist-induced changes in endometrial topography of immature and cycling rats. J. Cell Biol. 64:692–703.

Enders, A., and D. Nelson. 1973. Pinocytotic activity of the uterus of the rat. Am. J. Anat. 138:277–300.

Fathalla, M. A., R. M. Liptrap, and H. D. Geissinger. 1975. Combined scanning electron and light microscopy of biopsy samples of bovine uterus. Can. J. Comp. Med. 39:457–461.

Ferenczy, A., and R. M. Richart. 1973. Scanning and transmission electron microscopy of the human endometrial surface epithelium. J. Clin. Endocrinol. Metab. 36(5):999–1008.

Hafez, E. S. E., H. Ludwig, and H. Metzger. 1975. Human endometrial fluid kinetics as observed in scanning electron microscopy. Obstet. Gynecol. 122:929–938.

Hayashi, K., S. Hamanishi, and W. Lee. 1974. Scanning electron microscopic observations on the luminal surface of the rat uterus in the process of implantation and its hormonal regulation. Folia Endocrinol. Jap. 50(4):813–827.

Johannisson, E., and L. Nilsson. 1972. Scanning electron microscopic study of the human endometrium. Fertil. Steril. 23:613–625.

Johnson, E. C., and J. D. Hamer. 1975. Scanning electron microscopy of the luminal epithelium of the mouse uterus. J. Reprod. Fertil. 42(1):95–104.

Macdonald, A. A. 1976. Uterine vasculature of the pregnant pig: A scanning electron microscope study. Anat. Rec. 184(4):689–697.

Motta, P. M., and P. M. Andrews. 1976. Scanning electron microscopy of the endometrium during the secretory phase. *J. Anat.* 122(2):315–322.

Nilsson, O. 1974. Changes of the luminal surface of the rat uterus at blastocyst implantation. Scanning electron microscopy and ruthenium red staining. *Z. Anat. Entwicklungsgesch* 144(3):337–342.

Nilsson, O., and K. Hagenfeldt. 1973. Scanning electron microscopy of human uterine epithelium influenced by the TCU intrauterine contraceptive device. *Am. J. Obstet. Gynecol.* 117(4):469–472.

Richart, R. M., and A. Ferenczy. 1974. Endometrial morphologic response to hormonal environment. *Gynecol. Oncol.* 2:180–197.

Sheppard, B. L., and J. Bonnar. 1974. Scanning electron microscopy of the human placenta and decidual spiral arteries in normal pregnancy. *J. Obstet. Gynaecol. Br. Commonw.* 81(1):20–29.

Steger, R. W., H. Huang, G. Kuppe, J. Meites, E. S. E. Hafez, and H. Ludwig. 1976. Effect of age on uterine surface ultrastructure. In O. Johari and R. Becker (eds.), *Scanning electron microscopy/1976.* Chicago: IIT Research Institute, pp. 359–366.

White, A. J., and H. J. Buchsbaum. 1973. Scanning electron microscopy of the human endometrium. I. Normal. *Gynecol. Oncol.* 1:330–339.

White, A. J., and H. L. Buchsbaum. 1974. Scanning electron microscopy of the human endometrium. II. Hyperplasia and adenocarcinoma. *Gynecol. Oncol.* 2(1):1–8.

Cervix

Bonilla-Musoles, F., J. Hernandez-Yago, and J. V. Torres. 1974. Scanning electron microscopy of the cervix uteri. *Arch. Gynäk.* 216:91–97.

Chretien, F. C., C. Gernigon, G. David, and A. Psychoyos. 1973. The ultrastructure of human cervical mucus under scanning electron microscopy. *Fertil. Steril.* 24:746–757.

Ferenczy, A., and R. M. Richart. 1973. Scanning electron microscopy of the cervical transformation zone. *Am. J. Obstet. Gynecol.* 115(2):151–157.

Hafez, E. S. E., and H. Kanagawa. 1972. Ciliated epithelium in the uterine cervix of the macaque and rabbit. *J. Reprod. Fertil.* 28:91–94.

Murphy, J. F., J. M. Allen, J. A. Jordan, and A. E. Williams. 1975. Scanning electron microscopy of normal and abnormal exfoliated cervical squamous cells. *Br. J. Obstet. Gynaecol.* 82(1):44–51.

Murphy, J. F., J. A. Jordan, J. M. Allen, and A. E. Williams. 1974. Correlation of scanning electron microscopy, colposcopy and histology in 50 patients presenting with abnormal cervical cytology. *J. Obstet. Gynaecol. Br. Commonw.* 81:236–241.

Riches, W. G., R.E. Rumery, and E. M. Eddy. 1975. Scanning electron microscopy of the rabbit cervix epithelium. *Biol. Reprod.* 12:573–583.

Rubio, C. A. 1976. Cervical epithelial surface. 3. Scanning electron microscopic study in atypias and invasive carcinoma in mice. *Acta Cytologia* 20(4):375–380.

Wilbanks, G. D. 1975. In vitro studies on human cervical epithelium, benign and neoplastic. *Am. J. Obstet. Gynecol.* 121(6):771–788.

Williams, A. E., J. A. Jordan, J. M. Allen, and J. E. Murphy. 1973. The surface ultrastructure of normal and metaplastic cervical epithelia and of carcinoma in situ. *Cancer Res.* 33(3):504–513.

Williams, A. E., J. A. Jordan, J. F. Murphy, and J. M. Allen. 1975. The cervix uteri (man). In E. S. E. Hafez (ed.), *Scanning Electron Microscopic Atlas of Mammalian Reproduction* Toyko: Igaku Shoin, p. 223.

Zaneveld, L. J. D., P. F. Tauber, C. Port, D. Propping, and G. F. B. Schumacher. 1975. Structural aspects of human cervical mucus. *Am. J. Obstet. Gynecol.* 122:650–654.

Vagina

Parakkal, P. F. 1974. Cyclical changes in the vaginal epithelium of the rat seen by scanning electron microscopy. *Anat. Rec.* 178(3):529–537.

Rubio, C. A. 1976. The exfoliating cervico-vaginal surface. II. Scanning electron microscopical studies during the estrous cycle in mice. *Anat. Rec.* 185(3):359.

APPENDIX C TECHNIQUES OF SPECIMEN PREPARATION

Anderson, T. F. 1951. Techniques for the preservation of three-dimensional structure in preparing specimens for the electron microscope. *Trans. N. Y. Acad. Sci.* 13:130–134.

Batson, O. V. 1955. Corrosion specimens prepared with a new material. *Anat. Rec.* 121:425 (abstract).

Boyde, A. 1972. Biological specimen preparation for the SEM: An overview. In O. Johari (ed.), *Scanning Electron Microscopy/1972.* Chicago: IIT Research Institute, p. 257.

Boyde, A. 1974. Freezing, freeze-fracturing, and freeze-drying in biological specimen preparation for the SEM. In O. Johari and I. Corvin (eds.), *Scanning Electron Microscopy/1974.* Chicago: IIT Research, pp. 1043–1046.

Boyde, A. 1976. Do's and don'ts in biological specimen preparation for the SEM. In O. Johari and R. Becker (eds.), *Scanning Electron Microscopy/1976.* Chicago: IIT Research Institute, pp. 683–690.

Cohen, A. L. 1974. Critical point drying. In M. A. Hayat (ed.), *Principles and Techniques of Scanning Electron Microscopy* (Vol. 1). New York: Van Nostrand Reinhold Co., pp. 44–112.

Echlin, P. 1974. Coating techniques for SEM. In O. Johari and I. Corvin (eds.), *Scanning Electron Microscopy/1974.* Chicago: IIT Research Institute, p. 1019.

Echlin, P. 1975. Sputter coating techniques for scanning electron microscopy. In O. Johari and I. Corvin (eds.), *Scanning Electron Microscopy/1975.* Chicago: IIT Research Institute, pp. 217–224.

Fujita, T., and T. Murakami. 1973. Microcirculation of the monkey pancreas with special reference to insulo-acinar portal system. A scanning electron microscope study of vascular casts. *Arch. Histol. Jap.* 35:255.

Humphreys, W. J., B. O. Spurlock, and J. S. Johnson. 1974. Critical point drying of ethanol-infiltrated, cryofractured biological specimens for SEM. In O. Johari (ed.), *Scanning Electron Microscopy/1974.* Chicago: IIT Research Institute, pp. 275–282.

Ingram, P., N. Morosoff, L. Pope, F. Allen, and C. Tisher. 1976. Some comparisons of the technique of sputter (coating) and evaporative coating for scanning electron microscopy. In O. Johari (ed.), *Scanning Electron Microscopy/1976.* Chicago: IIT Research Institute, pp. 75–81.

Kelley, R. O., R. A. F. Dekker, and J. G. Bluemink. 1973. Ligand-mediated osmium binding: Its application in coating biological specimens for scanning electron microscopy. *J. Ultrastruct. Res.* 45:254–258.

Malick, L. E., and R. B. Wilson. 1975. Evaluation of a modified technique for SEM examination of vertebrate specimens without evaporated metal layers. In O. Johari and I. Corvin (eds.), *Scanning Electron Microscopy/1975.* Chicago: IIT Research Institute, pp. 259–264.

Murakami, T. 1971. Application of the scanning electron microscope to the study of the fine distribution of the blood vessels. *Arch. Histol. Jap.* 32:445.

Murakami, T. 1972. Vascular arrangements of the rat renal glomerulus. A scanning electron microscope study of corrosion casts. *Arch. Histol. Jap.* 34:87.

Murakami, T. 1975. Pliable methacrylate casts of blood vessels: Use in a scanning electron microscope study of the microcirculation in rat hypophysis. *Arch. Histol. Jap.* 38(2):151–168.

Murakami, T., M. Miyoshi, and T. Fujita. 1971. Glomerular vessels of the rat kidney with special reference to double efferent arterioles. A scanning electron microscope study of corrosion casts. *Arch. Histol. Jap.* 33:179.

Murakami, T., M. Unehira, H. Kawakami, and A. Kubotsu. 1973. Osmium impregnation of methyl methacrylate vascular casts for scanning electron microscopy. *Arch. Histol. Jap.* 36:119.

Nowell, J. A., and C. L. Lohse. 1974. Injection replication of the microvasculature for SEM. In O. Johari and I. Corvin (eds.), *Scanning Electron Microscopy/1974.* Chicago: IIT Research Institute, p. 267.

Page, R. B., and R. M. Bergland. 1977. The neurohypophyseal capillary bed. I. Anatomy and arterial supply. *Am. J. Anat.* 148:345–358.

Winborn, W. B. 1976. Removal of resins from specimens for scanning electron microscopy. In M. A. Hayat (ed.), *Principles and Techniques of Scanning Electron Microscopy* (Vol. 5). New York: Van Nostrand Reinhold Co., pp. 21–35.

INDEX